T0221180

CALIFORNIA NATURAL HISTORY GUIDES

FIELD GUIDE TO BUTTERFLIES
OF THE SAN FRANCISCO BAY
AND SACRAMENTO VALLEY
REGIONS

California Natural History Guides

Phyllis M. Faber and Bruce M. Pavlik, General Editors

Field Guide to
BUTTERFLIES
of the San Francisco Bay and Sacramento Valley Regions

Text by Arthur M. Shapiro
Illustrations by Timothy D. Manolis

UNIVERSITY OF CALIFORNIA PRESS

In memory of Clyde Wahrhaftig (1919–1994),
David Graves (1938–2002), and C. Don MacNeill (1924–2005)

———————

University of California Press, one of the most distinguished university presses in the United States, enriches lives around the world by advancing scholarship in the humanities, social sciences, and natural sciences. Its activities are supported by the UC Press Foundation and by philanthropic contributions from individuals and institutions. For more information, visit www.ucpress.edu.

California Natural History Guide Series No. 92

University of California Press
Oakland, California

Library of Congress Cataloging-in-Publication Data

Shapiro, Arthur M., 1946–
    Field Guide to Butterflies of the San Francisco Bay and Sacramento Valley Regions / Arthur M. Shapiro ; illustrations by Timothy D. Manolis.
        p. cm. — (California natural history guide series ; 92)
    Includes bibliographical references.
    ISBN-13: 978-0-520-24469-6 (cloth : alk. paper)
    ISBN-13: 978-0-520-24957-8 (pbk. : alk. paper)
1. Butterflies—California—San Francisco Bay Area. 2. Butterflies—California—Sacramento Valley. I. Title.

QL551.C3S36 2007
595.78'909794—dc22   2006024978

27 26 25 24 23 22
10 9 8 7 6 5 4 3

The paper used in this publication meets the minimum requirements of ANSI/NISO Z39.48–1992 (R 1997) (*Permanence of Paper*).

Cover: Sonoran Blue *(Philotes sonorensis)*. Painting by Timothy D. Manolis.

The publisher gratefully acknowledges the generous
contributions to this book provided by

the Gordon and Betty Moore Fund
in Environmental Studies
and
the General Endowment Fund of the
University of California Press Foundation.

CONTENTS

THINGS TO DO WITH BUTTERFLIES   251

Plates follow page 218

PREFACE

Several years ago I began teaching a University of California extension course for senior citizens entitled "Butterflies Demystified." (The title was brazenly adapted from an excellent book on mushrooms by David Arora.) The course grew out of twenty-some years of leading public butterfly walks and giving butterfly talks to service clubs, garden clubs, and so forth. In a sense, this book has been incubating for a long time. But, unexpectedly, it has acquired a certain urgency, because our butterfly fauna appears to be in crisis. As crises often do, this one crept up on us undetected until it was nearly full-blown.

I began monitoring the butterfly fauna of northern California in the early 1970s, establishing a permanent set of field sites along the Interstate 80 corridor. In 2003 my research group received a substantial grant from the National Science Foundation that enabled us to begin sophisticated statistical analysis of the mountain of data that had accumulated by then. We began by examining my low-elevation sites, for which the data are densest and most continuous. We almost immediately discovered that a substantial portion of the Sacramento Valley butterfly fauna is emerging earlier now than it did 30 years ago—in a couple of cases, nearly a month earlier. (How do you spell "global warming"?) But then we began to realize that much of the emerging news was not good. Declines in our fauna had been gradual, so slow as to be imperceptible without statistics. But these had accelerated dramatically in the last few years. Formerly common species such as the Purplish Copper, Sylvan Hairstreak, Large Marble, Lorquin's Admiral, and Mourning Cloak were clearly in trouble. And the cause could not be purely local—the declines were occurring at the regional level. Studies are under way to try to identify what is going on and why. In the meantime, it is plain that tomorrow's butterfly faunas may be dramatically different from yesterday's. This book, then, is in part an attempt to describe the recent fauna for posterity—not exactly its intent when first conceived.

Let's try to put this in some kind of context. As far as we know, no one ever collected butterflies in our area before the Gold Rush. Though nearly all of our fauna had been named and described by the early 1900s—and there is a solid historical tradition of butterfly study in the Bay Area beginning in the nineteenth century— ecological and biogeographic information on that fauna is very scarce before the 1960s. We know nothing of the butterfly fauna before Europeans came to northern California, and nothing of the fauna of the Sacramento Valley before the valley's ecology was drastically altered in the nineteenth and early twentieth centuries. We are confronted with a recent fauna whose characteristics suggest it is largely a product of human activity, and our ability to reconstruct its history is very limited. That is a recurrent theme of this book. Few of our butterflies are living today as they must have lived even 150 years ago.

On a time scale of a few thousand years or longer, the much-touted "balance of nature" is an illusion. The physical environment (especially climate) is constantly changing, and as it does so it reshapes the rules governing biotic interactions. Paleoclimatology and paleoecology have made tremendous strides in recent decades and can give us hints about potential futures. Biotic communities as we know them are freeze-frames taken from a very long movie. The changes humans have wrought on the lives of butterflies are merely the most recent of the many changes they have gone through in their history. We have no hope of restoring communities to some hypothetical pristine state on any but a miniature scale. At best we create gardens that more or less resemble what we think those pristine communities looked like at some arbitrary time in the past. Like all gardens, they require constant effort to keep them from becoming what today's conditions drive them to become—conditions dominated by what we characterize as "weeds." We can, however, try to protect the bits of nature that have survived relatively unchanged despite us, cognizant that larger forces than we control may override our efforts.

Our butterfly fauna, then, has always changed and will continue to change. In 50 years perhaps the Marine Blue will be as much a fixture in the Bay Area as it is now in coastal southern

California. The Large Marble may be but a memory. My advice to you is to love and enjoy your butterfly fauna and to learn as much as you can so you can approach questions of environmental change and conservation as an informed, responsible citizen. The Bay Area is one of the few places on earth where one can elicit political passion by talking about the impact of weed control on the urban butterfly fauna. If there's any place where a butterfly field guide can and should engage with broad questions of global change, it's right here. Okay, here's your fauna and here are the challenges. What do we do now?

A final note: This is not an academic text, and I have not loaded it down with bibliographic citations. However, throughout this book, I *have* deliberately used the best word I know when introducing or explaining scientific ideas. I try to define it the first time I use it, but since one does not read a book like this the way one reads a novel, you may encounter it anywhere, and be baffled. There is a glossary at the end of the book. Please use it!

Davis, California
July 20, 2004

ACKNOWLEDGMENTS

Thirty-odd years ago I gave my first expert testimony on an environmental matter in California to a Sacramento County agency. Picking up some (linguistic?) cue, one of the commissioners asked me if I was a native Californian. No, I explained; I was born in Maryland, raised in Pennsylvania, and educated on the East Coast. She let me know in no uncertain terms that native Californians (like her) didn't cotton to carpetbaggers (like me) telling them how to manage their resources. Well, I am a carpetbagger. I had never set foot in northern California until I tried out for a job at the University of California at Davis in 1971. I had been offered a Miller Postdoctoral Fellowship at U.C. Berkeley but turned it down because under the rules then in place, it was equal to a ticket to Vietnam. Had I been a Miller Fellow I would no doubt have been inducted into the Bay Area fellowship of the net and learned the lore passed down by word of mouth from the elders to their younger apprentices. As it was, I ended up in Davis, pretty much on my own. It took a number of years until J.W. "Bill" Tilden of San Jose State University told me I had earned my stripes. After 34 years, I'm still not sure I have, but I'll let you judge that for yourself.

At any rate, this book represents an unconventional "take" on our fauna, since it was not conditioned by being steeped in Bay Area lore from childhood. Still, the fundamental data are all ultimately derived from the old-timers by way of their apprentices. John Steiner performed an invaluable service by amassing much of the data in his master's thesis at California State University at Hayward (1990). Some of this, in turn, had been brought together by Harriet Reinhard in an unpublished article on the butterflies of San Francisco. Between them, they made it much easier to track down old records vaguely alluded to in later literature. Over the years Jerry Powell, Paul Opler, Ray Stanford, John and Tom Emmel, Sterling Mattoon, and Barbara Deutsch have taught me a great deal, but I claim sole credit for all the errors you may find.

Special thanks are due field companions too numerous to list; my grad students—especially Matt Forister, Erik Runquist, and Kayce Casner—who had to bear with the metastasizing mess associated with this project; my wife Adrienne for bearing with me at all; Kathy Merk for typing my chaotic manuscript; and Doris Kretschmer, Jenny Wapner, Scott Norton, Kate Hoffman, and Matthew Winfield at University of California Press for shepherding it through to book form.—A. S.

I would like to thank Cheryl Barr (Essig Museum); Lynn Kimsey, Steve Heydon, and John DeBenedictis (Bohart Museum); and Rosser Garrison, Gillian Watson, and Stephen Gaimari (California Department of Food and Agriculture Entomology Laboratory) for providing access to collections in their care and space in which to work on the color plates. I would also like to thank Bruce Webb and his wife, Annette, for providing many and various kinds of support as well as companionship in the field.—T. M.

INTRODUCTION

What Are Butterflies?

Butterflies are members of the order Lepidoptera that belong to the superfamilies Papilionoidea and Hesperioidea. All other members of the order Lepidoptera are called moths by default. Usually butterflies are contrasted with moths, as shown in table 1. Every one of these generalizations about the differences between butterflies and moths has many exceptions, even in our own fauna. Although the most butterflylike moth family (Castniidae) has no representatives here, we have plenty of diurnal, brightly colored moths, some of which have somewhat-clubbed antennae (they taper to a club, rather than being abruptly clubbed). Most people take them for butterflies, just as many people call *Pieris rapae* a Cabbage Moth because a plain white animal just doesn't register as a "butterfly" to them. The Papilionoidea (true butterflies) and Hesperioidea (skippers) are defined technically on anatomical grounds that need not trouble most users of this book. In recent years a small family of dull-colored "moths" from the New World tropics (Hedylidae) has been found to be nocturnal butterflies after all. (Go figure!)

If you are new to butterfly study, simply familiarize yourself with the look of the various families and read the introductory matter about each. Different books divvy up the families differently—the Heliconiidae may be a family of their own or a subfamily of the Nymphalidae, for example. Don't worry too much about this. The evolutionary (ancestor-descendant) relationships among the groups are real, but the rankings we assign to them are human constructs applied for our convenience. There seems to be no evidence that the butterflies themselves care what we call them.

However you define them, butterflies have been perceived as charismatic by many cultures, past and present. In Hindu tradition, Brahma watched the metamorphosis of a butterfly and was filled with great peace as he looked forward to his own reincarnation and perfection. The ancient Greeks objectified the soul as a butterfly. Associations of the soul with butterflies can be found in folklore from Germany and the Balkans to New Zealand and Assam. Butterflies are prominent in pre-Columbian Mesoamerican art (usually swallowtails!). The Pima Indians say the Creator, Chiowotmalki, flew over the world as a butterfly, looking for the

TABLE 1	Key Differences between Butterflies and Moths	
BUTTERFLIES	**MOTHS**
Active by day	Active by night
Brightly colored	Dull colored
Clubbed antennae	Simple or feathery antennae
Relatively small body	Relatively large body
Rest with wings held over back	Rest with wings open, or rooflike at sides

right place to put mankind. Although butterfly life histories were known to peasants around the world, the first scientific descriptions of them date only to the seventeenth century. Extremely accurate portrayals of adult butterflies can be found in many late Medieval and Renaissance illuminated manuscripts. They are somewhat mysterious, insofar as the butterfly net had not yet been invented—where did the models come from?

It is surprising but true that we have no record of butterfly study or collection in California before the Gold Rush. At that time a Frenchman, Pierre Joseph Michel Lorquin, collected extensively and sent material to the distinguished French entomologist J. B. A. de Boisduval who, in turn, named and described many of our species (see the sidebar "California's First Lepidopterist"). Although California has produced many butterfly workers since, the early efforts were mostly taxonomic. In fact, the first really *biological* treatments of California butterflies only appeared in the middle decades of the twentieth century. The first butterfly book in the California Natural History Guides series was *Butterflies of the San Francisco Bay Region* by J. W. Tilden (1965), followed by Garth and Tilden's *California Butterflies* (1986). We have learned a lot since then, and the emphases in field guides have changed—away from collecting and toward nonconsumptive activities and conservation.

Richard Vane-Wright, a distinguished lepidopterist at the British Museum (Natural History), has written that scientists should devote themselves to studying the origins and dynamics of life in all its manifestations, in the hope that this understanding will help us to "coexist with nature in an intelligent way." That's the slant of this book.

Regional Butterfly Geography

This book covers the traditional 10 Bay Area counties (Marin, Sonoma, Napa, Solano, Contra Costa, Alameda, Santa Cruz, Santa Clara, San Mateo, and San Francisco), the Sacramento–San Joaquin Delta (including part of San Joaquin County), and the Sacramento Valley portions of nine more counties (Sacramento, Yolo, Sutter, Butte, Colusa, Glenn, Tehama, Shasta, and Placer). This only *seems* odd.

The Bay Area is defined rather arbitrarily. It incorporates very diverse landscapes, vegetation, and climates. In fact, no similar-sized chunk of real estate anywhere else in the United States commands such diversity. The Sacramento Valley and the Delta occupy a larger chunk of real estate but are much more uniform physiographically, climatically, and ecologically. In fact, the butterfly fauna of Turtle Bay near Redding is pretty much identical to that along the American River Bikeway in Sacramento, 165 miles away.

Anyone who has driven Interstate 80 between Sacramento and the Bay Area knows that drawing a line between the Sacra-

CALIFORNIA'S FIRST LEPIDOPTERIST

The first known butterfly collector in California, Pierre Joseph Michel Lorquin (1800?–1877), was a French forty-niner. Unsuccessful in the gold fields, he nonetheless remained in California into the 1860s, collecting insects for entomologists in Europe and the eastern United States. His most famous patron was Jean Baptiste Alphonse D. de Boisduval (1799–1879), the most distinguished French lepidopterist of the time. Boisduval published a series of papers naming and describing the species Lorquin sent him, producing the first monograph of the Californian fauna. A great many common California butterflies were thus named by Boisduval. Unfortunately none of Lorquin's correspondence or field notes seems to have survived, so what we know of his travels is largely at second hand, through references in Boisduval's works. We know he ascended the Sacramento and Feather Rivers and was frequently in the Bay Area, where he befriended the early San Francisco lepidopterist Herman Behr. Boisduval praised his courage in "braving the tooth of the bear [grizzlies were still abundant] and the fang of the rattlesnake." He

mento Valley and the Bay Area is an arbitrary exercise. Putting aside purely human artifacts (county lines, zones for automobile insurance rates, the limits of irrigation districts, or Catholic dioceses), one can draw lines based on various natural criteria that may not agree among themselves. The eastern edge of the Inner Coast Range can be used as a natural boundary. But the Coast Range breaks between Vacaville and Fairfield and doesn't reappear until Antioch—while the east-west trend of the Potrero Hills further clouds the picture. Vegetation is largely a function of climate, with significant effects of soil type. Many of you already know that the cold California Current is responsible for the clammy gray summers of the coastal fog belt, and that the Central Valley is clear and blazing hot in the summer. Heating inland causes the air to rise and the barometric pressure to fall. This draws in cold, dense marine air ("nature abhors a vacuum") which rushes inland through the gap at the Carquinez Straits. This is the "Delta wind" or "sea breeze" that makes the area near the Delta so much more livable than areas farther north or south. The Delta and Suisun Marsh are cooled substantially but see little coastal fog or low cloudiness. They thus form a climatic transition between the Bay Area and the Sacramento Valley. In summer, Vaca-

---

said Lorquin reached "the glaciers of the Sierra Nevada," but this is doubtful, and there are no alpine species among the material he sent back.

Because it is important to fix the type-localities of Boisduval's names, J. F. and T. C. Emmel and S. O. Mattoon (1998), after carefully reviewing everything that was known about Lorquin and his travels, examined the extant type-specimens and compared them to recent material from throughout the species ranges. On this basis they designated type-localities they believe to be good matches. According to them, quite a few common, widespread California butterflies were probably described from specimens Lorquin collected in or near San Francisco or Sacramento.

Lorquin also traveled widely in the Orient, collecting as he went, and was about to leave for the Philippines when he died in 1877.

Until someone invents a time machine, or Lorquin's diary turns up in an attic somewhere, there is no way to prove the inferences by the Emmels and Mattoon to be true. But isn't it fun to imagine what it would be like to be a French-speaking butterfly collector in the rough-and-tumble days of Gold Rush California?

ville is usually about 5 degrees C (10 degrees F) warmer in the afternoon than downtown Fairfield, and 8 degrees C (15 degrees F) warmer than Cordelia. The pivot point is typically at Lagoon Valley.

The same gap produces a precipitation anomaly in the Vaca Hills just west of Vacaville, where winter storms are amplified as moist air rises upslope on the *east* side of the range. This is reflected in the vegetation and in butterfly distributions. Many unusual butterfly distributions reflect such local situations. For well-informed lepidopterists, it's great fun to search them out. In this case, the Tailed Copper *(Lycaena arota)* and the Crown Fritillary *(Speyeria coronis)* are beneficiaries of the anomaly.

However, you do not need to go that far inland to find dramatic climatic gradients. The cold, gray marine layer is usually less than 600 m (2,000 ft) thick, which means the higher hills in the Bay Area poke well above it and into the hot sunshine. Slight topographic features can have dramatic effects on microclimate, as every San Franciscan knows. Every San Francisco neighborhood has one or more microclimates all its own. For critters the size of butterflies—*cold-blooded* critters—that kind of variation can mean a lot. (See Harold Gilliam's *Weather of the San Francisco Bay Region* [2002] in the California Natural History Guides series.)

So what butterflies should you expect to see where?

Butterflies are heavily dependent on plants—as adult food sources and as larval hosts. The mere fact that a species' host plant occurs in a place does not guarantee that the butterfly will—but if the plant is *absent,* the butterfly will surely be (except as a stray). Butterflies are also responsive to vegetation structure (layering, shade, distribution and geometry of light gaps—remember that these move during the day) and to the overall landscape. It is a truism that to know the butterflies of an area, you should learn the plants. Experienced butterfly workers can often go to a place they have never been before, perhaps even in the dead of winter, and successfully predict what species will be found there. But life is full of surprises. Things that "should be" in a place may *not* be because of some past catastrophe, such as fire. If sources of colonists are available, the species may eventually return. But it takes time—years, decades, perhaps even centuries.

Because of its complex geography, the Bay Area has many—at least 20—major types of plant communities. (Plant communities are classified for our convenience, and most grade into others along their borders, but it is often useful to act as if they are "real.") They are classified in two ways—based on species composition (floristics) or growth form (physiognomy)—and all of this gets to be a bit much for users of this book. Relatively few of our butterflies are tied rigidly to one or only a few communities. The ecological determinants of what occurs where can be simultaneously both coarser and more subtle than plant community classification. Rather than review the vegetation in depth, I encourage you to use this book in conjunction with the plant community information in *Introduction to California Plant Life,* by Robert Ornduff et al. (2003) in the California Natural History Guides series. In this book I have tried to be as precise as possible in defining vegetational association species by species, rather than using standard classificational formulas. In many cases the physiognomy is more informative than community species composition. Usually both factors come into play.

About Serpentine

You will find frequent references to serpentine in this book. Aside from being California's state rock, serpentine (really serpentinite) has fascinated ecologists for decades because its unusual chemistry selects for unusual floras and plant communities. Species found only on serpentine are called serpentine endemics (leather oak *[Quercus durata]* is a good example). Some species occur on serpentine and other unusual soils, but not on "normal" soils; MacNab cypress *(Cupressus macnabiana),* which also occurs on gabbro, is an example. The native herbaceous genus *Streptanthus,* collectively known as jewel flowers (in the mustard family [Brassicaceae]) has speciated dramatically on serpentine soils; several species have very restricted ranges. In general, serpentine vegetation tends to be open, with much bare soil (hence, serpentine sites are often referred to as barrens); tree oaks are excluded; and community boundaries can be razor-sharp where the serpentine contacts other rock/soil types. Serpentine is common in the Coast Ranges and occurs within the city of San Francisco. It has a characteristic butterfly fauna. Two species (John Muir's Hairstreak *[Mitoura gryneus muiri]* and the Sleepy Duskywing *[Erynnis brizo lacustra]*) are serpentine endemic in our region, and about six more species occur preferentially on serpentine. (See *California Serpentines* by Arthur R. Kruckeberg

[1984] and *Introduction to California Soils and Plants,* also by Kruckeberg, both in the California Natural History Guides series.) There is no serpentine in the Sacramento Valley.

The Sutter Buttes

There is one big chunk of bedrock smack in the middle of the Sacramento Valley: the volcanic Sutter Buttes, sometimes called "the world's smallest mountain range." The Buttes are mostly privately owned. Public access, including guided tours, is provided through the Middle Mountain Foundation in Chico. The natural history of the Buttes is documented in the book *Inland Island* by Walt Anderson (2004). This book includes a preliminary butterfly list, but it is probably far from complete. The Middle Mountain Foundation is interested in having more butterfly work done in the Buttes. As might be expected, the fauna of the Buttes is more a (depauperate) foothill fauna than one characteristic of the Sacramento Valley floor.

To Be Continued

Several dozen unusual or unique botanical sites are within the area covered by this book. Some have special soils; others, special microclimates; some have both. Any area that has endemic and/or relict plants is of interest. Although they are well known to native plant enthusiasts, most of these areas have never been surveyed for butterflies. Every one of them could easily become a pet project for a reader of this book—or even a senior honors thesis or master's thesis for a student. Some 150 years after Lorquin first collected in the Bay Area, our ignorance is a sad and shameful state of affairs.

We know the isolated Antioch dunes have a special butterfly—Lange's Metalmark *(Apodemia mormo langei).* But what about the isolated sandhills at Bonny Doon in the Santa Cruz Mountains, home of various rare plants including the endemic Bonny Doon manzanita *(Arctostaphylos silvicola)?* We know John Muir's Hairstreak has followed its host, Sargent cypress *(Cupressus sargentii),* to Cedar Mountain Ridge near Livermore in southeastern Alameda County. But what other serpentine "goodies" are there? When will someone do a butterfly fauna of Ring Mountain?

Talk to your California Native Plant Society chapter. Scrutinize the geologic maps of your area. Not all ecological "islands" will bear fruit lepidopterologically, but you never know until you try. When Bruce Gervais and I looked for edaphic-endemic butterflies in the Sierra foothills, we were successful beyond our wildest dreams. On the other hand, vernal pools, with their fleeting and unpredictable seasonality, never have special butterflies.

Don't just sit there. *Do It.* (Good luck!)

Where Are Our Butterflies From?

The science of historical biogeography tries to interpret the origins and evolution of floras and faunas in geographic context. The species found in a given area may be derived from very different sources and can be assigned to geographic "elements" reflecting the historical influences affecting that area. The butterflies covered in this book reflect a long and complex history of geologic and climatic change. The climatic preferences shown by the species today may be clues to their origins in the distant past. California has been getting drier for millions of years, but the mediterranean climate we have now, in which nearly all the precipitation falls in the winter, is relatively young. Most, if not all, of our butterflies evolved in climates with summer rain. Adapting to present-day California meant adapting to that long, hot, dry summer.

Before the Pleistocene ice ages (roughly two million to 15,000 years ago), during the Tertiary Period, rich and diverse forest floras dominated the midlatitudes of the Northern Hemisphere. Tectonic (mountain-building) activity combined with climatic change to fragment this flora, leaving species that had once grown side by side eventually isolated in widely different parts of the globe. A substantial part of our butterfly fauna belongs to genera or species groups whose contemporary distributions strongly suggest that they were involved in this process. (We have abundant plant fossils, but very few of butterflies—none from California.)

Our California Tortoiseshell *(Nymphalis californica),* for example, is very closely related to two Old World species, the Large Tortoiseshell *(N. polychloros)* and Yellow-legged Tortoiseshell

(N. xanthomelas). Our Gray-veined White *(Pieris "napi")* belongs to a very complex group of populations found all around the Northern Hemisphere in cool, moist, mostly forested habitats; no one is sure how many biological species there are. (The true *P. napi* is European.) Our orange-tips are related to species on the East Coast and in Europe, the Middle East, and the Far East. Our Anise Swallowtail *(Papilio zelicaon)* and the skippers of the genus *Ochlodes* have very close Old World kin. Our tailed blues are outliers of a Eurasian group, and so on.

A few species are found naturally (not by human intervention, as far as we know) in both Eurasia and North America. The Arctic Skipper *(Carterocephalus palaemon)* is one. The Painted Lady *(Vanessa cardui),* Red Admiral *(V. atalanta),* and Mourning Cloak *(Nymphalis antiopa)* are others. In the nineteenth century the great naturalist Louis Agassiz, the "father of the Ice Age," suggested (on religious grounds!) that the last three must have been accidentally introduced in commerce in the early days of European colonization. No hard evidence exists pro or con, but most biologists today assume the distributions of these migratory butterflies are natural.

The Cabbage White *(Pieris rapae),* was, however, introduced to North America from Europe in the mid-nineteenth century. The great American lepidopterist Samuel H. Scudder, who had been a student of Agassiz, mapped its spread over the continent from its point of introduction in southern Canada. As discussed in its species account, there remains a possibility that it was introduced even earlier in California, perhaps by the Spanish during the Mission Period.

The history of our biota also involves a tropical element. During the earlier Tertiary, climates were moist and warm, and vegetation characteristic of Central America extended up into our area. As the climate dried, these plants and communities disappeared. We have two native butterflies, the Great Purple Hairstreak *(Atlides halesus)* and the Pipevine Swallowtail *(Battus philenor),* of unambiguously tropical origin, but neither is likely to have persisted through millions of years of climate change. I explain below why the Pipevine Swallowtail is almost certainly a recent arrival. We have one more tropical species, the Gulf Fritillary *(Agraulis vanillae),* in the Bay Area, but it is an extremely recent arrival and probably an introduction.

The Mormon Metalmark *(Apodemia mormo)* represents a

faunal element derived from Mexico. Many plants characteristic of California chaparral and oak or pine-oak woodland entered our area from Mexico, where they evolved in a climate with cool, dry winters and warm, wet summers. The plants appear to have made the transition to the mediterranean climate spectacularly well, while Mexican butterflies by and large have not. The butterflies of our Southwest deserts, which enter southeastern California, are still dependent on the summer monsoonal rains. One of them, the Marine Blue *(Leptotes marina)*, is a frequent visitor to our area but at present does not persist. Its own lineage goes back much further, rooted in the Old World tropics.

One of our strangest butterflies is the Golden Hairstreak *(Habrodais grunus)*. It belongs to an Old World group and is not closely related to our other hairstreaks. This may seem surprising, since it is so well adapted to our summer drought, but it seems historically akin to the Eurasian element, not the Mexican. Our Pine White *(Neophasia menapia)* is not closely related to our other whites. It has one living congener in northwestern Mexico and southeastern Arizona. The genus *Neophasia* is related to pierid genera in the New World tropics and to the genus *Delias* from the Indo-Australian region. It is a real enigma in our fauna.

Molecular Keys to the Past

A scientific breakthrough that is helping us work out the history of our butterfly fauna is called molecular phylogeography. This is the study of gene distributions among populations in space. Relatively easy and inexpensive DNA sequencing now allows us to do precise, quantitative genetic comparisons among populations and species. Sometimes we can use a technique called the polymerase chain reaction to obtain usable amounts of DNA from dead, dry museum specimens—we might even sequence the DNA of extinct species! These data are then analyzed using computerized procedures that reconstruct ancestor-descendant relationships and historical movements. For example, if we found a population of a foothill butterfly isolated in the Sutter Buttes, it might be possible to tell whether it got there from the Coast Ranges or the Sierra Nevada, even with no morphological evidence.

In the case of the Pipevine Swallowtail, phylogeographic studies over its entire range reveal that our populations have very little genetic variation, compared to those from the southeastern United States and especially Florida. We infer that our populations went through a genetic "bottleneck," probably due to their having been founded relatively recently by only a few individuals. We believe that the Pipevine Swallowtail had an Ice Age refuge in or near Florida and spread out after deglaciation, attaining its current range only recently. The genetic evidence strongly contradicts the hypothesis that it is an ancient relict from the Tertiary Californian tropical biota. Its host plant, the endemic California pipevine *(Aristolochia californica),* belongs to a genus found in both the Old and New World tropics. We still don't know if the plant is old or new in our biota. One of the curious facts about the Pipevine Swallowtail is that, as a poisonous and warningly colored species, it has mimics everywhere else in its range, but not in California where it is actually most abundant. This also suggests it hasn't been here very long. (See the sidebar "Mimicry" later in this introduction.)

Molecular phylogeography is being applied to more and more groups of butterflies and promises to make testable many ideas that have generated much heat and little light in biogeography for decades.

Regional Affinities of Our Fauna

In 1890 the Bay Area lepidopterist H. H. Behr pointed out that the butterfly fauna of the coast from the Bay Area north was largely a subset of that of the Sierra Nevada. The same is true of the flora. The summer fog markedly reduces the physiological stress imposed by the rainless regime. Without it there would be no redwoods, no Douglas fir *(Pseudotsuga menziesii)*, and no "montane" element in the Bay Area butterfly fauna—species such as the Great Arctic *(Oeneis nevadensis)*, Blue Copper *(Lycaena heteronea)*, Two-banded Skipper *(Pyrgus ruralis)*, Arctic Skipper *(Carterocephalus palaemon)*, and the now-extinct Strohbeen's Parnassian *(Parnassius clodius strohbeeni)* from the Santa Cruz Mountains. Many of these are hanging on in very special microclimates. We have abundant fossil plant evidence that shows repeated exchange between the Coast Ranges and Sierra Nevada during the Quaternary (Pleistocene plus Holocene). However, such exchange should not be taken for granted. For example, molecular phylogeography shows that Coast Range (including the Bay Area) and Sierra Nevada populations of the Colorado Skipper *(Hesperia colorado)* are distinct and perhaps not all that recently separated. It would be very interesting to compare molecules from populations of, say, the Two-banded Skipper at sea level in the Bay Area and at tree line in the central Sierra, since we think we know the time frame in which their distribution developed.

Taking a L-O-N-G View

Butterflies first appear in the fossil record in the early Tertiary, some 60 million years ago. By the Oligocene, in the mid-Tertiary (about 35 million years ago), we find most of the extant families—and some modern genera, for example, the painted ladies. It seems strange to think of such delicate creatures preserved in stone, and indeed butterfly fossils are rare—too rare to be very useful in reconstructing historical biogeography, though they can and do present us with astonishing surprises. Because plant fossils are much more numerous (both macrofossils such as leaves and cones, and microfossils—pollen grains, which can often be identified under a strong microscope), we can reconstruct ancient vegetation and then ask ourselves what butterflies might have inhabited it. To do this we have to assume that host-plant and community associations have been stable over the time frame involved—a risky assumption, as we will see. But it is a stimulating intellectual exercise. The great Berkeley paleobotanist Ralph Chaney showed, for example, that a redwood forest similar to Muir Woods grew in the Miocene (some 20 million years ago) in an area of central Oregon that supports sagebrush steppe today. Imagine how different its butterfly fauna must have been then!

Butterfly Life Histories

Metamorphosis

Like all Lepidoptera, butterflies are holometabolous—that is, they have a life cycle encompassing four distinct stages: egg, larva (caterpillar), pupa (chrysalis), and adult. Another term for such a life cycle is "complete metamorphosis."

Metamorphosis is regulated by the endocrine system. Several glands and hormones are involved, forming a "chain of command" with built-in checks and balances (feedback loops). Stretch receptors in the caterpillar tell the nervous system to secrete eclosion hormone, which then stimulates the prothoracic gland to produce ecdysone, or molting hormone. Another hormone, juvenile hormone (JH), is produced by structures in the head called corpora allata. So long as circulating JH is high, each molt leads to another larval stage. As the caterpillar matures, JH levels drop off while ecdysone levels do not. This situation triggers the critical events of metamorphosis. As the adult forms, JH production may resume, stimulating reproductive maturation. In butterflies with seasonal reproductive diapause, or dormancy, JH production is triggered by changing day length or other environmental stimuli and causes ovulation and sexual behavior to begin. Sperm production (spermatogenesis) begins in the late larval stages and seems to be independent of JH. Much of the research on the physiology of metamorphosis has been done with moths, but the principles seem to apply broadly to butterflies as well.

Most people are familiar with a number of butterfly species, but far fewer know their early stages. In fact, the early stages of some of our butterflies have still not been described scientifically. The characteristics of the early stages are just as specific as those of the adult and can be used in classification. Field guides to American caterpillars are just beginning to appear; perhaps one for California is in the cards.

Eggs

Butterfly eggs may be spindle or milk-bottle shaped (Pieridae and Danainae), turban shaped (Lycaenidae), or more or less hemispherical or dome shaped (Hesperiidae and Papilionidae; some Nymphalidae). They are usually white, yellow, or pale green, but those of sulphurs and several whites and orange-tips whose larvae feed on flowers and fruit of the host plant are red or orange. Eggs of the Pipevine Swallowtail *(Battus philenor)* are also rusty red. Seen under an electron microscope, butterfly eggs display intricate patterns of sculpturing or meshwork in the shell, or chorion. Often they have vertical "ribs," which may be connected by thinner horizontal ones. The ribs converge to the upper end of the egg where there are openings, called micropyles, through which sperm enter to achieve fertilization. The micropyles are often surrounded by a rosette of "cells" resembling the petals of a flower. Some butterfly eggs are modified to admit and hold air (through holes called aeropyles), functioning like little diving bells when the host plant is submerged in water or buried in snow. We rarely find empty butterfly eggshells; most larvae eat them immediately after hatching. In species that lay masses of eggs, the first larvae to hatch often nibble at the chorion of nearby eggs, accelerating their hatching in the process.

Larvae

Butterfly larvae undergo a number of molts. At each molt the caterpillar sheds its old skin, which is then often eaten to recycle the nutrients it contains. The head capsule, which is heavily reinforced with a material called sclerotin, is completely inelastic as well as inedible. When rearing larvae, one may collect the head capsules to determine reliably how many molts (usually five) occur. A few butterfly larvae actually molt once inside the eggshell before hatching! The number of molts is usually fixed but may vary in some species, especially those with larval dormancy. The period between molts is called an instar, and larvae can conveniently be classified by instar using the abbreviations L_1, L_2, and so forth. A turn-of-the-century entomologist named Harrison G. Dyar discovered that the sizes of successive head capsules of the same larva are related in a nearly constant ratio, each being about 1.2 to 1.3 times the previous one. This is known as Dyar's law.

Some larvae change their appearance dramatically during development. The Anise Swallowtail *(Papilio zelicaon)*, for example, looks like a bird dropping as a first or second instar but then develops a characteristic green ground color with black rings on each segment, incorporating yellow or orange spots. Anise Swallowtail larvae are blacker in cool and humid environments; a few are nearly all black. In hot, dry environments they have much more light green.

The Pipevine Swallowtail larva is normally purplish black with soft red "horns," but development in intensely sunny, hot conditions induces mostly red or all red coloration. Some lycaenid larvae may, to a degree, match the color of their host. Many larvae change color prior to pupating. The green larvae of the Small Marble *(Euchloe hyantis)* and the Western Tiger Swallowtail *(P. rutulus)* turn a livid purple color at this time. Male larvae of light-colored or green species may often be recognized, as the testes can be seen on either side of the dorsal midline about two-thirds of the way aft. Actually, larvae can usually be sexed at hatch by examining their eighth and ninth abdominal segments below, under a microscope, but this usually requires sacrificing the individual because first instars are so fragile.

When larvae first hatch they bear bristles, called primary setae, that may be useful for classification. After the first molt these become obscured by numerous secondary setae. So it is important to preserve a few neonate larvae when rearing for study. Many larvae are nearly colorless at hatch and "color up" only after beginning to feed, so the primary setae and other anatomical features may be very easy to see in neonates.

Larvae may be solitary or gregarious. Young larvae of the Variable Checkerspot *(Euphydryas chalcedona)* feed together in a silken web on the host plant. After overwintering, they are solitary. Tortoiseshells and the Mourning Cloak *(Nymphalis antiopa)* feed gregariously and conspicuously on the host but spin no web. Oddly, when reared in captivity, larvae of the large fritillaries—which are always solitary in nature—aggregate and feed together.

The solitary larvae of the West Coast Lady *(Vanessa annabella)*, the Painted Lady *(V. cardui)*, and the Red Admiral *(V. atalanta)*, as well as most hesperiid larvae, construct shelters by rolling the leaf of the host into a tube or tying it into a tent. The half-grown larva of Lorquin's Admiral *(Limenitis lorquini)* attaches a leaf with silk, constructs a tent, and lives in it over winter. We call such a structure a hibernaculum. The tent of the West Coast Lady on a mallow leaf is open at the top with a light silken cover in summer, but during the rainy season the larva lives underneath the leaf and pulls it downward like an umbrella. The larva of the American Painted Lady *(V. virginiensis)* produces an untidy web incorporating the chaffy flowers and seeds of its host plants, species of everlastings *(Gnaphalium,* Asteraceae). The brightly colored larva is completely concealed within.

A few lycaenid larvae begin feeding by eating through the leaf surface and mining out the soft tissue (parenchyma) below. The Sonoran Blue *(Philotes sonorensis)* larva spends its entire life inside one leaf of its succulent host, dudleya *(Dudleya,* Crassulaceae). The larva and its frass can be seen through the translucent leaf surface. Larvae of our tailed blues feed on seeds inside the host's pods; after boring into the pod they turn around and seal up the hole with silk, presumably denying entrance to natural enemies.

Larvae that feed in the open are usually camouflaged (cryptically colored); think of the Cabbage White *(Pieris rapae)* larva hidden in plain sight on a leaf. The larva of the Large Marble *(Euchloe ausonides)*, which feeds on buds and fruit of mustards (Brassicaceae), is striped lengthwise in yellow and purplish gray and sits stretched out among the fruits, appearing for all the world to be one of them. Brightly colored, very conspicuous larvae—especially if gregarious—are usually "warningly colored." Their colors and behavior advertise that they are inedible for some reason. They may be spiny (Mourning Cloak, tortoiseshells) or chemically defended by toxic or distasteful compounds (Monarch *[Danaus plexippus]*, Pipevine Swallowtail) or even both (Variable Checkerspot, Buckeye *[Junonia coenia]*). The technical term for warning coloration is "aposematism," from roots meaning "a sign at a distance." The chemicals used for defense may be gathered from the host plant and sequestered or (less frequently) manufactured by the insect itself. Such defenses may be carried over into the adult and form a basis for mimicry. Mimicry also occurs in larvae. Large larvae of the Clodius Parnassian *(Parnassius clodius)* apparently mimic warningly colored toxic millipedes. The large larva of the Spring White *(Pontia sisymbrii)* is strikingly similar to that of the Monarch and quite unlike the larvae of its relatives, but it is not known if this is a true case of mimicry or merely convergent evolution. An orange-tip from the desert *(Anthocharis cethura)* also has a very Monarch-like larva. The larva of our race of the Indra Swallowtail *(Papilio indra)* is perhaps our most striking caterpillar, gaudily banded in black and shocking pink. We do not know if it is edible.

Larvae of papilionids have an eversible organ, called an osmeterium, normally concealed in a slit behind the head. When disturbed, they inflate the osmeterium and discharge a defensive secretion. In some species this is barely noticeable, but in the common Anise Swallowtail it is butyric acid (rancid butter!) and it may linger in the air. If you rear this species, you have been warned!

Prior to pupating, most larvae leave the host plant and wander for several hours. It has been suggested that parasitoids "key

MIMICRY

Mimicry is a form of protective coloration in which organisms escape attack by predators because they resemble other organisms that the predators avoid — for example, because they are toxic or bad tasting, or bite, sting, or itch. The two best-known types of mimicry are named Batesian and Mullerian, after nineteenth-century naturalists who described them in the American tropics. In Batesian mimicry there is an edible mimic that resembles an inedible model. Typically, the mimic departs significantly from the normal coloration in its lineage. In Mullerian mimicry, multiple inedible species in the same region resemble one another; thus the predator need learn only one color or pattern and will then avoid all of them. Truly spectacular mimicry situations are common in tropical butterflies and formed an important test case for Darwin's theory of evolution. Although mimicry is less common outside the tropics, one of the most famous cases occurs in the eastern two-thirds of temperate North America, where the Viceroy butterfly *(Limenitis archippus),* which is moderately distasteful, mimics the Monarch *(Danaus plexippus).* This is actually an unusually complex situation because the Monarch acquires its chemical defenses (cardenolides) from its host plants, milkweeds *(Asclepias).*

in" on the odor of the host plant to increase the efficiency of their search for specific hosts; by leaving the host the mature larva makes itself harder to find. In the case of swallowtails whose pupae may be either brown or green, the larva is sensitive to the backgrounds it encounters in its wandering, and this usually determines the ultimate pupal color.

Pupae

Larvae attach themselves to a substrate when about to pupate. The way they do this is family specific. Pierids and papilionids spin a button of silk at the tail end and a girdle around the middle, in which they hang at an angle. Lycaenids are similar but usually tightly appressed to the substrate. Hesperiids are attached at the posterior end but do not hang vertically, head down, as nymphalids and their relatives do.

Since the caterpillar skin is shed at pupation, how does the pupa manage to attach itself to the button of silk? The pupa bears

Different milkweeds vary widely in their nastiness, so individual Monarchs are only as nasty as the plants they ate were. Monarchs thus can be Batesian mimics of their own species, a situation called automimicry. We do not have Viceroys in our area. The Pipevine Swallowtail *(Battus philenor)* gets its protection (aristolochic acids) from its hosts, pipevines *(Aristolochia).* California Pipevine Swallowtails are just as nasty as those from elsewhere in the range but have no mimics here. In the East they are mimicked by several different butterflies and the day-flying male of the Promethea Moth (*Callosamia promethea,* Saturniidae), and they have one mimic in Arizona as well. This situation, as noted in the text, supports the inference that Pipevine Swallowtails have not been in California very long.

The California Sister *(Adelpha bredowii californica)* has been shown to be moderately distasteful and is apparently mimicked here by Lorquin's Admiral *(Limenitis lorquini),* which is edible. We also have diurnal noctuid moths, the Blue Moth *(Caenurgina caerulea)* and species of *Euclidea,* which appear to be very effective mimics of "blue" butterflies (e.g., Boisduval's Blue *[Plebejus icarioides])* and of duskywings *(Erynnis),* respectively. No one has demonstrated the advantage conferred by such mimicry. Why don't you?

a structure at its tail end called a cremaster, covered with more or less recurved hooks. In a deft and nearly instantaneous maneuver, having worked the cast skin back to its rear end, the pupa simultaneously flings it away and embeds the cremaster and its hooks in the silk pad. In captivity this maneuver often fails, leading to death or deformity for the hapless pupa. (There is actually a moment when it is not connected to the substrate!) When rearing, it is important to provide a proper substrate for pupal attachment and to cushion the blow if the new pupa falls.

Most people seem to think butterflies emerge from cocoons. A cocoon is a silken structure produced by the larva and enclosing the pupa. It may incorporate leaf material or larval hairs. Many moths make cocoons, but none of our butterflies makes a true cocoon. Hesperiids and parnassians may make a crude web of silk threads, and some hesperiids as well as the species of *Vanessa* occasionally pupate in their last larval shelter.

The new pupa commonly changes color as its cuticle hardens. On its surface we can see cases, within which compound eyes, mouthparts, antennae, legs, and wings will form. We can also see the reproductive structures, so pupae can be sexed. What happens inside the pupa is truly remarkable. The larval body is broken down into a nutrient broth. (A few tissues, notably the nervous system, persist.) Special clusters of cells, called imaginal discs, are activated and begin proliferating rapidly, feeding on the nutrients released from the larval body to construct the new body of the adult. During this process, found in all holometabolous insects, specific genes are turned on and off at specific times; the whole sequence is under hormonal control. As discussed in the section "Dormancy and Diapause," the building of the adult may progress very rapidly, or it may be postponed, sometimes for years! When it is nearly complete, the adult, including the wing pattern, can be seen clearly through the pupal cuticle. If you observe this process carefully you will notice that different wing pigments and pattern elements are laid down at different times in a predetermined sequence. The adult encased within the pupa is called a pharate adult. In a few butterflies, but apparently none of ours, mature males are attracted to pharate females and may break through the pupal cuticle to mate with them.

Emergence

When the adult butterfly emerges its wings are soft and crumpled; they must expand to full size through a combination of gravity and the pumping of blood, or hemolymph, into the wing veins to help extend them. If a butterfly cannot hang freely with its wings drawn out by gravity, it will be permanently crippled. If you rear in glass containers, line them with paper towels or the like so the newly emerged (teneral) adult can climb freely! Once the wings are hardened and ready for flight there is no further blood circulation in the veins.

You may notice that the proboscis is contained in two cases on the pupa. Upon emerging the butterfly must join the two halves, fitting them together like a zipper. It does this by repeatedly coiling and uncoiling the proboscis even as its wings are hardening. If the joining maneuver fails, the butterfly cannot feed and will starve. One occasionally finds such defective specimens even in nature.

Shortly after emerging, butterflies void one or more large drops of liquid (meconium), which contains the waste products of metamorphosis. Meconium is often colored bright red due to the presence of compounds called ommochromes. It may be squirted under pressure at predators (teneral adults are otherwise defenseless). Gregarious nymphalids have red meconium. In the Middle Ages in Europe, showers of "red rain"—taken to be blood by the credulous—inspired superstitious dread and sometimes led to riots, pogroms, and public displays of religious fanaticism.

Little butterflies do not grow into bigger ones. Adult butterflies may get fatter or thinner, but once they get their wings, their overall dimensions are fixed for life.

Voltinism and Seasonality

Butterfly life cycles are strongly seasonal, and for good reason: their vital resources are too. The study of biological seasonality is called phenology (from a Greek root meaning "appearance"). Phenology was a necessary preoccupation of early humans, be they hunters, gatherers, or primitive agriculturalists. As an applied scientific discipline it has served agriculture and public health. At a more basic level it helps us understand how both individual species and ecological communities function in nature.

The term "voltinism" (from a root meaning "time" or "instance") means the number of adult emergences a species has per year. Strictly speaking, voltinism need not equal the number of generations—there can be "split flights" as described below—but usually the two are interchangeable. Univoltine species or species populations have only one flight per year; bivoltines have two, and in this book the term "multivoltine" refers to three or more. All our univoltine butterflies are strongly seasonal, occurring at a specific time of year. Some bivoltine species, such as the Echo Blue *(Celastrina ladon echo)*, Large Marble *(Euchloe ausonides)*, Sara Orange-tip *(Anthocharis sara)*, Gray-veined White *(Pieris "napi")*, and Woodland Skipper *(Ochlodes sylvanoides)* have two successive flights in the same half of the year and are absent in the other half. Others, such as the Columbia Skipper *(Hesperia columbia)*, Yuba Skipper *(H. juba)*, and Yuma Skipper *(O. yuma)*, have widely separated spring and late-season flights. Multivoltine species may be present most of the year as adults.

The West Coast Lady *(Vanessa annabella)* and Red Admiral *(V. atalanta)* fly all winter; the Orange Sulphur *(Colias eurytheme)* and Cabbage White *(Pieris rapae)* typically fly about 44 weeks of the year in the Sacramento Valley, but in dry years may be nearly continuous. Multivoltine species may have well-defined flights, or these may overlap broadly. The more diverse the local topography and climates, the less obvious the flight sequence is likely to be, as animals will move among areas with local differences.

Because butterfly life cycles are adaptations to resource availability, different climates and vegetation types in our area select for quite different phenologies. For example, in chaparral and foothill woodland communities, the mediterranean climate produces a burst of new plant growth in late winter and early spring. By midsummer growth has ceased; many plants have senesced or become dormant, and the foliage of evergreen species has usually become tough, dry, chemically defended, and indigestible. (Plants cannot afford to lose photosynthetic tissue in the mediterranean summer, because replacing it requires water, which is just not available.) There is only a brief window of time when the weather is sunny and warm enough for adult activities *and* the vegetation is in good enough condition for larval feeding and development. Thus we see a burst of butterfly activity at that time, with most chaparral and foothill woodland species either spring-univoltine or spring-bivoltine. Numbers of species and individuals drop off precipitously in summer, and very few species emerge only in the second half of the season.

In riparian (riverine) forest the proportion of multivoltine species is high because the high water table allows many plants to put on additional growth in summer. Riparian tree-feeders such as the Lorquin's Admiral *(Limenitis lorquini)*, California Sister *(Adelpha bredowii californica)*, Western Tiger and Two-tailed Swallowtails *(Papilio rutulus* and *P. multicaudatus)*, and Mournful Duskywing *(Erynnis tristis)* are able to breed repeatedly. The Pipevine Swallowtail *(Battus philenor)*, which has only one host plant, California pipevine *(Aristolochia californica)*, is a riparian species with a twist. Its larvae must begin feeding on young tender shoot tips, and females lay only on them. The plant stops growing by the 4th of July—in dry years as early as May!—yet the butterfly has a very long flight season, from February to November. How is this possible? Its voltinism is difficult to characterize because it routinely produces "split flights." Among the progeny of a single female, some develop directly to the adult with no dormancy; others reared under identical conditions enter pupal dormancy. Some of these will eclose later the same season, while others will lay over until the following spring. In most places the Pipevine Swallowtail is basically spring-bivoltine, and nearly all reproduction occurs before July. But some adults are always flying in warm weather. Normally their reproductive success should be zero, but if pipevine burns—or is cut down to the ground—it regenerates quickly, even in late summer or fall. Any inseminated females in the neighborhood quickly find and oviposit on it. Any year's early-spring flight consists of individuals representing all of the previous year's generations and partial generations. We still do not understand how the butterfly makes its developmental "choices."

In urban and suburban areas and in the Central Valley generally, the phenology of the butterfly fauna is markedly different. Here, most species are multivoltine, often with very long adult flight seasons. These butterflies feed for the most part on herbaceous plants that depend on summer water supplied by humans. Much of this fauna is probably recruited from the tule marshes that once lined the east side of the valley (see "Butterflies in the Anthropic Landscape," later in this book). Most of them undergo a population bottleneck over winter, so they start off scarce and local in the spring, building up their numbers with successive generations and peaking in September and October. Fall is the best season for urban and suburban butterfly gardeners in our area.

The coastal fog belt has a major impact on butterfly phenology. Coastal fog eases the water stress on the vegetation in summer, but it also reduces the number of hours of sunlight per day, discouraging adult activity. It also holds down daytime temperatures, reducing larval growth rates. As a result, even univoltine species experience prolonged flight periods at the coast; species that disappear by May inland may still be active at Point Reyes in July. Multivoltine species generally have fewer generations near the coast. As all coastal residents know, inland and coastal temperatures equilibrate in September and October, producing usually the finest sunny, warm weather of the year. Butterfly diversity is at its peak in Berkeley in late September and on into October.

Some bi- and multivoltine butterflies rely on a seasonal succession of host plants as the generations go by. The first generation of the Echo Blue typically lays eggs on buds of wild-lilac

(Ceanothus). By the time the second generation is flying there are no wild-lilac buds, and females lay on flower buds of California buckeye *(Aesculus californica)*. The Acmon Blue *(Plebejus acmon)* can breed continuously where the perennial host plant white buckwheat *(Eriogonum nudum)* occurs, but elsewhere it has to use a series of seasonal annuals, some of which it shares with the Eastern Tailed Blue *(Everes comyntas)* and the Purplish Copper *(Lycaena helloides)*. A seasonal host succession requires constant dispersal; local "populations" last only a generation or two. In fact, there may be no real populations at all. (See "Population Dynamics and Metapopulations," below.)

The impacts of plant phenology on butterfly phenology are striking. Butterflies that feed on tannin-producing plants (which include most chaparral shrubs as well as oaks *[Quercus]* and some other trees) need to feed as larvae early in the season, when the foliage is most nutritious and palatable. Tannins are compounds that bind nitrogen, making it virtually indigestible. We tan leather to protect it against fungi and bacteria that might attack it. Tannins are also feeding deterrents; they are bitter. Native Americans, who often relied heavily on acorns as a dietary staple, treated them to leach out the tannins. Caterpillars do not have that option! We find that larvae of tannin-plant-feeding species are all active in spring. But the adults fly at various times of the year in response to other factors, arranging their dormancy to assure proper larval timing. (Butterflies such as the Mournful Duskywing manage to have multiple broods on oak by ovipositing only on tender young growth, which, in turn, reflects a reliable water supply.) The California Tortoiseshell *(Nymphalis californica)*, Hedgerow Hairstreak *(Satyrium saepium)*, and Pacuvius Duskywing *(Erynnis pacuvius)* all feed on wild-lilac foliage in spring, but they overwinter as an adult, an egg, and possibly a larva, respectively, in the same localities.

Dormancy and Diapause

All our butterflies undergo some kind of dormancy in the adverse season, but what constitutes the adverse season varies for different species. "Hibernation" refers to winter and "estivation" to summer dormancy. For spring-univoltine and spring-bivoltine species, dormancy means estivation followed by hibernation, with no active interval between! Such species may be dormant 9 months of the year. The technical term "diapause" refers to dormancy characterized by developmental arrest: in the case of early stages, growth and metamorphosis come to a halt, and in diapausing adults reproduction is delayed. Diapause is a genetically controlled syndrome but subject to environmental cues as discussed below. Diapausing individuals enter an alternative physiological state characterized by elevated resistance to environmental stress and negligible aging. True diapause is not a simple function of temperature. Once triggered, it requires specific conditions to be reversed—more on this in a moment. Inactivity caused directly by low temperature is referred to as torpor, or chill coma. All insects have a temperature at which growth grinds to a halt. This is known as a developmental zero. If temperatures remain below this level long enough, the insect will use up its reserves of nutrients and die. Of course, still lower temperatures can be directly lethal. The physiology of diapause is a way to counter such thermal threats, as well as others such as desiccation.

Any butterfly species, if it is capable of dormancy at all, has a specific stage in the life cycle in which this can occur; the stage varies from group to group. Very few species can undergo dormancy in more than one stage, except in the most benign climates. Every individual is, as it were, in a race to reach its dormancy-capable stage in time to meet the challenges of the season. And no two years are identical. What to do?

Some species—univoltines—always enter dormancy. If they have true diapause, we say it is obligate. For multivoltines, life is more complicated: every individual has the potential to develop directly or enter dormancy, and to make the correct "decision," it must in effect predict the weather weeks or months ahead. It does this in response to environmental cues, and generally there is a well-defined sensitive period when it is responsive to such cues.

What are the cues? Temperature seems obvious, but think about it. Short-term weather phenomena can produce a dramatic difference in temperature on the same date in different years. Consider the impact of offshore winds in the East Bay hills in fall—classic fire weather. Temperatures may reach into the nineties Fahrenheit in October under such conditions, while with the more usual light onshore winds they would be in the sixties and seventies. Yet October is still October, and winter is still coming in either case. In temperate midlatitudes in both the Northern and Southern Hemispheres, insects and plants rely primarily on day length (photoperiod) as a seasonal indicator or predictor. The length of the day is the same on the same calendar date year after year regardless of weather, except for the infinitesimal effect of the precession of the equinoxes. Species with obligate diapause show no sensitivity to photoperiod when reared under experimental regimes in growth chambers. In species with optional (facultative) diapause, the decision to diapause or not is determined by photoperiod during the sensitive period, and you can induce both conditions among the offspring of a single mating by rearing under different regimes, a so-called split-brood experiment. As we have already seen, the Pipevine Swallowtail (*Battus philenor*), which is indifferent to photoperiod, somehow splits all its broods in nature.

Of course, things are rarely so simple. Photoperiod is often, perhaps usually, modulated by temperature. Temperature alone cannot induce diapause, but it can alter the photoperiod threshold at which it is induced. That should not be surprising. For many multivoltine species, warm nights shift the diapause threshold a little later in the year, and cold nights do the reverse. Thus in years with warm falls multivoltine butterflies continue emerging and breeding later in the year, perhaps even getting in an extra generation. Applying game theory to this, we can see that the system juggles benefits (additional opportunities for reproduction) against risks (how likely is a hard freeze before one's offspring reach the diapause stage?).

This line of argument also points up a seeming paradox. Butterflies occur in the Arctic and in the subantarctic (e.g., Tierra del Fuego), where the best summer weather is comparable to our best winter weather in California. Butterflies breed there under those conditions, but ours shut down for the winter. Why don't they breed here in winter, since they almost certainly could? The Darwinian answer is that in high latitudes the only chance to breed is under harsh conditions—either you try or you don't exist there. But here, on average, you will leave more descendants if you postpone breeding until spring than if you try to breed in winter. So you don't breed in winter.

Diapause is induced hormonally and has multiple effects on the animal. Diapausing pupae may keep their spiracles (breathing pores) closed to conserve water, breathing only once or a few times a day. Winter diapausers typically secrete chemical "antifreeze" into their hemolymph, thus lowering the freezing point and hardening themselves against cold weather. This is less important in our mild climate than elsewhere. In fact, nondiapausing pupae of the Cabbage White *(Pieris rapae)* and the Anise Swallowtail *(Papilio zelicaon)* can and do overwinter successfully some years in the Bay Area.

Once in diapause, many species must accumulate a predetermined number of hours below some threshold temperature in order to break diapause and resume development. This is known as a chilling requirement. Photoperiod is rarely a factor in diapause termination. In our area most species have satisfied their chilling requirement in January and subsequent (postdiapause) development is at the mercy of the temperature.

In some cases factors other than photoperiod may control diapause induction. These include food quality, humidity (either directly or through the water content of the food), population density, and host-plant chemistry. When the Anise Swallowtail is reared in split-brood experiments on different hosts under identical regimes, it is more likely to enter diapause on poison hemlock *(Conium maculatum)* than on sweet fennel *(Foeniculum vulgare)*. It is also more likely to diapause if reared on the flowering tops of poison hemlock than on basal rosette leaves.

As noted before, the ability to respond to these environmental cues is genetically determined. The Anise Swallowtail has several ecological races or ecotypes adapted to the phenology of different hosts. When breeding on native hosts on serpentine in the Coast Ranges, it has obligate diapause. These plants, while perennial, are available for only a few weeks in spring. On the other hand, races breeding on both fennel and cultivated citrus *(Citrus)* (!) in the Sacramento Valley are multivoltine, with little tendency to diapause, and a weak diapause at that. Such races can be crossed, allowing us to study how diapause is inherited. But such experiments take a long time, for obvious reasons!

Adult reproductive diapause occurs in hibernators such as members of the genera *Vanessa, Nymphalis,* and *Polygonia;* at least some of these may mate in fall but store the sperm until spring, when they begin to ovulate. The Monarch *(Danaus plexippus)* is in reproductive diapause in its winter clusters along the coast. Reproductive diapause also occurs in estivators such as the second brood of the California Ringlet *(Coenonympha tullia cal-*

ifornia) and at least some populations of the Crown and Zerene Fritillaries *(Speyeria coronis* and *S. zerene).* In such species, reproductive diapause can be broken early by injecting the proper hormone.

In at least some photoperiod-sensitive species it has been shown that the critical cue is night length, not day length (scotophase rather than photophase). This is done by interrupting the phases with very brief pulses of light or dark. The Cabbage White, for example, reads a 14-hour scotophase as a long night, but if interrupted in the middle by 5 seconds of light it reads two short (7-hour) ones and does not diapause. In this species the sensitive period is the half-grown larva, but it is the pupa that actually diapauses.

Once adopted in a group, a given diapause strategy tends to persist across species boundaries and evolutionary time. You will notice in the species accounts that related species usually spend the adverse season in the same stage—hairstreaks and coppers as eggs, whites and orange-tips as pupae, and so on. Most sulphurs overwinter as larvae but the California Dogface *(Zerene eurydice)* apparently hibernates as an adult. The Pine White *(Neophasia menapia)* is the only pierid in our fauna that overwinters as an egg, but it is of tropical origin. Despite its seeming vulnerability, the larva is often the overwintering stage, especially in nymphalids. In the fritillaries *(Speyeria* and *Boloria)* the egg hatches quickly; the larva eats its eggshell and immediately diapauses. It may remain dormant for as long as 7 months! The Great Arctic *(Oeneis nevadensis),* is unique in our fauna in having an obligatory 2-year life cycle, diapausing twice as a larva. It is even odder in that most populations fly as adults only in even-numbered years. This phenomenon is widespread among alpine butterflies and moths, but the Great Artic is not alpine, and no one really understands the synchronization in the alpine ones either.

The univoltine serpentine race of the Anise Swallowtail, already mentioned, faces a very difficult seasonal challenge. If the adults emerge and breed early, the larvae are guaranteed their food supply, but there is a high risk of severe weather, including snow, which can wipe out all the reproduction at a site in a given year. If they emerge and breed later the probability of catastrophic weather is low, but the plants are prone to dry up and die before the larvae are full-fed. Perhaps this dilemma explains why a high proportion of the pupae diapause for more than 1 year—

up to at least 5 years. This can be interpreted as a way of spreading the risk, akin to the "banking" of dormant seed in the soil by plants. We do not know how the system is controlled, either genetically or physiologically. The same phenomenon occurs in a related swallowtail, the Desert Black Swallowtail *(Papilio coloro),* in the deserts of southeastern California. Serpentine populations of Spring White *(Pontia sisymbrii)* and Small Marble *(Euchloe hyantis)* also display multiyear pupal diapause. The diapausing larvae of both the Variable Checkerspot and Edith's Checkerspot *(Euphydryas chalcedona* and *E. editha)* can also lay over at least one extra year in drought years. Some insects, including yucca moths (Prodoxidae), leave our butterflies in the dust by having indeterminate diapause that may last 20 years or more!

Natural Enemies

Populations of butterflies are held in check by a variety of mortality factors. Some of these, such as severe weather, are essentially unpredictable and will kill the same proportion of the population whether it is dense or sparse. We call these density-independent factors. Others, however, vary in their effectiveness depending on whether few or many animals are around. An easy one to visualize is infectious disease: the more potential victims present, the more efficiently it can be transmitted. Such factors are called density dependent and include not only disease but other biological factors such as parasitism and predation. The most effective control occurs when the proportional impact of the factor increases as the population density increases, producing negative feedback on numbers. This is a basic principle of population biology.

All stages of the life cycle are vulnerable to mortality factors. Among the biological factors—"natural enemies"—some are generalized while others are highly specific. Some attack only particular stages in the life cycle; others are far less picky.

Predators

When thinking about predators of butterflies, most people name birds first. Have you ever seen a bird eat a butterfly in the wild? Few people have; it seems to be an uncommon event. Most bird predation on adult butterflies probably occurs early in the morning or on cool, cloudy days, when the butterflies are at rest and too cold to escape. Birds rarely pursue healthy butterflies flying normally, but they will readily go after an injured one that is flying clumsily. During mass migrations of the California Tortoiseshell *(Nymphalis californica)* in the mountains, Steller's Jays *(Cyanocitta stelleri)*—which rarely bother butterflies—are sometimes hit and killed by cars while trying to harvest injured torties flopping in the roadway. It has been suggested that butterflies in general are difficult to catch on the wing and on average may not repay the energetic cost of chasing them. Of course, a significant percentage of our fauna is protected by chemical defenses against predators (see "Butterflies and Plants," below). Most bird predation is on immature butterflies.

Other predators of adult butterflies include dragonflies and damselflies (Odonata), ambush bugs (Phymatidae, Hemiptera), spiders (especially crab spiders [Thomisidae]), mantids (Mantodea), toads, lizards, and small mammals. Dragonflies and damselflies take mostly small butterflies such as blues, hairstreaks, hesperiids, and crescents, but I have seen the Green Darner *(Anax junius)* eat an Anise Swallowtail *(Papilio zelicaon)*, and once saw one catch a California Tortoiseshell, eat the abdomen, and drop the rest, which flew away. Robber flies (Asilidae, Diptera) take significant numbers too. I once saw one eat a migrating adult Painted Lady *(Vanessa cardui)*. Ambush (sit-and-wait) predators such as crab spiders, ambush bugs, and mantids can be very effective. Small butterflies killed by crab spiders can often be found hanging limp from flowers. They do not dry out for days due to the spider venom and can be harvested as specimens! Thomas Eisner at Cornell, using a special apparatus to see in the UV spectrum, demonstrated that our common crab spiders match their backgrounds in the visible spectrum but not in the UV. Since butterflies can see in the UV spectrum, the implication is that the spiders are hidden not from their prey but from their own (vertebrate) predators! Wasps occasionally attack adult butterflies, almost always in the just-emerged (teneral) state. They will bite off the abdomen and carry it away, leaving the butterfly to bleed to death. Mammals—mice and shrews—eat butterflies at rest on or near the ground at night. In our motorized age, the automobile may be a very significant predator of adults.

Evidence of bird predation exists in the form of V-shaped

beak marks on the wings of butterflies—either indentations or actual missing wedges of wing membrane. Every beak-marked butterfly bears testimony to an unsuccessful bird attack—but why was it unsuccessful? Either the bird dropped the butterfly voluntarily (perhaps it tasted bad), or it fumbled it (remember, birds have no hands!), or the insect broke free. Breaking free usually causes real wing damage. Careful examination may allow you to discriminate among these situations. Even though predation is seldom observed, up to 10 percent of field-collected specimens may be beak-marked, especially at seasons when alternative prey are rare.

Several lycaenids, especially the Common Hairstreak *(Strymon melinus)*, have apparent "false heads" at the anal angle of the hindwing, consisting of a brightly colored (usually red) eyespot with a black "pupil," combined with a short hairlike tail. Many lycaenids rotate the hind wings when at rest. This causes the tails to move up and down, resembling antennae. The "false head" apparently draws a predator's attention away from the real head (and body). The result is that the predator gets a scrap of wing, and the would-be prey gets away. Up to 20 percent of field-collected specimens have the anal angle missing, usually on both hind wings, indicating this ruse actually works. Up to 10 percent of Buckeyes *(Junonia coenia)* show evidence of predator strikes at the large hindwing eyespots, which seem to function in the same way.

Butterfly eggs are eaten by ants, predatory bugs (Hemiptera), and ladybird beetles ("ladybugs" [Coccinelidae, Coleoptera]). The introduced Asian Ladybird *(Harmonia axyridis)* may be a significant mortality factor for butterfly eggs in some areas. Most density-dependent egg mortality is probably due to parasitoids. Birds actually eat the egg masses of butterflies that are batch layers, if not chemically defended, but they rarely if ever bother with singletons.

Butterfly larvae are eaten by many things. Birds and wasps are particularly important. Yellowjacket wasps (Vespidae, Hymenoptera) are very efficient caterpillar-eating machines. Small larvae are eaten by ants, lacewing larvae (Chrysopidae, Neuroptera), ladybirds and ladybird larvae, and various true bugs (Hemiptera). Larger larvae are eaten by carabid beetles (Carabidae, Coleoptera), hunting spiders (non-web-weaving), and stinkbugs (Pentatomidae, Hemiptera). Little is known of predation on pupae, which are usually well concealed and difficult to study in the field. Pupae on and near the ground are probably eaten often by mice and shrews.

Parasitoids

Parasitoids, commonly called simply "parasites," are probably the most frequent density-dependent controls on butterfly numbers. Parasitoids are insects with free-living, nonparasitic adults (generally wasps or flies) that seek out and oviposit on, in, or near living host insects on which their larvae will feed from within. The most important parasitoids attacking butterflies belong to the dipteran family Tachinidae and the hymenopteran families Chalcididae, Braconidae, Ichneumonidae, and Pteromalidae. There are extremely tiny hymenopterans (Trichogrammatidae) that are important parasites of butterfly eggs. Most parasitoids attack larva or pupae. They vary greatly in their degree of specificity for host species and life-history stage; some attack caterpillars of many families (including moths) while others may be species specific. Many restrict the universe they need to search for hosts by keying in on the chemistry of their host's host plant; some parasitoids of pierids, for example, are attracted to mustard oils. Some are adept at detecting chemical cues produced by damaged foliage, or released from the host's feces. Many, often most, larvae collected in the wild have been found by parasitoids before being found by us. When rearing, it is wise to segregate wild material to avoid introducing parasitoids into the culture, and larvae and pupae reared from eggs should be properly screened to keep parasitoids out. Because relatively little is known about butterfly parasitoids, any that you rear should be saved along with host material and given to a specialist for identification if at all possible. Some may be of potential economic importance, or even new to science.

Parasitoids may inject their eggs into the host (most hymenopterans), lay them on the host (many tachinids), or even broadcast them on the host's host plant. In the latter case the parasitoid egg may be ingested with leaf material and hatch inside the host, or the parasitoid larva may "ambush" a passing host and bore into it. Each species is quite consistent in these matters. The parasitoid larva eats nonessential parts of the host first, carefully avoiding killing it until it itself is full-fed and ready to pupate.

The parasitized host may appear perfectly normal right up until the end, or it may become listless and stop feeding. The braconid wasp *Cotesia glomerata* is a major mortality factor for the Cabbage White *(Pieris rapae)* and occasionally attacks other pierids. It lays up to 40 or 50 eggs in a small host larva, which then develops normally. When ready to pupate, it spins the usual pad of silk and a button to anchor the tail end. It would then normally spin a silken girdle around its midsection—but at that exact moment all the parasitoid larvae bore out through its skin at once and proceed to spin yellow cocoons on its back. The caterpillar soon dies and shrivels, concealed by the mass of parasitoid cocoons. This precision timing has to be regulated by the hormonal condition of the host. Several common icheumonids as well as solitary species of braconids (one per host) also pupate on the dead or dying host's back. *Cotesia lunata*, a very common parasitoid of the Anise Swallowtail, lays one egg in a first-instar larva and kills it in the second instar. Some parasitoids lay only one egg per host, but it subsequently undergoes repeated cycles of twinning, giving rise to many embryos. This is known as polyembrony. In a polyembryonic species all the offspring produced from one host are normally of the same sex, while in species that lay many eggs, both sexes may be produced.

Some parasitoids oviposit in the host larva but wait until it pupates to eat and kill it. *Pteromalus puparum* is a beautiful little wasp that attacks many lepidopterans but seems to prefer pierids. It oviposits in the prepupa or the newly formed, unhardened pupa right after the molt. The tachinid fly *Phryxe vulgaris* attacks pierids also, ovipositing in the larva. When the host pupates, its hormonal condition dictates the behavior of the maggot. If the pupa is not in diapause, it is eaten immediately. But in a diapausing host the maggot remains quiescent—probably itself in diapause—until development resumes. It then quickly eats and kills the host. The adult fly is thus synchronized with the next brood of butterflies! *Cotesia* larvae do not read their host's diapause condition in deciding whether to diapause as pupae after emerging from the caterpillar. Studies in Russia suggest they read day length directly, through the host caterpillar's skin!

Oviposition on the same host by more than one parasitoid could easily result in too many parasitoid larvae for one host to sustain, a condition called superparasitism. This may result in all the parasitoids dying, or metamorphosing prematurely. We should not be surprised that many parasitoids mark their hosts chemically as an advertisement to later comers that "this one is taken."

Butterfly larvae are not defenseless against parasitoids. Aside from thrashing behavior that may deter them from laying eggs, many are able to surround the eggs with blood cells and prevent their hatching, a process called encapsulation. Parasitoids have parasitoids of their own, which are called hyperparasitoids. They may be highly specific and may greatly reduce the ability of the primary parasitoid to control the numbers of its host. Some parasitoids are highly seasonal, and some change hosts opportunistically from generation to generation, much like multivoltine butterflies. Some parasitoids, especially the hymenopterans, are very long-lived as adults (they feed on nectar and pollen; we have kept some of them alive for 4 to 6 months in the lab) and thus may overlap several generations of hosts. The diversity of tachinid parasitoids is highest in our area in fall, when many adults feed at flowers of coyote brush *(Baccharis pilularis)*.

Diseases

Caterpillars are vulnerable to devastating viral diseases. Almost everyone who rears will encounter them. There are two major groups of inclusion-body diseases, granulosis and nuclear polyhedrosis, each with numerous host-specific strains. In these diseases the virions (virus particles) are released into the environment embedded in protein crystals (the inclusion bodies), which are almost indestructible in natural environments. If ingested by a susceptible caterpillar, they dissolve in the gut, liberating the infective virions. Granulosis is a pierid disease and is relatively uncommon in our area. Polyhedroses are common in nymphalids (especially the Buckeye) and *Colias* species. All these diseases cause the larva to liquefy into rancid goo laden with millions of inclusion bodies. Polyhedrosed larvae typically climb to the top of the container, die while clinging with their abdominal prolegs, and then drip from the head end. Granulosed larvae simply turn to mush wherever they are. Rearing equipment contaminated with the goo should either be discarded or sterilized with a strong bleach solution. The infection can be spread on the feet of flies and other insects visiting the cadavers. Viral disease in caterpillars is strongly density dependent and is rarely seen outdoors except during population outbreaks. It is, however, common in crowded and unsanitary laboratory culture. The viruses may be carried asymptomatically and transmitted from generation to generation: it is not certain whether infection can occur inside the egg, but it is known that virions can be ingested when the larva eats its way out of the chorion, so some labs carefully sterilize the eggshells. An apparently "clean" line carrying inapparent infection can develop a severe outbreak under crowded, dirty, humid conditions or if reared on poor-quality food. These same conditions predispose to bacterial and fungal disease, much of it nonspecific. In the wild, entomopathic fungi may be rather common in caterpillars, but they are rarely seen in the lab. The Monarch *(Danaus plexippus)* is susceptible to a protozoan parasite that may be a significant factor in its notoriously episodic population dynamics. *Note:* No butterfly disease is known to infect humans—and conversely.

It seems likely—we don't really know—that a lot of the heavy mortality of early stages during overwintering, which results in a seasonal bottleneck for multivoltine species, is due to nonspecific bacterial and fungal disease thriving in the moist environment near the soil surface with temperatures mostly well above freezing.

Population Dynamics and Metapopulations

Butterflies lend themselves very well to research at the population level, as was recognized by Paul Ehrlich and his associates at Stanford University back in the 1960s. Many hundreds of such papers have been published in scientific journals. However, only a handful of species can be considered well studied, and then at only a few places. We are far from understanding what controls the numbers and distribution of most butterflies, and there is lots to be done.

To study population dynamics, ecologists often construct life tables, which are concise quantitative statements of the vital statistics of a population. The technique was pioneered for human populations in the insurance industry. Life tables may record age-specific fecundity or mortality or (preferably) both.

As might be imagined, getting the requisite data for the whole life cycle of a holometabolous insect such as a butterfly is a lot of hard work. Adults are traditionally studied using mark-release-recapture methods. Individuals are captured, marked in a distinctive way to allow for recognition, and re-released. We usually use ordinary marking pens for the purpose. Sometimes the marks can be sighted (as with binoculars) so the individual need not be recaptured to be recorded. The population is resampled regularly, and its numbers are estimated from the ratio of new captures to recaptures—assuming the marked individuals assimilate randomly into the population and that their probability of recapture is unaffected by having been marked. These are tricky assumptions. If the marked individuals are injured or traumatized, their probability of recapture or resighting goes down (because they die, they leave the area, or their behavior is changed). Not seeing a given individual again is not, of course, evidence that it is dead. We rarely find dead butterflies—all sorts of scavengers, especially ants, see to that. We are measuring not mortality but "residence time," how long an individual remains available for recapture. Still, we know that most common small butterflies, even the Western Pygmy Blue *(Brephidium exile)*, have the physiological capacity to live about a month, but hardly any do. Life is full of perils, and most individuals disappear within a few days, never to be seen again.

Survival of the early stages can be studied by setting out groups of eggs or neonate larvae and tracking them regularly until the last one has metamorphosed or disappeared. This is called a cohort study. Such studies may have little relation to reality, even if conducted outdoors, if the local distribution or density of the animals is dissimilar to what would normally occur. Apparent mortality is usually very heavy, but usually we cannot tell what is causing it. For monitoring levels of parasitization, we can set out a cohort of the susceptible stage for a few days, then bring them in and rear to see how many were attacked during the experiment.

Butterfly populations may fluctuate greatly in absolute numbers or density (number of individuals per unit area), or both. For long-term monitoring a method familiarly known as a "Pollard walk" was developed in England and is widely employed. Analysis and interpretation of population data constitute complicated and highly technical subjects, and new methods are con-

WHAT IS A BUTTERFLY FAUNA?

A fauna is a list of animal species recorded from a defined area. Comparisons among faunas help us to understand what controls the species richness (biodiversity) of an area, a question of immense importance in conservation.

The simplest thing to do with faunas is to compare their species richness. An initial assessment reveals that some places seem blessed with exceptionally rich faunas, while others seem impoverished. Can we say anything more rigorous about this? Yes, but such exercises are fraught with complications.

One potential complication is area. If we say San Francisco County (117 sq km [45 sq mi]) has fewer species than San Bernardino County (59,909 sq km [20,131 sq mi]), how much of the difference can we attribute to area alone? San Francisco might have more butterfly species than we would predict based on area alone, but how could we tell? If we have enough comparable faunas, we can spot "outliers" easily and ask why they deviate from expectations. But there is far more habitat (and climate) diversity per unit area in some places than in others. The relationship is not simple and straightforward.

The quality of coverage is another major issue. Conservation biologists are often called upon to do rapid biodiversity evaluations. How reliable are faunas generated on a "quick-and-dirty" basis compared to those established through long-term monitoring? This is a problem of diminishing returns. The first time one visits a site, every species encountered is "new." The percentage of "new" species should diminish with each subsequent visit. Eventually one can assume the fauna is well documented. But can one, really? Each species has its own seasonality; if one stops looking when the number of "new" records drops off, one may miss a whole set of species that only appear later in the year. Also, we routinely find that adding a new habitat (say, a serpentine barren) causes a significant blip in the species-over-time curve.

One more problem: What about "strays"? In 35 years in Yolo County, I have seen one each of the Variegated Fritillary *(Euptoieta claudia)* and the Mexican Yellow *(Eurema mexicana)*. Both are accidental visitors from the desert. It would be absurd to consider them members of the Yolo County fauna. More frequently, we see the Cloudless Sulphur *(Phoebis sennae)* and the Dainty Sulphur *(Nathalis iole)*, also nonbreeding visitors; more frequently still, the Marine Blue *(Leptotes marina)*, which sometimes breeds here but does not persist. How do we decide who is "in" and who is "out"?

So—given all these difficulties—should you try to generate local butterfly faunas? Yes! Well-documented faunas are absolutely necessary for intelligent conservation action. Theory is wonderful; occasionally it is even useful. But it can never substitute for real data.

stantly being developed. All of this is beyond the scope of this book but can be found in recent texts on ecological and population-biology techniques. Because of increasing concern about the status and management of endangered and threatened species, a set of techniques called population viability analysis has been developed, using past fluctuations in numbers to try to estimate the degree of risk facing a given population. These techniques are often applied to butterflies.

The factors causing the most mortality in a population are often not the factors driving population dynamics. This seems counterintuitive or even paradoxical but is in fact true. Using a technique called key-factor analysis, for example, it was found that failure of females to lay most of their eggs (say, due to bad weather during the spring flight period) may be much more important than more obvious factors such as predation or parasitism. In fact, we might not recognize it as a form of mortality at all. Various studies suggest that such density-independent factors may suppress butterfly populations to a level where their natural enemies can then hold them in check. Alternatively, such factors may free them from density-dependent control to a degree that it will take many generations for the natural enemies to catch up. Most long-time collectors believe they have observed qualitative correspondences between weather and local butterfly dynamics. Have you? Sophisticated statistical analyses often support such observations, if good quantitative data are available.

A few butterfly species are reputed to have population cycles. A cycle requires that numbers fluctuate in a more or less predictable way, with a relatively constant return interval for the peaks (a particular number of years or generations). Identifying cycles rigorously, again, requires statistical analysis of good quantitative data; often we have only broad qualitative observations. The Painted Lady *(Vanessa cardui)* spends the winter in northern Mexico and in the desert Southwest, breeds there in late winter, and migrates into our area—and all of temperate North America—sometime between January and April. Statistically, big Painted Lady years in our area are associated with years of heavy (often early) winter rains in the desert, and thus with El Niño. But there does not seem to be any regular periodicity, and the great Painted Lady year of 2005 was very wet in the desert without benefit of El Niño. The California Tortoiseshell *(Nymphalis californica)* was thought to be cyclical, but in recent years the pattern has broken down. Its movements are seasonal and predictable in general terms, but the areas where it breeds in summer vary greatly. The factors governing its population dynamics are not understood. The Checkered White *(Pontia protodice)* had massive outbreaks in our area in the 1970s but not since. And so it goes.

Most butterfly populations are very transitory. In the case of weedy, multivoltine species it may be uncommon to see even two successive generations in the same place. This is because resources, such as host plants and nectar sources, may come and go seasonally. Butterflies with long-lived perennial host plants are, by and large, much more likely to have stable, long-term populations than those feeding on annuals. In the case of the Painted Lady and California Tortoiseshell and some other species, there are no local populations—only one big gene pool whose geographic range is constantly changing. The Buckeye *(Junonia coenia)*, Fiery Skipper *(Hylephila phyleus)*, and other species that die out over most of their ranges in the winter and then recolonize from local foci of survival cannot really be said to have discrete populations.

Ecologists have coined the term "metapopulation" ("meta-" means "beyond") for these ever-shifting sets of transient local "populations" that are interconnected by individual transfers through space and time. Think of an amoeba on a microscope slide, constantly changing shape as it pushes out projections (pseudopodia) on one side and withdraws them on another! To a degree this model applies to all butterflies, but for some the turnover rate is so slow that we can treat local populations as effectively freestanding. Populations stranded by climatic change in isolated favorable localities (relicts)—for example, the Arctic Skipper *(Carterocephalus palaemon)* and the Blue Copper *(Lycaena heteronea)* in the North Bay—were once part of a dynamic metapopulation but have been cut off from other local populations of their species for perhaps thousands of years. Contrast this with the Marine Blue *(Leptotes marina)*, which may colonize widely in our area some years but then disappears completely over winter.

A special case of metapopulation dynamics involves sources and sinks. Source populations are persistent and relatively long lasting. They may occupy only a small fraction of the species' total range. The remainder is occupied by sink populations, which are incapable of long-term persistence without occasional renewal by immigrants from source populations. A classic example is our Western Pygmy Blue. Its source populations are associated with evergreen, perennial host plants (semisucculent members of the goosefoot family [Chenopodiaceae], especially sea-blites *[Suaeda]*) that grow in alkaline soils that do not flood in winter—a very special habitat. Populations on flood-prone hosts are wiped out over winter. This butterfly can be very abundant in salt or brackish marshes but often does not persist there; in the Suisun Marsh, for example, its usual host is sea purslane *(Sesuvium)*, which may either flood or freeze, depending on the year. Either way, it becomes unavailable in winter, and the butterfly dies out! Most populations of this tiny butterfly breed on Russian thistle *(Salsola)*, also called tumbleweed, along roadsides and railroad tracks. When the plants die in late fall, these populations—however dense—go extinct; they are true sinks and need to be reestablished every year. This species probably overwinters in less than 10 percent of its range in our area and must recolonize the remainder each year, relying on its high dispersal capability, high reproductive rate, and short generation time.

Most such species in our area typically go through a bottleneck in winter. Each generation is larger and occupies a higher percentage of the range than the last, peaking in fall when the range is most fully occupied. At the end of the season females disperse very widely, scattering eggs over the landscape in what amounts to a risk-spreading strategy in which by pure chance survival will be high in a few localities. These then serve as foci for recolonization the next year.

Butterflies and Climate Change

Monitoring studies in the past 30 years or so suggest that the butterfly fauna covered by this book is in decline. Especially since 1999, even once-common and once-widespread species (Large Marble *[Euchloe ausonides]*, Purplish Copper *[Lycaena helloides]*, Mourning Cloak *[Nymphalis antiopa]*, Lorquin's Admiral *[Limenitis lorquini]*, Sylvan Hairstreak *[Satyrium sylvinum]*, etc.) have suffered dramatic declines. Even a weedy species such as the Common Sootywing *(Pholisora catullus)* has lost a significant part of its former range. Are these declines related to the much-discussed pattern of global climate change? Answering that question requires high-powered statistical methods. In particular, one must try to disentangle effects of climate change from those of changing land-use patterns. And as usual our ability to identify environmental correlations is limited in part by our own imaginations; if we cannot identify potential causes we cannot try to quantify their importance!

As we saw under "Where Are Our Butterflies From?," geographic ranges have changed in response to changing climate in the past and can be expected to do so in the future. In general, as climates warm, organisms in the short term would be expected to track their preferred climates by shifting northward, upslope, or (in the case of the California fog belt) coastwise. Because butterflies are limited by the distributions of their host plants, and plants can disperse in general only much more slowly than flying insects, such shifts could take much longer than one might imagine. (Paleobiogeographers often use certain beetles to track climate change. Not only do they fossilize well, but generalized predators and scavengers can relocate without waiting for changes in the slow-moving vegetation.) The ability of butterflies to track climates could also be hampered by patterns of land use, creating barriers to free dispersal that did not exist during previous episodes of climate change. Our ability to predict future ranges is further limited by the potential ability of butterflies to adapt to novel conditions, even thriving where one would never have expected them to. The ability of the Umber Skipper *(Poanes melane)* to thrive on Bermuda grass *(Cynodon dactylon)* in lawns in the East Bay is a good example; no one would have predicted it from its biology elsewhere. In addition to shifting to cooler sites, we might expect species to die out in sites that warm beyond their ranges of tolerance. Some of the riparian species in the Central Valley might end up restricted to coastal valleys and the fog belt.

One thing we can say with confidence—because the statistics have been done—is that some species in our area are emerging significantly earlier in spring now than they did 30 years ago. In two cases (the Field Skipper *[Atalopedes campestris]* and the Red Admiral *[Vanessa atalanta]*) they are emerging nearly a month earlier! For multivoltine species this can be either a good thing or a bad thing. In general, increasing the number of generations per year is good if it increases the number of progeny able to overwinter successfully. But if survival in the added generation is poor (resources become scarce or decline in quality, or there isn't time for most individuals to reach the stage in which they can overwinter successfully), the results can be catastrophic. Most species are conservative in that they shut down earlier in the season than they "need" to in most years, suggesting that there is strong natural selection exerted in the occasional very bad year. So far there is little if any evidence to suggest that multivoltine species in our area have benefited from longer warm seasons, but there are hints that some have been hurt. At least one species, the Anise Swallowtail *(Papilio zelicaon)*, seems to be entering dormancy earlier now than it used to! We can model phenology and generate predictions about these things, given various assumptions. It looks as if we will get to test our models against a rapidly changing reality.

In 1992 unusual atmospheric-circulation patterns allowed butterflies from the desert Southwest (the Cloudless Sulphur *[Phoebis sennae]* and the Dainty Sulphur *[Nathalis iole]*) to enter our area in significant numbers, although, of course, they could not breed—there are no host plants here. The Sierra Nevada had been considered a major barrier to such species. The future promises to be full of surprises.

Butterfly Behavior

Adult Feeding and Pollination

Adult butterflies feed mostly on flower nectar. Some, especially nymphalids and especially in the tropics, visit sap or gum oozing from trees, damaged or rotting fruit, honeydew deposited by aphids or scale insects, and much less elegant materials such as dung, urine, dead animals, decaying mushrooms, and the like. One of our species, the Golden Hairstreak *(Habrodais grunus)*, is not known to feed at all.

Some flowers are specialized for specific pollinators. Hummingbird flowers, for example, are typically tubular and red in color. Bee flowers tend to be white or whitish with shallow tubes and strong fragrances. There is no specific "butterfly pollination syndrome," and butterflies—at least in our area—are rarely if ever critically important as pollinators. Butterflies with short proboscides (most lycaenids) visit flowers with short corolla tubes, while flowers with deep tubes can accommodate only longer-tongued, generally larger species. To visit a given flower, a butterfly must have a place to put its feet. Small butterflies, especially hesperiids, may crawl inside deep tubular flowers, but many cannot. Vinegar weed (*Trichostema lanceolatum,* Lamiaceae) is a native late-summer annual with an elaborate balancing mechanism that affords access only to certain bees of the proper shape and weight. It is rarely visited by butterflies, the vast majority of which cannot reach the nectar, but the Woodland Skipper *(Ochlodes sylvanoides)* does so and visits it routinely while other, similar-sized hesperiids do not. This skipper also feeds successfully on the flimsy flowers of *Epilobium brachycarpum* (Onagraceae). In both cases it must hang upside-down to feed. (The Cabbage White *[Pieris rapae]* is the only other butterfly that routinely feeds successfully on *E. brachycarpum.*)

Flowers that do not produce nectar, no matter how showy, do not attract butterflies (except the occasional naïve individual, often on its maiden flight): poppies, buttercups, and many members of the rose family (Rosaceae) are examples. Among native plants, California buckeye *(Aesculus californica)* is an extremely important nectar source in late spring, attracting virtually all species that are flying then. In fact, given that most species have some flexibility in adult emergence times and can adjust them in response to resource availability, the large number of univoltine species that emerge while buckeye is in flower (hairstreaks, checkerspots, Farmer *[Ochlodes agricola]*, and others) may be no accident. Other particularly good native butterfly flowers include many wild buckwheats *(Eriogonum)*, especially sulphur flower *(E. umbellatum)*; rabbitbrush *(Chrysothamnus)* and similar yellow-flowered sunflower family (Asteraceae) shrubs (e.g., goldenbush *[Ericameria]*); asters *(Aster)*; goldenrods *(Solidago)* and grass-leaved goldenrods *(Euthamia)*; coyote brush *(Baccharis pilularis)* and other species of *Baccharis*; lippia *(Lippia)*; giant hyssops *(Agastache)*, lemonade-bush *(Rhus trilobata)*, buttonbush *(Cephalanthus)*, and all species of milkweeds *(Asclepias)* and dogbanes *(Apocynum)*. The latter two have an unusual pollination mechanism. Their flowers are built in such a way that a visiting pollinator must place its feet where they can slip easily into a slot. There a pair of pollen sacs (pollinia), resembling saddlebags, are clamped onto them. These slip off when the insect visits the next flower and inserts its feet. You can often find these sacs adhering to the feet of hairstreaks and other small butterflies. You may also find the occasional specimen hanging dead, having failed to pull its foot free from the trap. The large lilies *(Lilium)* are very attractive to swallowtails, which can often be found plastered with their extremely sticky orange pollen—especially striking on a male Pale Swallowtail *(Papilio eurymedon)!*

Butterflies have been shown to be capable of short-term learning. They will often revisit productive nectar sources again and again over a period of days and may concentrate for a while on a single floral species. These behaviors are especially well developed in long-lived species.

Flowers of papilionaceous ("butterfly-shaped," referring to the arrangement of the petals, not adaptation to pollinators) legumes, except the small clovers *(Trifolium)*, are seldom visited by butterflies. Two exceptions are redbud *(Cercis)*, which may be heavily visited by early-spring lycaenids when little else is in bloom, and some species of the genus *Lotus*, including the common naturalized alien birdfoot trefoil *(L. corniculatus)*. Some of the annual vetches *(Vicia)*, however, are heavily visited by many species. Lycaenids that feed on species of the large genus *Eriogonum* routinely fly when their hosts bloom and use them as nectar sources—sometimes exclusively.

The Great Copper *(Lycaena xanthoides)*, whose larvae eat dock *(Rumex*, Polygonaceae, a wind-pollinated plant), is unusually picky as to nectar sources, and its absence from many seemingly suitable sites seems to depend on the lack of such plants. The preferred plants are dogbanes, gumweed *(Grindelia camporum*—not all species will do!), horehound *(Marrubium vulgare)*, and perennial peppergrass or tall white-top *(Lepidium latifolium)*. Except for gumweed, these are all preferred nectar plants for various hairstreaks too. Woolly sunflower *(Eriophyllum lanatum* "complex") is a bit of a mystery. In some places it is very attractive to the Gorgon Copper *(Lycaena gorgon)* and Nelson's Hairstreak *(Mitoura nelsoni)*, among other species, while in other places it attracts no butterflies at all and is pollinated primarily by flat-headed borers (Buprestidae, Coleoptera).

Among naturalized (nonnative) plants in our area, the following are especially good butterfly flowers: heliotropes *(Heliotropium)*, mints *(Mentha)*, tall blue verbena *(Verbena bonariense)*, various thistles *(Carduus, Cirsium*, and *Silybum)*, star-thistles *(Centaurea)*, and last but not least, dandelion *(Taraxacum officinale)*.

Butterflies often behave as nectar "thieves"—they take nectar but fail to pollinate. Hesperiids often insert the proboscis from the side of the flower in a way that does not bring the wings or body into contact with the anthers. The Tailed Copper *(Lycaena arota)* has been observed exploiting the behavior of carpenter bees *(Xylocopa,* Hymenoptera). These large bees often bite a hole in the base of a tubular flower and take nectar through it, never contacting the anthers. The Tailed Copper may then insert its proboscis after the bee leaves, taking nectar from flowers that are normally not visited by butterflies.

Puddling and Roosting

Adults of many butterfly species aggregate at mud puddles. Though several species may be present, they typically form single-species clusters; even the very similar-looking blues usually assort in this way. Moreover, on examination one finds all the animals are young males. What is this all about?

Among species in our fauna that puddle regularly are the Cabbage White, Orange Sulphur *(Colias eurytheme),* many blues including the Boisduval's and Silvery Blues *(Plebejus icarioides* and *Glaucopsyche lygdamus)* and the Echo Blue *(Celastrina ladon echo),* all of our swallowtails, the Common Checkered Skipper *(Pyrgus communis),* all of our duskywings (especially the Propertius Duskywing *[E. propertius]),* the Woodland Skipper *(Ochlodes sylvanoides),* and the Colorado Skipper *(Hesperia colorado),* and plenty more. It is thought that they are collecting dissolved mineral salts that seem to be important for male reproductive success, but the phenomenon is still not well understood. It is easy to attract puddling species by setting out decoys, even dead specimens or good photographic reproductions. "Mud puddle clubs" offer outstanding photographic opportunities and may often be approached very closely, even to the point of removing individual specimens with forceps without causing general alarm.

Many butterflies roost gregariously for the night, once again typically in single-species clusters (and once again attracted to decoys). These roosts typically occur at the top of herbaceous vegetation between knee- and waist-high. They are often placed facing east to catch the first rays of the rising sun; at first light the
text continues on page 48

A LITTLE ABOUT HOW BUTTERFLIES FLY

Biomechanics is the application of physical principles to biological structures and processes. Studies of animal locomotion, including butterfly flight, are carried out for their own sake, but also because they regularly yield insights applicable in biomedical or aerospace engineering. We know a great deal about how butterflies fly and how this is related to their structure, wing shape, behavior, and physical environment. The subject is highly technical, but a few fairly simple generalizations can be made.

There are two sets of opposing flight muscles in the thorax. The thorax acts like a relatively rigid box, such that forces applied in one area must be compensated in another. When the dorsal-ventral muscles contract, the sides of the box bulge outward. Due to the way the wings articulate with the thorax, this causes the wings to go up — the upstroke. Then these muscles relax, and the longitudinal muscles contract. The sides of the box move inward, the top of the thorax bulges up, and the wings go down — the downstroke. In addition to these muscles, which are called indirect flight muscles, there are direct flight muscles that control the precise orientation of the wings in flight. The complete sequence of wing movements in a beat cycle is rather similar to the motions of a swimmer's arms during the breaststroke.

Flight is all about lift. The combined wings act as airfoils (note that the forewing and hindwing overlap sufficiently to function as a unit). The front margin, or costa, of the forewing is thickened, giving a slight curvature in flight — like an airfoil. That curvature creates a pressure differential between air flowing over the wing and beneath it; the air above the wing is forced to travel farther — and faster. This differential generates wing-tip vortices, or whirlpools of air, which further accelerate the air above the wings and enhance the lift. The morphology of the wing clearly enhances this effect. Tails on the hindwings typically have a half-twist in life (eliminated in spread specimens, of course), which helps to "catch" the vortex "braids" trailing off the wings and reduce performance-reducing drag.

During the wingbeat cycle, various kinds of vortices known in the trade as "starting" and "stopping" vortices are generated and shed into the wing's wake, which under certain circumstances permits a process called "wake capture" to extract energy from them and use it to enhance lift. This is particularly important at takeoff and for complex in-flight maneuvers. Studies carried out with wind tunnels and high-speed videography have permitted extraordinary insight into these mechanisms, which are difficult to explain in words (and best modeled mathematically).

The most energy-efficient mode of flight is gliding, but it is practical only in a nearly wind-free environment. Most of the butterflies that rely heavily on gliding are found in forests, well beneath the canopy. The various long-winged butterflies of the tropics, such as the passionflower butterflies (Heliconiinae) and the family Ithomiidae, are essentially gliders, with low "wing loading" (weight to wing-area ratio) and high "aspect ratio" (span to chord ratio, where "chord" refers to the distance between the leading and trailing edges of the airfoil). Wind exerts drag on any flying object, and the demands of controlled flight in windy environments produce very different wing proportions. This can be seen even in comparisons of the wing shapes of seasonal forms of the same species, especially pierids. A very triangular forewing with relatively low aspect ratio is a formula for very powerful flight and is characteristic of territorial males with outstanding aerobatic capabilities; paradoxically, this kind of flight also depends on there being little wind resistance.

The surface characteristics of the wings also contribute to flight abilities. To minimize drag, ideally there should be no turbulence over the wing surface (pure laminar flow). The tightly overlapping scale rows create a smooth surface that approximates this condition. Damage to the scales not only can create a weight imbalance between wings, but also potentially serious problems with drag. The lower surface of the wings is characteristically less smooth, creating turbulence beneath the wing, which reduces airspeed relative to the upper surface and thus enhances lift.

Even species which normally exercise strong control over their flight, such as the Mourning Cloak *(Nymphalis antiopa)* or Monarch *(Danaus plexippus)*, will at times go into their best gliding mode (i.e., soar) in order to exploit rising currents of warm air (thermals). This may be a significant mechanism in facilitating migration.

For strong fliers—not gliders!—the fastest wing speed that can be sustained for long is around 32 km/h (20 mi/h), though some hesperiids are capable of short spurts at much higher speeds. Migrating Painted Ladies *(Vanessa cardui)* on their single-minded trip north from the desert can apparently cover 320 km (200 mi) in 3 days. If we allow them 8 hours' flying time per day, that works out to only 13.3 km/h (8.3 mi/h). But many of us have seen them at least briefly pacing our car on the road at three times that speed.

animals go into their usual thermoregulatory mode. Both sexes participate, and there is no sexual interaction in roosts. Typically, roosting sites are on plants too slender or fragile to support a climbing rodent or shrew; to catch the butterflies a bird would have to hover in the air while pecking at them. Roosting can thus be interpreted as protective behavior directed against nocturnal or eocrepuscular (dawn/dusk) predators. If you wait until dark and then place the butterflies directly on the ground, most are usually eaten overnight, and you find only detached wings in the morning. Orange Sulphurs will roost on the ground, however, and many nymphalids *(Vanessa, Polygonia, Nymphalis)* shelter overnight in ivy or other dense ground cover, under shingles or loose bark, or even between stones in rock piles. In the tropics, long-lived, often inedible and warningly colored butterflies such as heliconiids (Heliconiidae) may form single-species clusters that endure for months. Of course, overwintering Monarchs *(Danaus plexippus)* along the coast cluster in eucalyptus, pines, and other trees.

Migration

Migration is systematic dispersal. Butterflies differ greatly in their tendency to disperse. Some—such as many blues—seldom stray more than a few feet from their birthplace. Others, such as the Monarch and Painted Lady *(Vanessa cardui)*, undergo regular seasonal movements of hundreds or even thousands of miles. General appearances can be deceiving: our smallest butterfly, the Western Pygmy Blue *(Brephidium exile)*, is a great wanderer, and much of its range is recolonized every year.

Migration, like diapause, is usually a seasonal adaptation and is often controlled by photoperiod in temperate latitudes. In fact, migration is often associated with adult reproductive diapause: butterflies are as a rule uninterested in sex while migrating. In some species the physiology of the migratory generation(s) may be radically different from others—just as different as in the "phases" of migratory locusts (which are grasshoppers [Orthoptera]). The generation of Painted Ladies that migrates into our area from the desert in late winter or early spring not only looks different (small, dull, pale) but is different hormonally and behaviorally, too. You may have noticed that migrating Painted Ladies fly in a set direction (generally toward the north-

northwest) like the proverbial bats out of Hell; they stop for nothing—neither food nor sex—and, faced with an obstacle such as a building, they go over the top rather than reset their bearing to go around it. All of this is preprogrammed hormonally and associated with the presence of a large supply of yellow fat carried over from the caterpillar stage, which is the fuel for their flight. It is also the material that makes the distinctive splotch on your windshield when you intersect a Painted Lady's trajectory. When the fat supply is depleted they stop migrating and begin reproducing, but it is not sure which is cause and which effect. Painted Lady migration is the same in the Old World, where late-winter breeding occurs in North Africa and the Levant, and the resulting adults cross the Mediterranean, breeding in northern and central Europe and even reaching Britain and Scandinavia. In both North America and Europe the next generation heads still farther north (or upslope) to breed. There is a reverse migration southward in fall. These generations, however, are not provisioned with fat and are more leisurely in their movements. The northward migration seems to be a way of tracking host plants (mainly annuals) as the spring moves northward; if they tried to breed in their wintering grounds in summer, they would usually find no suitable vegetation available. The Painted Lady almost never survives the winter in the north—on either continent. The Red Admiral *(Vanessa atalanta)* is a seasonal migrant in Europe and eastern North America; it is not certain that it migrates here, and we know it overwinters routinely in our area.

The California Tortoiseshell *(Nymphalis californica)* is another seasonal migrant whose movements appear to track the availability of young, tender foliage of the host plants (wild-lilac *[Ceanothus]*). The butterfly larvae must start feeding on new growth, which is only briefly available at the start of the growing season (February to March in the foothills, June to July in the high mountains), so normally it is impossible to rear two successive generations in the same place. However, the close relatives of the California Tortoiseshell—the Mourning Cloak *(N. antiopa)* and Milbert's Tortoiseshell *(N. milberti)*—migrate too, and their cases cannot be rationalized on similar grounds since their host plants remain available all season. Perhaps they simply are poorly equipped to deal with summer heat at low elevation inland. The Mourning Cloak may have three broods a year in the coastal fog belt, where it is seemingly not migratory, but only one from Fairfield inland, where it is.

Many common, multivoltine species may have had to disperse elevationally on a seasonal basis to track available host plants. The need to do so is reduced today due to irrigation and the availability of exotic hosts that stay green in summer at low altitudes, but many of these species still disperse upslope in late spring. As noted earlier, most of these species become very dispersive in late fall, spreading eggs all across the lowland landscape. At this time they also show up at rabbitbrush along roadsides in the high country, where any attempt at reproduction is probably doomed.

In our area the Buckeye *(Junonia coenia)* has large directional flights in late spring some years, moving in a generalized south or southwest to north or northeast direction. In 2003 and 2005 such flights triggered unprecedented reproduction in northern California. The sources of these apparent migrations are unknown. They may have played a role in the recolonization of our area after the Buckeye was eradicated in the great freeze of December 1990, but the sequence of first observations in our area argues instead for spreading out from multiple sites of very local survival.

Quite a few southern California butterflies, including some from the deserts, show up as rare "strays" in our area. Only one, the Marine Blue *(Leptotes marina)*, breeds here with any regularity, but it seems unable to overwinter. The fact that we see desert species so far out of range is important in the context of global change, because it tells us something about potential mobility of species and the pool of species that might be available to colonize northern and central California under a greenhouse climatic regime.

Our best-known migrant is of course the Monarch. Monarchs east of the Rockies overwinter in the mountains of central Mexico, but ours have a shorter journey—to the California coast. Monarchs begin leaving the western Great Basin and Sierra Nevada as early as August, and a gradual movement westward can be observed in late summer. Huge fall concentrations like those often seen on the Atlantic seaboard are rare here but have been observed in the Suisun Marsh in October, and in some years the effect of topography in funneling migrating Monarchs through Lagoon Valley Regional Park, between Vacaville and Fairfield, is conspicuous. While in their winter clusters along the coast, Monarchs are sexually inactive, but just before the clusters begin to break up (in late February to early March) there is a burst of copulation. The butterflies then move inland to oviposit, often arriving before their host plants (milkweeds) have broken ground. In some years, when hosts on the floor of the Central Valley are late, there may be a backwash westward into the Coast Range to breed on purple milkweed *(Asclepias cordifolia)*, which grows on warm rocks and is usually well advanced.

Both sun-compass navigation and the Earth's magnetic field have been implicated in butterfly migration, but there is much we do not yet understand.

What Are Wing Patterns "For"?

When most people think of butterfly wing patterns they think of defense against visual predators: camouflage or crypsis, warning coloration or aposematism, or mimicry. Actually, that's only part of the story. The other important factors are thermoregulation and sex.

In the 1960s it was first suggested that butterfly wing patterns and colors might play a role in thermoregulation (regulation of body temperature). Butterflies are cold-blooded. Except for some nymphalids *(Vanessa, Nymphalis, Polygonia)*, which can generate body heat by muscular activity or shivering, butterflies rely on a combination of color and pattern, insulation (fur), and subtle postural adjustment to maintain a thoracic temperature suitable for flight: between 25 and 44 degrees C (75 to 112 degrees F). But many butterflies can fly in bright sunlight at ambient temperatures as low as 13 degrees C (55 degrees F). Dark colors absorb incoming solar radiant energy and convert it to heat. Because there is no blood circulation in the wing veins, only the basal parts of the wing, which are in contact with the body, can deliver heat directly to the thorax (by conduction). But some butterflies, particularly the whites, have a reflective sheen which redirects light toward the body when the wings are held in basking attitude, acting much like a tanning collar.

Behavioral thermoregulation is directly responsive to the immediate energetic environment. Different butterflies exploit the sun in different ways. *Colias* species, for example, engage in lateral-basking—closing the wings over the back, tucking the forewings inside the hindwings, and turning one side so it is perpendicular to the sun's incoming rays. *Pieris* species engage in dorsal-basking, with the wings partly open over the back. Most branded skippers bask with the forewings slightly open and the hindwings more so. Checkered skippers and duskywings hold the wings open at their sides, as do some of the world's highest-altitude butterflies—tiny whites from the Andean highlands and the Himalaya and Pamir, where they reach 5,500 m (about 18,000 ft) and can fly at ambient temperatures slightly below 10 degrees C (50 degrees F) in full sun.

In very hot sunshine butterflies become inactive or seek shade. Some, for example, the Buckeye, remain in their male territories but close the wings over the back and align them parallel to incoming solar radiation to minimize the absorptive surface. Due to the need to thermoregulate, butterflies are active under overcast only if it is quite warm.

There is a substantial literature of butterfly thermoregulation, employing state-of-the-art physiological instrumentation to take the temperatures of living butterflies as they go about their business. Thermoregulation is discussed further under "Species, Subspecies, and 'Forms'" elsewhere in this book.

Communication

Most butterflies are cryptic on the lower wing surfaces—the surfaces normally exposed when the animal is at rest. Any bright colors are on the upper surface, where they are displayed in flight. As noted earlier, attacks by birds on flying butterflies are rarely observed. The underside crypsis is almost certainly directed at visual predators. But what are the upperside colors and patterns "for"? *Their* targets appear to be other butterflies, of the same or other species.

The compound eyes of butterflies contain between 2,000 and about 20,000 tiny facets or ommatidia, each of which is an eye in its own right, focused on a tiny portion of the overall visual field. Butterflies see the world very differently from humans. They cannot adjust for distance, and their ability to integrate complex patterns is poor, but they are excellent at detecting motion; for them the world appears as a mosaic of dots. Males often have larger compound eyes than females, which probably reflects the role of vision in mate location and courtship.

The wings are instrumental in telling other butterflies "I am a member of species __ and my sex is __." Since initial recognition of potential mates is almost always visual, this is very useful information. Males are almost constantly in competition for females, and females may use visual information to choose among suitors. This is Darwinian sexual selection: any traits preferred by females that have a genetic basis will increase in frequency in the population. Many butterflies have a pronounced difference in color and pattern between the sexes; this is known as sexual dimorphism. Think of the Great Purple Hairstreak *(Atlides halesus)* or the Checkered White *(Pontia protodice)*. As a rule the male is the more brightly colored sex. In the southwest Pacific occur the birdwing butterflies *(Ornithoptera* and *Troides)*, relatives of our Pipevine Swallowtail *(Battus philenor)*, in which "runaway sexual selection" has led to spectacular and sometimes bizarre males, while the females are dark and drab. They are the butterfly equivalents of birds of paradise.

On the other hand, some butterflies, notably the Painted Ladies, tortoiseshells, and Mourning Cloak, have almost no sexual differences in color and pattern.

Of course, we humans cannot see everything butterflies see. Butterflies, like bees, can see in the UV end of the spectrum, where we cannot. Many flowers have patterns reflecting UV light that guide bees (and other pollinators) to the nectar. Thus it should not be all that surprising that many butterflies have patterns visible to them but not to us. Often these are useful in sexual identification. Male and (orange) female Orange Sulphurs look the same color to us, and it takes a while to learn to distinguish them in flight. But they look very different to their own kind, because the orange part of the wings reflects UV light in the males, making it look "bee purple," while the females absorb UV light and are just plain orange. (If you rotate a large, richly colored pinned male through a variety of angles in sunlight, you can just pick out a pinkish sheen at the edge of the near-UV spectrum and get a hint of what it would be like to see like a *Colias*.) Behavioral researchers developed a video apparatus that makes it easy to visualize UV patterns and look for them in different species. Butterflies probably can detect polarization of light, but we are not sure what they do with that information.

The little details in complex butterfly wing patterns may appear irrelevant in mate selection. But altering the wing color or pattern artificially may have dramatic impacts on sexual attractiveness. In one recent study, the experimenters used computer software to alter minor pattern elements in various ways and generate lifelike decoys, which were set out in the field to attract males. In fact, these subtle changes did produce significant decreases in male approaches. Thus wing patterns may be under conservative "stabilizing selection" much of the time.

Courtship and Territoriality

How to find a mate? This is a significant question for a great many people in the Bay Area, and it is for butterflies too. In a complex landscape, it will not do to rely on random encounters—especially if you are a short-lived, small insect with relatively limited mobility. Butterflies have a number of mate-location strategies, which lead to some of their most conspicuous behaviors.

The fundamental problem is to narrow down potential encounter sites so that only those most likely to produce potential mates need be searched. Virgin females and males in mating condition are genetically programmed to seek out such sites, often at specific times of the day. Once in a suitable location, males have stereotyped behaviors that maximize the likelihood of sexual encounters. (Does that sound rather familiar?)

The character of such sites depends on the landscape. To be effective, sites cannot be very numerous but must be easily recognized at a distance, and readily accessible. In mountainous terrain, the preferred sites for many organisms—not just butterflies—are on exposed, rocky summits. This behavior is called hilltopping. Although it was long suspected that hilltops are mating rendezvous sites, early researchers were fooled by the rarity of females there. This turns out to be a function of sexual appetite. Many female butterflies mate only once in their lifetime and thus visit the hilltop only once, but a given male will return day after day. As a result, you may perceive the situation as an all-male aggregation, like what you see on mud puddles (where no sex ever takes place). In detailed studies the sex ratio is commonly about 20:1, males to females, on any particular day. Multiple species will use the same hilltop at the same time, producing a bewildering swirl of multicolored butterflies that must be seen to be fully appreciated. We know from mark-release-recapture studies that a single summit may draw from a surrounding area of several square miles, that individual males may cover several kilometers a day commuting to and from hilltops, and that males may move among hilltops over a period of days. The whole system promotes gene flow and works against genetic differentiation of local populations. Hilltopping is readily observed on serpentine barrens. One of the most dedicated hilltoppers is the Anise Swallowtail *(Papilio zelicaon)*, which will hilltop even on piles of debris at construction sites! (Hilltopping butterflies, interestingly, show no interest in the roofs of tall buildings.)

In forested areas, small sunlit clearings or sunflecks are used as rendezvous sites—often by the same species that hilltop elsewhere. A walk in the riparian forest in the Sacramento Valley often reveals male Anise Swallowtails occupying such sunflecks.

The rendezvous sites for our four species of *Vanessa* are particularly well defined. They are places open to the late-afternoon sun (i.e., to the west or southwest) with a well-defined vertical backdrop, be it trees, a wall, or whatever. We occasionally see sexual behavior in these butterflies shortly after sunrise in sites facing east or southeast, but only on warm mornings with no dew. Normally our male *Vanessa* take up their perches when the sun is low in the western sky—earlier on cool days, later if hot. It should be emphasized that sexual activity is completely dictated by context. A male and a virgin female may jostle each other on flowers at noon with no evidence of recognition—but the story is very different at a rendezvous site a few hours later.

Males typically perch in a conspicuous location affording a wide view of the area. Some species habitually perch on the ground (the Sandhill Skipper *[Polites sabuleti]* and Buckeye, for example) while others perch on relatively low vegetation (the Anise Swallowtail and Umber Skipper *[Poanes melane]*) and a few on leaves or twig tips high in the trees (the Tailed Copper *[Lycaena arota]* and Golden Hairstreak *[Habrodais grunus]*). From the perch they sally forth to investigate passing insects. Are they potential mates? Potential rivals for mates? Other species (whose presence might in fact interfere with courtship)? Some species always perch with wings closed over the back (hairstreaks), partly open (coppers, branded skippers), or fully open (checkered skippers, Common Sootywing *[Pholisora catullus]*). The species of *Vanessa, Nymphalis,* and *Polygonia* are more versatile, modulating their postures to reflect the need to regulate body temperature. They also may sit on tree trunks or on the ground, with wings fully or partly open or closed, as the thermal environment dictates.

Male hairstreaks and Lorquin's Admirals (*Limenitis lorquini*) usually perch on or very near the host plant, and most hairstreaks do so only for a limited time in late afternoon; the Golden Hairstreak is active into twilight.

An alternative to perching is called patrolling: males "fly a beat," typically up and down a linear habitat such as along a stream or road. The Western Tiger (*Papilio rutulus*), Two-tailed (*P. multicaudatus*), Pale, and Pipevine Swallowtails, and the California Sister (*Adelpha bredowii californica*) are conspicuous patrollers. So are several pierids including the Sara Orange-tip (*Anthocharis sara*) and the California Dogface (*Zerene eurydice*), and some hesperiids such as the Propertius Duskywing. For both perchers and patrollers, there seems to be a dominance hierarchy or "pecking order": if the resident males are removed, others generally move in quickly to take their place—suggesting that they had been excluded from the "best" sites.

Hilltopping males may be perchers or patrollers or a combination of the two. Careful observers often note that males usually arrive at and leave the hilltop in groups rather than singly; they look like they are chasing one another. They appear to be in a race, which is easily explained by a familiar analogy. Imagine that you are trying to catch a bus at a street corner. You cannot see whether a bus is coming, but you find yourself speeding up as you approach the corner, just in case one is. Hilltopping males are seeking mateable females. A male that sees another male ahead of him going up the hill speeds up to overtake him, "just in case" a female actually is there. Hence the apparent tag teams!

Many insects, including both diurnal and nocturnal moths, use airborne sex-attractant chemicals (pheromones) to facilitate mate location. The common Sheep Moth (*Hemileuca* or *Pseudohazis eglanterina*) is a familiar diurnal example in our area. As far as we know none of our butterflies does this, though there are hints that the Clodius Parnassian (*Parnassius clodius*) might. Although the initial phase of butterfly sexual interaction is typically visual, pheromones come into play at close range during courtship and may be absolutely critical to the outcome.

Much ink has been spilled over whether butterflies can truly be considered territorial. The problem is a semantic one. Butterflies have no "weapons" to fight with, so active "defense" of territory carrying a risk of injury is not normally at issue. But in fact, in vertebrates most territorial behavior is symbolic, and actual aggression is relatively rare despite lurid popularizations. The argument is also made that vertebrate territories represent resources such as food for the young. Can a similar claim be advanced for butterflies?

The territory defended by male butterflies is space at a sexual rendezvous point; the resource it represents is a potential mating opportunity. The familiar male-male chases—in which the previous occupant almost always returns to his perch or beat—are best seen as investigative rather than aggressive. This is best illustrated by the dynamics of such chases. If the intruder is another male or the wrong species, the resident male simply returns. If it is a mateable female a courtship typically ensues, usually away from the rendezvous site. The length of male-male or interspecific chases seems to reflect how easy it is to identify the intruder: strongly sexually dimorphic species have very short chases, while species in which the sexes look alike, such as those in the genus *Vanessa*, take quite a bit longer.

A lek (the word is of Swedish origin) is an area where males of a species gather to display and compete for females, which come to the site to meet mates. The word has been generalized from vertebrates, such as grouse, to insects. Hilltop aggregations and *Vanessa* mating sites can fairly be called leks. So can the sites where male hairstreaks perch in numbers, generally on or near the host plant. Lekking behavior on the host plant can facilitate the differentiation of host races, which may in turn become full species in time.

Females of some species, especially blues, often sit passively on the host plant and let males come to them.

Males of most butterfly species produce aphrodisiac pheromones that they direct at females at close range. Most of these are undetectable by the human nose, but many people (in my experience, mostly women!) can smell the pheromone of the male Cabbage White if held just beyond the nose and allowed to flap. (It is usually described as resembling the scent of apple blossoms— try it.) Females may then respond with chemical signals of their own, each step in the courtship serving to trigger a specific response. Any error in the sequence can end the courtship then and there. Males often have conspicuous pheromone glands: in the middle of the dorsal forewing (branded skippers and the satyrines), along the costal edge of the forewing above (hairstreaks); along the anal margin of the hindwing (the Pipevine

Swallowtail); in a rolled-over flap along the forewing costa (duskywings and some checkered skippers); in a rounded or oval patch at the base of the hindwing above (in our area, the California Dogface and Dainty Sulphur *[Nathalis iole]*); and so forth. The male Monarch has a gland on one of the hindwing veins above. Some species (such as many blues) have scent scales (androconia) scattered over the wing surface; some have hair pencils that can be extruded from the legs (some hesperiids) or abdomen (Monarch) and are used like a powder puff.

Courtship may lead to almost instantaneous mating, especially if the female is newly eclosed. Or it may be a very long, drawn-out affair of a half hour or more. Unreceptive females may simply leave the area, but some (the Cabbage White is a good example) have a distinctive "rejection posture" familiarly known as a "tail in the air": the female alights, spreads her wings, and raises her abdomen perpendicularly, such that the male cannot make genital contact. In sulphurs an unreceptive female rises vertically in a spiraling motion, followed by the male (often out of sight), then drops suddenly, almost always losing him in the process. Both of these conspicuous behaviors are routinely misinterpreted by observers as invitations to mate.

Interspecific courtships are surprisingly common but very rarely successful. Even interfamilial matings have been recorded on occasion, but no offspring result in such cases. Such mismatches may be more frequent when few conspecifics of the opposite sex are available.

Mating

Butterflies have internal fertilization. During copulation the female's abdomen is held by a pair of clasping organs and centered by a hooklike structure above, the uncus (sulphurs have a larger, more "modern" structure called a superuncus). Sperm are delivered through an intromittent organ called the oedeagus. Copulation is initially achieved in a side-by-side position, with the male curving the tip of his abdomen toward the female until his claspers can grasp her genitalia. The male then typically walks around until he is facing directly away from her, and copulation proceeds tail to tail. Successful sperm transfer requires at least 20 minutes and often substantially more. A typical undisturbed mating will last 45 minutes to an hour or two, but pairs formed in late afternoon may remain joined (in amplexus) right through the night. Occasionally a male will mate a second time within a short period (a day, say) but is not prepared to ejaculate yet. In such cases he may hold the female, even overnight, until he is recharged and able to complete the act.

In some species—notably *Colias* species—males will often copulate with teneral females before their wings harden. There is no evidence that they are attracted to them pheromonally, however, unlike some tropical butterflies in which males may mate with pharate females, breaking through the pupal case in order to do so.

In cages, male Clodius Parnassians make a beeline for newly emerged females they cannot see, suggesting a pheromonal attractant. This may explain why virgin females of this species are almost never encountered afield.

In addition to sperm, the male delivers a package of material known as a spermatophore, along with various accessory-gland secretions. Using radioactive labels, it has been shown that these materials provide nutrients that can be used by the female in making eggs. They therefore represent additional "paternal investments." The spermatophore is squeezed out by the male like toothpaste from a tube and is sometimes extruded by males when dying in a killing jar. Inside the female it is deposited in an organ called the bursa copulatrix, whose shape it takes (resembling a vase with a neck). One side of the bursa typically has a roughened surface resembling the business end of a meat-tenderizing tool. This may abrade the surface and assist in its enzymatic digestion. By dissecting a female and counting spermatophore residues we can tell how many times she mated; each new mating pushes the remains of the last spermatophore back into a crumpled heap. The sperm leave the spermatophore quickly after deposition and migrate across the oviduct to an organ called the spermatheca, where they can be stored, even for months. (Some hibernating nymphalids may mate in October but do not ovulate until, say, February; butterflies with estivation—summer dormancy—also may store sperm. The sperm are presumably fed in the spermatheca!) Once ovulation begins the sperm leave the spermatheca one by one and meet the individual eggs as they come down the oviduct. This happens just before they are laid, which is why it is not possible to salvage mature eggs from a dead female and rear them—alas!

If males are numerous and females can afford to be picky, they should, in theory, hold out for the best mate they can get—in terms of genetic quality, sperm quality and quantity, and nutrient contributions. Genetic quality can be assessed only indirectly through such traits as size, color, and so on. At hilltops, as noted earlier, the sex ratio is invariably highly skewed, and females can and should be "coy." The male is effectively being tested; the assumption is that his persistence and vigor reliably indicate his fitness (this has been called the "honest salesman hypothesis"). Typically a mated female will not mate again unless her supply of sperm and/or nutrients is depleted; females seem able to sense the condition of the spermatophore, perhaps employing stretch receptors in the bursa. Most females mate only once or twice in a lifetime, while males can mate every day or two. Hence, there is usually a large surplus of males (despite a 1:1 sex ratio at birth), and most courtships are doomed to fail. We do not yet know whether (as in many vertebrate species) a small proportion of males accounts for most of the matings, with most males never mating at all, or whether things are more equitable among butterflies.

If a female remates, any remaining sperm from the preceding mating are somehow inactivated, so the latest male will father all the subsequent offspring. This phenomenon is known as sperm precedence, and it clearly serves to intensify competition for mates among the males. It can be demonstrated in the lab by mating the same female sequentially to males with distinctive genetic markers that will show paternity in the offspring (the red-eyed mutant in the Anise Swallowtail, for example). It is in a male's genetic interest that he prevent the female from remating, if he can. In the Clodius Parnassian and some swallowtails (not ours) the male secretes a "chastity belt," called the sphragis, during the act of mating. It plugs her copulatory opening, rendering it inaccessible to subsequent males. (Her egg-laying organ is on a different segment and is unaffected.) Other males may court vigorously but are unable to dislodge the plug. It is the presence of this large, conspicuous structure that enables us to say with confidence that virgin female Clodius Parnassians are almost never encountered in nature! In some other butterflies a chemical remating deterrent is deposited, and in some moths the male deposits an anaphrodisiac pheromone that repels males.

At high population density, males frequently interfere with

courtships and mating pairs. They are almost never successful in displacing the first male, but I have seen a frenzied Pipevine Swallowtail grasp the side of a mating female's abdomen and hang on for dear life! Any disturbance of a mated pair may cause it to fly. Contrary to popular belief, butterflies cannot initiate copulation in the air, but they can sustain it. It is impossible for both partners to flap their wings simultaneously, however, due to interference. One must carry the other, which hangs limp with wings tightly closed and legs folded against the body. In a given species the carrying partner is always the same—the male in papilionids and pierids, the female in nymphalids and hesperiids, and so on. In species whose females carry, mating with tenerals is impossible. It is usually thought that copulation entails a predation risk above and beyond solo activities, but we have no hard data on that.

Egg Laying (Oviposition)

A given butterfly species lays eggs singly or in clusters. Most butterflies lay singly, with a mandatory interval of flight (however short) between ovipositions. The location on the plant is often very specific. The Large Marble *(Euchloe ausonides)* always oviposits on the terminal flower bud on a shoot of a tall mustard (Brassicaceae) plant. The Pipevine Swallowtail deposits eggs in small groups—rarely up to 25—on young, tender, growing shoot tips of the host. It is notoriously picky; a female may examine and reexamine a plant for an hour or more before deciding to lay. One Rocky Mountain butterfly has been shown to select its oviposition sites to maximize exposure to morning sun. Thermal considerations may well play a role in the oviposition behavior of some of our butterflies, too.

The crescents and checkerspots, as well as the members of the genus *Nymphalis,* lay eggs in clusters up to 100 or more, and the larvae—at least initially—feed gregariously. Crescent and checkerspot females hatch with their first egg batch ready, and subsequent egg production depends on their nutritional status.

Some large true fritillaries fly after their host plants (violets *[Viola]*) have dried up; they can apparently tell where there will be hosts next spring, and lay there. (The larvae eat only their eggshells upon hatching, then remain dormant until spring.) Some populations of Lindsey's Skipper *(Hesperia lindseyi)* ovi-

EGGS THAT COUNT

Most butterfly eggs are inconspicuous or even hidden, but pierids that lay on fruit, flowers, or buds of the mustard family (Brassicaceae) lay very conspicuous red or orange red eggs (sometimes the eggs are pale green when first laid but then change color). Newly hatched larvae of many of these species are known to be cannibalistic. Shortly after hatching they walk all over the plant and eat any eggs they encounter, whether of the same or different species. Both the shared egg color and the cannibalism have been interpreted as evolutionary responses to competition for a limited food resource.

Many ecologists argue that because "the world is green," foliage is rarely in short supply, and therefore competition is rare or nonexistent among foliage-feeding insects. But the reproductive structures of plants are typically much more limiting; not only do they constitute less biomass, but they are usually highly seasonal and transient. Experiments have shown that the presence of red eggs on hosts deters subsequent visitors from laying on them. (One Old World pierid definitely has an oviposition-deterrent pheromone associated with its eggs. We do not know if any of ours do.) Many parasitoids mark their hosts chemically to

posit on lichens growing on fence posts and tree trunks. When the larvae hatch they drop to the ground and forage for their (bunchgrass) hosts. This may be a way of evading egg parasitoids that key in on the host plant. The Great Copper often oviposits on or near the dead, dry tops of the current year's growth of its perennial host plants, docks *(Rumex),* and some butterflies (not in our area) systematically oviposit not on their hosts but in litter or soil near them. Butterfly eggs normally adhere strongly to their substrate, but some alpine species drop off the host and overwinter in the soil nearby. Had they adhered strongly they might have blown away in the fierce gales that regularly rake their habitat. Might the same phenomenon have evolved on the coastal strand?

Some butterflies lay eggs readily in captivity and may not even require the host plant as a stimulus. Most are pickier, however, and a few, such as the Pine White *(Neophasia menapia),* rarely if ever lay in captivity. Care should be taken to maintain eggs at moderate relative humidity, as they may be in water equilibrium with the substrate and are at risk of dying if the latter is too dry.

deter superparasitism (all the larvae would end up starving), as do fruit flies on their host berries, so this situation should not be surprising.

Now consider the situation from the plant's point of view. One pierid larva can destroy all the reproductive output of one medium-sized or several small hosts. The fitness of such hosts is zero. There should be intense natural selection for defense, and there is. In our area several species of native mustards of the genus *Streptanthus* (jewel flowers) produce orange-pigmented callosities on their leaves that mimic pierid eggs. Again, field experiments have shown that these actually deter oviposition by pierid females! Most cases occur on serpentine soils, where the cost of replacing tissue lost to herbivory may be especially high. Strikingly, some tropical passion vines *(Passiflora)* have also evolved egg mimics that fool their herbivores, butterflies of the genus *Heliconius.* The phenomenon is known overall as egg-load assessment.

Not all red eggs are involved in oviposition deterrence. The eggs of the Pipevine Swallowtail *(Battus philenor)* are apparently warningly colored and may even attract subsequent females to lay nearby. The eggs of sulphurs are bright red too, but we don't know why. There is no evidence of egg-load assessment in these foliage feeders.

Butterflies and Plants

It's a truism that to really know your butterflies, you have to know your plants. The reason for this basically is that the butterflies themselves "know" and depend on plants. Their ecological and evolutionary ties to plants are profound—so much so that more than 50 years ago a distinguished lepidopterist wrote a paper titled "Butterflies as Botanists" in which he argued that by observing patterns of butterfly-plant interaction we could be led to new insights about the plants. He was right.

Host Plant Specialization

Plant-eating insects can be arranged along a continuum of specialization. Polyphagous species feed on plants of many families; oligophagous species feed on just a few (usually related) families; and monophagous species are narrowly specialized, sometimes on only a single host species. We have two truly polyphagous butterflies in our fauna: the Painted Lady *(Vanessa cardui)* and the Common Hairstreak *(Strymon melinus)*. Each is recorded (globally, not necessarily here) on 20 or more plant families in several orders. No insect species can eat all plants, or even all flowering plants, but some are remarkably tolerant. Our most extreme specialists are the Pipevine Swallowtail *(Battus philenor),* the Yuma Skipper *(Ochlodes yuma),* and the Least Checkered Skipper *(Pyrgus scriptura),* each with only one host species. Many lycaenids have impressive lists of hosts *as species,* but any given local population usually has only one host.

Even the most polyphagous species do not select hosts at random. Both the growth form (habit) of the plant and its chemistry affect host choice. Species that oviposit on trees will not do so on herbs, and vice versa; different life stages (phenophases) may be differentially attractive, even within a plant species. (The Checkered White *[Pontia protodice]* oviposits freely on small basal rosettes of mustard [Brassicaceae] species it will not lay on as large, mature adults.) A great deal of research has focused on the chemical component of host selection. The decision whether or not to lay on or feed on a plant is based on the *presence* of nutrients and, often, of specific chemical substances that are not nutrients but are required to release the behavior—more about these in a moment—and the *absence* of repellents, deterrents, or toxins. Both adult females and larvae are able to detect water, sugars, and amino acids in plant tissues. These, however, are present in all plants. It's the chemicals that are present only in certain plants and absent in all others that make discrimination possible.

The chemicals in question are often called "secondary plant substances" or "secondary metabolites" because they are not a part of the basic, or primary, metabolism of the plant. There are several different groups of such compounds, each associated characteristically with certain plant families. It can be shown that the insects feeding on these plants have specific neuroreceptors for the compounds and will not accept plants that lack them. (Often, though, they can be fooled into accepting other plants, or even filter paper, if coated with the compounds.) It's obviously not in the interest of a plant to stimulate its natural enemies to eat it, yet that is what it does when it produces these compounds. Is there a logical explanation consistent with Darwinian fitness?

Pieris and *Pontia* butterflies require chemicals called glucosinolates (or mustard oil glycosides) as egg-laying and feeding stimulants. These compounds are found in most plants of the mustard, caper, mignonette, and nasturtium families (Brassicaceae, Capparidaceae, Resedaceae, and Tropaeolaceae), all of which are used by the insects. Botanists used to put the first three families in one order and the nasturtium family in another and argued that the compounds arose independently in the two lineages. We now know that the nasturtium family really *is* closely related to the others, thanks in part to modern DNA evidence. The butterflies were better botanists than the botanists! It is striking that plants in these families lacking the proper compounds are not used, despite their relationships.

The Anise Swallowtail *(Papilio zelicaon)* requires essential oils shared (and in this case, apparently really independently derived) by the carrot family (Apiaceae) and the rue family (Rutaceae), which includes citrus *(Citrus)*. The Anise Swallowtail is one of the very rare cases where a single species will feed both on herbs (carrot family) and trees (cultivated citrus in southern California and in Butte County); the commonality is entirely chemical. But the butterfly will not feed on carrot family plants that lack these oils—we have several—nor will it feed on any member of the ginseng family (Araliaceae), which is the sister family to the carrots but never produces the oils.

The solution to the seeming Darwinian paradox is that the secondary compounds are protective against a variety of *other* enemies, such as polyphagous insects. The plants have, in effect, traded one set of (generalized) herbivores for another set (specialized). The specialists are insects that overcame the initial defense and subsequently adopted the essentially competitor-free plant as their primary host. The plants, of course, are evolving too. They may acquire a second line of chemical defense. The mustard family genus *Erysimum* (wallflower), for example, produces cardenolides—a novel class of chemicals in its family—and these seem to exclude pierine butterflies (the whites) from using it as a host. We might predict that very ancient plant lineages, "living fossils" such as horsetails or cycads, would have many different secondary chemical defenses and have outlived most of their enemies. And that does seem to be true.

We can use the distribution of secondary chemicals to understand the patterns of host utilization in different butterfly lineages. For example, the genus *Satyrium* in the Lycaenidae and the genus *Erynnis* in the Hesperiidae both have species feeding on wild-lilac *(Ceanothus)*, oaks *(Quercus)*, willows *(Salix)*, and papilionaceous legumes (members of one branch of the pea family [Fabaceae]). This is not an accident, but reflects the role of secondary chemistry in bridging the gap from one host plant to another in time and space. The term "coevolution" has been used often in this connection but requires qualification. For coevolution to occur, each species has to serve as an agent of natural selection on the other. There are relatively few cases of coevolution in our butterfly fauna and its host flora, at least in a rigorous sense. But in a more diffuse sense, they are broadly coevolved.

When a species can utilize more than one host in an area, the Darwinian expectation is that it will preferentially use the one on which it "does best." But there are multiple criteria for performance (such as growth rate, pupal weight, fecundity—but also such things as risk of being parasitized or eaten!), and such predictions are difficult to test. Often it is difficult to understand how a narrow specialization arose or why it is maintained. The Least Checkered Skipper, for example, feeds only on alkali mallow *(Malvella leprosa)*. The Common Checkered Skipper *(Pyrgus communis)* feeds on many species of the mallow family (Malvaceae), both native and introduced, including alkali mallow. Although the Least Checkered Skipper will not lay eggs on other mallows eaten by the Common Checkered Skipper, its larvae will not only eat them but, by some criteria at least, do better than they would on their own host! Clearly we have a lot to learn about butterfly–host plant relationships.

Butterflies may lay eggs on plants that their larvae cannot or will not eat. This occurs when the plant provides the chemical cues for oviposition but not for feeding, or when it is toxic to the larvae. There are several such cases on record. One of the most interesting concerns the Anise Swallowtail on species of bishop's weed *(Ammi majus* and *A. visnaga,* the latter also called toothpick weed), a carrot family genus, in the Central Valley. Bishop's weed has a spotty distribution but can be locally abundant. The two species are naturalized from the Middle East. They are summer annuals and are thus in good shape when hosts other than sweet fennel *(Foeniculum vulgare)* are rare, and females oviposit eagerly on them. However, all the larvae die within a few days. We can predict that natural selection will either remove from the Anise Swallowtail the propensity to lay on bishop's weed, or result in the evolution of tolerance to its toxic chemistry. For either to occur, however, the right genetic variations have to be available. Strikingly, montane Californian populations of the Anise Swallowtail, which do not have contact with bishop's weed, appear preadapted to eat it; apparently the chemistry of their native host plants is similar.

Butterflies in the Anthropic Landscape

The butterfly fauna covered by this book occupies habitats more or less altered by our species—an "anthropic landscape." Even seemingly intact plant communities have been altered by the naturalization of exotic species, resulting in changes in competitive and feeding interactions. The Central Valley and adjacent foothills have been profoundly altered by the replacement of native, perennial grasslands with a new kind of grassland made up of (mostly Mediterranean) annuals naturalized since the Mission Period. The rich golden brown of the California summer is a human artifact. Before Europeans came, the prevailing color was grayish green. The tule marshes were diked and tamed; the riparian forests were reduced from several miles wide to often a single rank of trees by the water's edge, and their continuity was often broken. Although no one collected butterflies in pre-European California, one overriding fact has emerged from studies of the fauna as it exists today: California butterflies, for better or worse, are heavily invested in the anthropic landscape. About a third of all California butterfly species have been recorded either ovipositing or feeding on nonnative plants. Roughly half of the Central Valley and inland Bay Area fauna is now using nonnative host plants heavily or even exclusively. Our urban and suburban, multivoltine butterfly fauna is basically dependent on "weeds." We have one species, the Gulf Fritillary *(Agraulis vanillae)*, that can exist here only on introduced hosts. Perhaps the commonest urban butterfly in San Francisco and the East Bay, the Red Admiral *(Vanessa atalanta)*, is overwhelmingly dependent on an exotic host, pellitory *(Parietaria judaica)*. And that's the way it is.

Most of our common multivoltine species rely on weedy hosts that can persist only due to the availability of summer water, namely, irrigation. There is no reason to think these species are recent arrivals to lowland California or that multivoltinism evolved only very recently (they are nearly all multivoltine everywhere). The explanation for this odd situation can be found in the history of California's wetlands. As recently as the early twentieth century there were extensive freshwater tule marshes in our area, especially along the east side of the Sacramento Valley. These wetlands stayed green in the summer and could support multivoltinism because native host plants were available. We can still find bits and pieces of this ecology in the Delta and the Suisun marshes and in a few relict wetlands elsewhere. The draining, diking, and agriculturalization of the wetlands corresponded in time to the widespread naturalization of exotic weeds related to native marshland plants. What did our Mylitta Crescent *(Phyciodes mylitta)* feed on before the various pestiferous annual Mediterranean thistles came into California? Native, mostly wetland thistles, just as it does in mountain bogs today. In fact, it can still be found using the endangered Suisun thistle *(Cirsium hydrophilum)* at Rush Ranch! The Common Checkered Skipper still uses checkerbloom *(Sidalcea)* in wetlands where it can find it, but thanks to the weedy species of mallow *(Malva)* it is now found in every garden and weedy lot in the northern part of the state. And the Anise Swallowtail still lays eggs on water hemlock *(Cicuta)* and on *Oenanthe* in the marshes, but percentage-wise, very, very few of them.

I once escorted a famous plant ecologist from abroad on his first field trip in the Sierra Nevada. Several grad students came along. On the way up, they loudly bemoaned the domination of the landscape by naturalized weeds. Finally our visitor demurred. "Why do you insist on considering this a disaster? Why don't you consider it an *opportunity*? Think of all the evolution that will go on in the next thousands of years!"

The butterflies are certainly treating the anthropic landscape as an opportunity.

Butterfly Classification

The Importance of Phylogeny

Taxonomy is the business of naming the living world. Systematics is the broader view, incorporating not only the process of naming things but also the philosophical rationale and biological basis for naming. According to Genesis, God told Adam to name all the animals—so taxonomy is, in fact, the oldest profession. (The word "butterfly," however, does not appear in the Bible.) All human cultures practice taxonomy, but the modern scientific practice dates from the mid-eighteenth century when a Swede named Carolus Linnaeus (Karl von Linné) standardized biological nomenclature. He fixed as the standard a genus (adjective: generic) name and a species (specific) epithet—what we call binomial nomenclature: *Homo sapiens, Pieris rapae.* (The generic name is capitalized; the specific epithet is not; the combination is the name of the species and, by convention, is italicized or underlined.) European taxonomists of Linnaeus's day wrote descriptions in Latin, a dead language except among scholars. The use of Latin (or Latinized) names provided universality and sidestepped national and linguistic rivalries. *"Pieris rapae"* is the scientific name for what we call the Cabbage White or European Cabbage Butterfly, the Swedes call Liten Kalfjaril, the Spanish Blanquita de la Col, the Germans Kleiner Kohlweissling, and so on. The rationale is the same as that for the use of Latin on the stamps and currency of polyglot Switzerland.

The idea of formal taxonomy was to provide a fixed identifier for every biological entity, recognized across all borders and for all time. To achieve such standardization a legalistic structure was needed—one that would specify how priority would be decided, what it meant for a name to be "available," and how to handle duplication (homonymy; you can't have two animal genera named *Platypus,* which is why *Platypus* is a beetle and the weird mammal is now named *Ornithorhynchus*). An International Code of Zoological Nomenclature grew up in the late nineteenth and twentieth centuries and is revised periodically. Its application is overseen by the International Commission on Zoological Nomenclature, a group of senior taxonomists who act as a sort of taxonomic supreme court. They have the power to set aside the code in specific cases when its application would be detrimental to stability—which is often the case.

Linnaeus was active a century before *The Origin of Species* came out, and like most people in Christian Europe, he accepted that God had created the biosphere pretty much as it is today. He based his classifications on similarities and differences and sought to make his system as straightforward and useful as possible. He believed every taxonomist should learn all the genera by heart and recognize them on sight. Darwin, in 1859, brought a new rationale for classification: rather than giving us a map of the mind of God as reflected in His creation, taxonomy would now mirror ancestor-descendant (phylogenetic) relationships among the organisms. Relationship was no longer merely a measure of similarity; it was a measure of the recency of common ancestry. But in practice, these two ways of looking at classification were not always compatible. For most of the modern era, taxonomy was conducted as a series of compromises and was heavily dependent on the judgment and prestige of its practitioners. One learned taxonomic judgment by apprenticing oneself to a master; the methods used could be communicated by osmosis, but in philosophical jargon they were not "operational"—they could not be set down in an instruction manual like directions for assembling a bicycle. This style of taxonomy is known as "evolutionary," because it at least paid lip service to the main ideas of the neo-Darwinian synthesis of the 1940s through 1960s.

In the 1960s a reaction set in—originally called "numerical taxonomy," later "phenetics." Pheneticists wanted to make classification more objective, repeatable, and operational. They sought to rid it of its "artistic" and judgmental components, seizing on the emerging availability of computers to make the process rigorously quantitative. Without being hostile to evolutionary theory, they sought to decouple classification from it—claiming that the same techniques could and should be used equally to classify sea anemones or doorknobs. Ultimately their movement failed (having made little impact on butterfly taxonomy), but it was quickly followed in the 1970s by another, originally called "phylogenetic systematics" and later "cladistics." Cladistics also sought to make classification more objective, repeatable, and operational, but in a different way—by making it dependent on and congruent with phylogeny. Overall similarity or difference were irrelevant; only those characters that were logically informative about phylogeny were to be used in classification. The broad principles of cladistics are now very widely accepted by biologists, although methodological quarrels still abound. The important thing is that there is now general agreement that taxa—named "things" in a taxonomic hierarchy—are meaningful only if they represent groups of organisms descended from a common ancestor. The jargon word for this is "monophyly" (adjective: monophyletic).

The cladistic approach gives us new tools for analyzing the history of life on Earth as well as the evolution of lineages, taxa, or faunas. In particular, cladistic reasoning combined with molecular technology allows us to reconstruct the deep relationships of higher taxa (phyla, classes, orders, and so on) and to trace the history of specific lineages in a geographic context (phylogeography). It even allows us to test hypotheses of herbivore-plant coevolution in a rigorous way! Because there are very few butterfly fossils, cladistics has been very useful in understanding the evolutionary history of the group. We have already seen how useful phylogeography has been in understanding the history of our own fauna.

Why Are the Names Always Changing?

Nothing is more discouraging to the beginner in butterfly study than finding him- or herself in a taxonomic morass. The thirteenth-century Sufi mystic Rumi warned us not to be overly preoccupied with the names of things. Many things have more than one name, he said, and obsessing over which is correct leads on to strife and even war.

Butterfly taxonomy is a mess. The aim of the system is supposed to be universal stability, but if you compare the butterfly books currently in print you find discrepancies on almost every page. Is our Western Tiger Swallowtail *Papilio rutulus* or *Pterourus rutulus*? Is our Echo Blue *Celastrina argiolus echo* or *Cyaniris ladon echo*? Is our Field Cresent *Phyciodes campestris* or *Phyciodes pulchellus*? Is Muir's Hairstreak a species or a subspecies or only a "host race"? Should *Speyeria atlantis* and *Hesperia comma* each be split into two species? Who decides and how? What do you label your specimens or your slides?

Let's get a few things straight.

The code and the commission are strictly neutral on matters of either biology or ideology. The code deals with what it means to be published, how to assign temporal priority, what is an adequate indication of the biological entity a name refers to, how to replace a lost type-specimen, and so forth. Decisions about what genus a species should be in, or whether something is a species or a subspecies, are biological decisions and outside its purview. *Papilio* versus *Pterourus* hinges on perceived limits of genera. The Muir's Hairstreak question depends on what criteria one is using to define species or subspecies. These cases resemble most instances of taxonomic instability in that they are grounded in honest differences in biological opinion, not in legalisms. You have probably heard of "splitters" and "lumpers." Splitters tend to emphasize differences; lumpers tend to emphasize similarities. These personality types have always been with us.

But the situation has been complicated by modern technology. Molecular methods give us new ways to look at relationships among populations: now we have multiple data sets that often do not converge to a straightforward taxonomic interpretation. For example, mitochondrial DNA is inherited only through the mother, while non-sex-linked (autosomal) genes in the nucleus pass through both parents. If females are either more or less likely to outbreed than males, the patterns of relationship suggested by the two data sets will be different. Then what? We have also discovered that molecules and wing patterns often give inconsistent results (see below). In a sense, we now know less than we thought we did a few years ago when we had only morphological data to work with. As the American humorist Artemus Ward said, it's not what we don't know that hurts us, it's what we know that ain't so.

There's a second front as well: common names. Ornithologists joke that the common names of American birds have been a lot more stable than the scientific names, and it's true. The common names of our butterflies are currently trapped in a rivalry between two competing lists, each seeking to be *the* standard. There has also been a lot of bad judgment in coining new common names, many of them quite unnecessary. For example, *Ochlodes agricola* had long been "The Farmer," which is what *agricola* means. Now in some books it is the "Rural Skipper." This is singularly unfortunate since there is a skipper whose scientific name actually is *ruralis (Pyrgus ruralis)*! Why "Variable Checkerspot" instead of the venerable "Chalcedon Checkerspot," when other checkerspots are at least as "variable" if not more so?

Anyway, in writing a book like this there is no way to please everyone. To avoid compounding the confusion one should pick an authority and stick to it unless there are compelling reasons to deviate, which should then be spelled out. That is what has been done here, and I apologize to anyone who thinks I should have picked a different authority. I would also argue that a book like this is not a proper place to introduce new names or combinations. So I won't, though inevitably some of them will *look* new to some readers. I have tried to include common alternative names in the more or less contemporary literature for cross-referencing purposes. And good luck.

A Bit about Genitalia

Users of this book are either unlikely to do genitalic dissections or already know how to do them. A word about why they appear so prominently in the butterfly literature does, however, seem warranted.

The male external genitalia of insects are typically heavily sclerotized, in other words, they are impregnated with the protein sclerotin, which makes them hard and rigid. They are also quite complex, with all sorts of bumps, hooks, spines, and such, all of which have names. (A diagram of the male genitalia has been compared to a map of San Francisco Muni Metro.) The female genitalia are mostly internal and soft. They are both harder to study and less rich in diagnostic detail. Diagnostic of what?

Since the mid-nineteenth century it has been recognized that male genitalia appear to vary morphologically at the level of species in insects. To put it another way, when wondering if two different-looking butterflies are species or subspecies, it's a good idea to check out the genitalia (known in the trade as "pulling tails"). Consistent genitalic differences are usually a good indicator of species status, as confirmed independently by other means (such as molecular genetics). Why should this be so? It has been suggested that genitalic differences act as a reproductive isolating mechanism—that is, interspecies hybridization is prevented because the parts don't fit together. This is known as the "lock and key" theory. But the variation in male genital morphology is generally not mirrored in the female; there seem to be many keys for the same lock. The lock and key idea is not dead, but it has not fared very well in studies to date. Whatever the reason for the variation (the current favorite is that the male morphology provides tactile stimulation to the female and this enhances the probability of successful reproduction), the generalization remains valid. There are cases where biological species are so similar that the genitalia *must* be examined for reliable identification. In this book the species pairs that may require tail pulling are *Everes comyntas/E. amyntula, Pyrgus communis/P. albescens,* and *Erynnis persius/E. pacuvius.*

Species, Subspecies, and "Forms," Oh My!

The word "phenotype" refers to the outward appearance of an organism: blue or green, smooth or hairy, tailed or tailless. Before Darwin taught us that speciation is a *process,* taxonomy was typological: it classified according to essence, that which all members of a taxon had in common by definition. Variation ("accidental properties") was a nuisance, if not an embarrassment. Imperfection in the reproduction of God's design could be read as a consequence of sin. Of course we know now that variation is the stuff of evolution and very important; for the butterfly watcher or collector, however, it can still be a nuisance. At least at first. Let's consider types of variation and their taxonomic implications.

Species have been defined in various ways. For purposes of this book we will use Ernst Mayr's biological species concept, in which the essential criterion is ability to interbreed. Members of the same species can exchange genes freely, while members of different species cannot. This criterion says nothing about phenotypes; things that look alike but cannot interbreed are different species, and things that look different but can interbreed belong to the same species. Simple, right? No. We know, for example, that Muir's Hairstreak and Nelson's Hairstreak can interbreed freely in the lab. But they sometimes co-occur in the field and don't interbreed. Are they one species or two? (Yes.)

In general, given such contradictions, it seems wisest to do what the organisms do in nature; if they behave as two species, treat them as such. Of course, if you have two entities, one in California and one in Tibet, they never have a chance to get together. Our Western White, *Pontia occidentalis,* is phenotypically nearly identical to *Pontia callidice kalora* from Tibet. But is it the "same species"? If we brought them together in the lab, could we answer the question definitively?

Sometimes we can get clear evidence about species status. To stay with whites for a minute, the Western White and the Checkered White *(Pontia protodice)* often occur together in the Great Basin and Sierra Nevada, but apparent phenotypic hybrids are quite rare. At the molecular level they have a fixed allelic difference at one enzyme locus (out of two dozen studied); in other words, all members of one species differ at that locus from all members of the other. Since it's a nuclear gene, any hybrid should have one of each allele. We haven't found such an individual yet; we've been looking. Hybrids certainly cannot be frequent.

We now know that named subspecies, however phenotypically distinct, cannot be assumed to be species in the making. Striking differences in color and/or pattern may not be mirrored at the molecular level, and there may indeed be very persuasive molecular evidence for gene flow. There is no generally accepted criterion for subspecies status; a subspecies is what someone claims it is— and others may disagree. In this book I have tried to minimize the use of subspecies, given their subjectivity and uncertain biological nature. However, sometimes one *has* to use them.

The U.S. Endangered Species Act of 1973 follows the International Code of Zoological Nomenclature in treating species and subspecies as more or less interchangeable (as, historically, many have been). Since subspecies qualify for legal protection, the law affords safeguards to things that may not be "real"! (A useful rule of thumb, the 75 percent rule—at least 75 percent of specimens must be nameable by eyeball—has no formal status or rigorous justification.) Because one is not allowed to collect or molest protected entities, such subspecies *must* be flagged in this book. And they are.

Subspecies are normally recognized at the level of populations or groups of populations occupying a defined geographic area. They are not normally sympatric, but in some lycaenids and one hesperiid (all outside our area) there are sympatric California populations that are isolated by having different flight seasons (allochrony). Molecularly, these do not seem to be full species (yet).

The butterfly literature is loaded with references to "forms" and "aberrations." These are nongeographic variants of various kinds. Many were originally described as new species. Forms can be broadly divided between polymorphisms and polyphenisms.

Polymorphisms reflect underlying genetic differences. The ongoing occurrence of different forms within populations reflects heterogeneity in selection pressures—that is, some forms are favored in some environments and others in others. Our most conspicuous example is the white form of the Orange Sulphur *(Colias eurytheme)*, which is expressed only in the female. (It is produced by a dominant sex-limited gene. Sex-limited genes are not on the sex chromosomes, as demonstrated by the pattern of inheritance, but they are expressed only in one sex.) The rare yellow female of the Sara Orange-tip *(Anthocharis sara)* and the buff female of the Spring White *(Pontia sisymbrii)* may be other examples. The female of the Northern Checkerspot *(Chlosyne palla)* occurs in three forms. One is orange like the male; one is predominantly black, with yellow and orange spots; and one is intermediate. The black form is probably a mimic of the Variable Checkerspot *(Euphydryas chalcedona)*. This situation cries out for study. You may have noticed that in all of these examples the female is polymorphic but the male is not. It is a general rule that if only one sex is polymorphic, it is always the female. Several hypotheses have been advanced to explain this. Can you suggest any?

In a polyphenism, genetically identical individuals may look radically different from one another due to environmental conditions during development. The regulation of seasonal polyphenisms is very similar to, and sometimes coupled with, that of diapause; the same cues (photoperiod, temperature, and photoperiod-temperature interactions) are generally used. Less frequently other factors may be implicated, such as population density, humidity, or food quality. Polyphenisms occur in larvae, pupae, and adults. Larvae of the Orange Sulphur and Anise Swallowtail *(Papilio zelicaon)* are darker in cool and humid conditions and lighter in warm and dry ones. The differences are probably adaptive for thermoregulation. Pupae of many pierids and papilionids and some other species may be green or brown. This has been shown to be inducible in some species by the color and/or texture of the substrates crossed by the larvae during the "wandering" period between cessation of feeding and settling down in a pupation site. However, the green/brown polyphenism is also partially linked to diapause, and in several species some individuals are genetically brown regardless of the environment— the story turns out to be quite complex.

Adult seasonal polyphenisms are best developed in our multivoltine pierids, in which the seasonal forms are so different looking as to have been originally described as species. Cold-season Gray-veined Whites *(Pieris "napi")* have gray veins, but the summer ones are often immaculate. Cold-season Orange Sulphurs are heavily dusted with gray green on the underside of the hindwing and have reduced or even obsolescent black borders on the upper surface. (The apparent gray green color is really a reflection from a high proportion of black to yellow scales; there is no actual green color.) This melanization—as noted earlier—assists in lateral-basking (thermoregulation). The wing shapes also differ seasonally in both species. Cold-season *Colias* are smaller and have narrower wings, reflecting a different mode of flight, as is obvious afield: the cold-season form flies near the ground out of the wind, while the broader-winged summer forms fly higher and exploit rising air (thermals) to get a free ride. There are many other differences as well. Seasonal polyphenisms also occur in the Cabbage White, Checkered White, Western White, Sara Orangetip, and Large Marble *(Euchloe ausonides)*. In other families they are found in the California Ringlet *(Coenonympha tullia californica)*, Buckeye *(Junonia coenia)*, Acmon Blue *(Plebejus acmon)*, Eastern Tailed Blue *(Everes comyntas)*, Field and Mylitta Crescents *(Phyciodes campestris* and *P. mylitta)*, Satyr Anglewing *(Polygonia satyrus)*, Sandhill Skipper *(Polites sabuleti)*, and Sachem *(Atalopedes campestris)*. The Least Checkered Skipper *(Pyrgus scriptura)* has an especially dramatic seasonal polyphenism in which the spring form is very similar to related species, but the summer form is very different. The Painted Lady *(Vanessa cardui)* has migratory and nonmigratory forms that look quite different as well.

The control of many of these polyphenisms is poorly if at all understood; their study requires strictly controlled rearing conditions, preferably in growth chambers that can be programmed for temperature, humidity, and photoperiod.

Finally, there are individual variations presumed to have no population-level significance; these are usually called aberrations. In the "old days" they were often named, but the names have no standing in formal nomenclature even if familiar and widely used. (Neither do names for polymorphs, like the *"alba"* white female of *Colias* or the *"ariadne"* cold-weather form of *C. eurytheme*.) Some aberrations have a genetic basis, some are developmental accidents, some are induced by environmental stress such as heat or cold shock, and many are just unexplained. Some are spectacularly bizarre and beautiful, and most are prized by collectors. Among the most interesting are sexual mosaics, in which part of the animal is male and part female. They are easy to spot in species with strong sexual differences in color or pattern. The mosaicism may be ragged or blotchy, or perfectly symmetrical left to right (one side male, the other female). The latter are called bilateral gynandromorphs and can occur in any species but are especially spectacular in the Orange Sulphur when the male side is orange and the female white. Bilateral gynandromorphs usually have male genitalia on one side and female on the other, one testis and one ovary. Sexual mosaics are possible because in insects the sex of each cell is determined by its chromosomes, not by circulating sex hormones as in mammals. In butterflies the female is the heterogametic sex (with two different sex chromosomes, like male humans), while the male is homogametic (with two like chromosomes). If a sex chromosome is accidentally lost, the cell and its descendants will be female. In bilateral gynandromorphs, either a sex chromosome was lost from one daughter cell at the first division of the fertilized egg (zygote), or else the egg accidentally had two nuclei—one chromosomally male, one female—both of which were fertilized (by different sperm). Bilateral gynandromorphs have been recorded in at least a dozen species in our area. By the way, forewing-hindwing gynandromorphs are virtually unknown. Intersexes, which are phenotypically intermediate between the sexes "all over," are extremely rare in butterflies, and their causation is generally not understood.

Plate 23 includes an aberrant West Coast Lady *(Vanessa annabella)* belonging to a series of recurrent aberrations known by the informal names *V. a.* "*muelleri*" and *V. a.* "*letcheri.*" Virtually identical aberrations occur in the Painted Lady and very rarely in the American Painted Lady *(V. virginiensis),* and obviously similar forms in the Red Admiral *(V. atalanta).* These can be produced experimentally by administering temperature shocks to the young pupa, but they occur spontaneously in nature at all times of the year, and there is some evidence that they may also have a genetic basis. (When both genes and environment can produce the same phenotype, we speak of the latter as a "phenocopy.") Dramatic aberrations also occur in all the tortoiseshells, including the Mourning Cloak *(Nymphalis antiopa),* which occasionally produces one called *N. a.* "*hygiaea*" in which the yellow border expands basally to cover nearly half the wings, obliterating the blue spots altogether. This again can be manufactured by pupal chilling. One distinctive aberration of the Buckeye has the white band on the forewing nearly or entirely covered with brown, closely resembling the usual form of one of the Buckeye's close relatives in the subtropics. All of these have been collected in our area more than once.

Checkerspots and fritillaries are prone to black suffusion—I once saw an all-black Variable Checkerspot!—and coppers and blues often have the complex underside patterns variously fused.

Historically, experiments on butterfly wing patterns in the late nineteenth century provided very important clues to the control of gene expression during development. This line of research has been revived with great success in our own time, when the molecular tools exist to really "get at" what happens in developmental genetics of wing pattern.

One more source of variation, albeit an infrequent one, is interspecific hybridization. In our area, hybrids between the West Coast Lady and Red Admiral occur very rarely—I have seen one in over 30 years of field work. Since these two species co-occur nearly everywhere and share their lek sites and mating behaviors, there must be strong barriers to hybridization. Other hybrids have been claimed, for example, between the Sara Orange-tip and Boisduval's Marble *(Anthocharis lanceolata)*, but these are even rarer. Despite their very similar appearance, the various species of fritillaries, hairstreaks, duskywings, and blues are not known to hybridize at all. There are several records of Checkered Whites in copula with Cabbage Whites, but nothing seems to come of it. The Orange Sulphur hybridizes with yellow relatives (the eastern and western populations of the Yellow Sulphur *[Colias philodice* and *C. eriphyle]*) wherever they co-occur, but they don't in the area covered by this book.

SPECIES ACCOUNTS

How to Use This Section

This book is specifically directed toward the Bay Area and Sacramento Valley. Every species that has been recorded from these areas is treated individually. The taxonomy broadly follows *Butterflies of North America* by Jim P. Brock and Kenn Kaufman (2003). In a few cases I have deviated slightly from their usage, but these are flagged in the text. Where current literature employs significantly different common or scientific names for the same biological entity, I have included them all in the headings. Where there is substantial uncertainty about what the limits of species are, I

THE FUZZY LOGIC OF BUTTERFLY NAMES

"**F**uzzy logic" was invented by Lotfi Zadeh in 1965 and was a critical development for the emerging computer revolution—it underlies how search engines work, for example.

Unlike "fuzzy thinking," which typically leads one into error, fuzzy logic leads one to insight. In ordinary logic things are either big or small, fat or thin, good or bad. But in fuzzy logic, as in the real world, they can partake of seemingly contradictory properties at once to varying degrees. Looking at the world through fuzzy glasses can help you navigate the seemingly illogical maze of butterfly names.

If common (English) names had all been coined at one time by one person—or even a committee—they could be rigidly hierarchical and thus "make sense" in the world of ordinary logic. Actually, most of them do, even though they have accumulated over as long as 300 years. All our coppers belong to one tribe. Taxonomists may argue about whether they can be accommodated in one genus *(Lycaena)* or several, but at least nothing outside that tribe is called a copper. But it isn't always that neat.

Consider the terms "marble" and "orange-tip." Marbles and orange-tips are closely related. Logically, all marbles should be in one genus *(Euchloe)* and all orange-tips in another *(Anthocharis)*. But *Anthocharis lanceolata* is an orange-tip with no orange in either sex, and it has never been called an orange-tip. Instead it has two common names in widespread use: Gray Marble and Boisduval's Marble. Why? Tradition—and logic at another level, in which names are descriptive of color, not phylogeny.

In this book, and most others, the satyrines (a subgroup of brush-

---

use the term "complex" and explain the nature of the problem in the text.

For each species you will find a general discussion, followed by **DISTRIBUTION** (by counties or regions), **HOST PLANTS** of the larvae, **SEASONS** of flight, and **SIMILAR SPECIES** sections. If you consult sources aimed at a larger area you will find many more host plants listed for most species, as well as information on other aspects of their biology that may contradict what you see here. All the information given in these accounts is based on the Bay Area and Sacramento Valley unless explicitly stated to be otherwise! The life cycle of almost any butterfly will be quite different in Richmond, California, and Richmond, Virginia.

---

foots [Nymphalidae]) are collectively referred to as "satyrs and meadow browns." You will not find any California butterflies called "Meadow Brown," however. The Meadow Brown is a common British and continental European species. Again, it's just a matter of tradition.

Or consider the word "admiral" as it applies to butterflies. It was originally "admirable" and had nothing to do with the Navy. Our Lorquin's Admiral belongs to the genus *Limenitis,* which also includes the Old World white admirals. But a similar *Limenitis* in the eastern United States is called the Banded Purple, a purely American name. Birders in this country more or less reluctantly adopted the British name "kestrel" for our sparrowhawk, but no such transplant of British names has occurred for butterflies. The result? Inconsistency in naming. The Red Admiral *(Vanessa atalanta)* is found in both countries and has the same name, but it is not related to the White Admiral or to Lorquin's Admiral. "Admiral" is not an indicator of phylogeny. The Gulf Fritillary *(Agraulis vanillae)* isn't a true fritillary *(Speyeria),* either.

There are ambiguities of a different kind in the use of scientific names. They reflect a different kind of history—evolutionary history.

If everything were clear-cut in taxonomy we would have little reason to believe speciation is a process. The messiness of nature is testimony to its being a work in progress. We know a lot more about butterflies than about most organisms, and that makes it that much harder to make decisions. It's a lot easier to convince yourself you know what the species of Bolivian bat fleas are, precisely because you've seen so few of them!

Don't fret. Go with the messiness. As a contemporary comic strip says, "get fuzzy."

Several species are included on the basis of vague old records that have been recycled through previous works. If the records are considered questionable, that is plainly indicated. Strays may be exquisitely rare or relatively frequent. All the strays known to me are included, with a frank discussion of the likelihood of seeing them, but most are not illustrated. If you have something—on a pin, on a memory card or slide, or just in your head—that doesn't match anything in this book, look in one of the larger-scale field guides. There's no reason why you can't add something new to the list.

Recent molecular genetic and other studies are causing major changes in plant classification. Because most users of this book are not professional botanists, I have chosen to stay with the familiar groupings found in currently available floras rather than risk wholesale confusion. So far, the greatest impact for our purposes is on the figwort family (Scrophulariaceae), which are hosts for checkerspots and the Buckeye *(Junonia coenia)*. Anyone interested in seeing what has happened to the figworts can look at an article by Olmsted et al. in the *American Journal of Botany* 88:348–361, 2001. Fortunately, despite the changes, these plants still "hang together" as an evolutionary unit.

A word about distribution: At the end of the book is a county checklist for the Bay Area and a regional checklist for the Sacramento Valley and Sutter Buttes. The valley checklist is not divided into counties because nearly all of them embrace foothill and montane areas, and their species lists thus do not properly reflect the valley as a physiographic and climatic unit. *Remember that most butterflies occur as discrete colonies.* The fact that a species is recorded in your county does not mean you are likely to find it in your backyard. We have seen many populations succumb to the advance of civilization, and even places that look pristine have often lost important species. Old-time collectors such as Robert Wind or Michael Doudoroff would be appalled to see how much of the North Bay and Santa Cruz Mountains fauna has been lost. In some cases, as with the Western Pine Elfin *(Incisalia eryphon)* in San Francisco or the Greenish Blue *(Plebejus saepiolus)* and Clodius Parnassian *(Parnassius clodius)* in Marin, we are not sure if a species is still around. Fortunately, most species can still be seen if you are willing to travel a little farther. But that's another book.

Swallowtails and Parnassians (Papilionidae)

Only about 600 species of papilionids exist worldwide, but this family, which is much richer in the Tropics than elsewhere, contains many of the world's most spectacular butterflies. Perhaps because they are so charismatic, perhaps because the numbers of species are manageable, the swallowtails have been very thoroughly studied from a phylogenetic standpoint—using both morphology and molecules—and we believe we understand the evolutionary history of the family pretty well. Many of the adaptive radiations in papilionid history appear to have followed a breakthrough into a new host plant "adaptive zone"—that is to say, by conquering new sets of secondary chemicals, they opened the door to all sorts of ecological experimentation. Our Pipevine Swallowtail *(Battus philenor)* is a representative of the pantropical "*Aristolochia* swallowtail" lineage, which includes the spectacular birdwings from Oceania *(Ornithoptera* and *Troides)*, as well as a bewildering set of Neotropical species often deeply involved in mimicry rings. The group to which our Anise and Indra Swallowtails *(Papilio zelicaon* and *P. indra)* belong is of evolutionary interest because it seems to be in great ferment since deglaciation, with all sorts of exciting things happening ecologically and genetically in both the Old World and the New. Rather bizarrely, given how big and showy these beasts are, new species are still being discovered in temperate North America! They are, of course, sibling species—genetically distinct but until recently hidden within one taxonomic entity. Nothing new has turned up in our area, however…yet.

Swallowtail and parnassian eggs are characteristically dome shaped or globular and smooth. They are laid singly, except in a few pipevine *(Aristolochia)* feeders, our Pipevine Swallowtail being one of them. The two basic larval types are sometimes called "red-tuberculate" and "smooth green," but of course these are oversimplifications. The red-tuberculate type is exemplified by the Pipevine Swallowtail and seems to be the primitive type in the family. All our other swallowtails are the smooth green type, but when young many appear to impersonate a bird dropping (some do so all the way through the larval stage), and the group that includes our Anise and Indra Swallowtails group are variously banded when large. The "tiger" group (Western Tiger, Pale, and Two-tailed Swallowtails [*P. rutulus, P. eurymedon,* and *P. multicaudatus*]) have inflatable eyespots on the thorax, which serve to deter predation. Swallowtail larvae have a characteristic eversible organ, the osmeterium, concealed in a slit behind the head. If the larva is disturbed it everts the osmeterium, which is usually brightly colored, and discharges a more or less foul-smelling substance. The Pipevine Swallowtail, which relies on toxic chemicals gotten from its host plant to render it inedible, seldom uses its osmeterium. Our others do, and he or she who would rear them indoors has been duly warned not to annoy them. The pupa is naked, commonly comes in both brown and green morphs that are not entirely seasonal, and is attached to the substrate by a girdle around the middle and a button of silk at the anal end, as in pierids. All of our swallowtails overwinter as pupae.

The parnassians are primitively associated with the pipevine family (Aristolochiaceae), but our species feeds on bleeding heart (*Dicentra formosa,* Fumariaceae, now in Papaveraceae; these plants are generally poisonous, too). The larvae are plain but not green, and the pupae are rounded (obtect) and usually enclosed in a rudimentary cocoon. Overwintering occurs as the egg. The parnassians, like the Anise Swallowtail group, seem to have been very active evolutionarily since the last glaciation and for their trouble have received a plethora of unnecessary names. Some 30 years ago the chemical ecologist Murray Blum commissioned me to collect the hitherto-unstudied osmeterial secretion of the Clodius Parnassian (*Parnassius clodius*) for analysis. I had trouble getting much, and he egged me on with a telegram reading "Keep those secretions coming! It's now or never!" He finally got enough.

Swallowtails are generally easy to rear, and most species have large genitalia and can be hand paired. Many hybridization studies have been done (bypassing behavioral barriers to crossing), and the transmission genetics of many color and pattern traits have been worked out.

CLODIUS PARNASSIAN *Parnassius clodius*
Pl. 3

This is a distinctive butterfly of mesic forest where the host plant, bleeding heart (*Dicentra formosa*), grows. Parnassians are a circumboreal group that includes some of the most cold-adapted butterflies on Earth. They reach the High Arctic and to about 6,000 m (20,000 ft) in the Himalayas and other high Asian ranges, making them among the highest-altitude butterflies on the planet. Our species is near the warm edge of its climatic tolerances and seems to be in retreat. Formerly, a very lightly marked population, called Strohbeen's Parnassian *(P. c. strohbeeni),* inhabited the Santa Cruz Mountains, but it has not been found since 1958 and is certainly extinct. The nominate subspecies, *P. c. clodius,* has disappeared from Sonoma County, so that the only known populations in our area at this writing are in Marin—if *they* are still extant. The species is still common farther north, in the high inner North Coast Range, the Klamath Mountains, and on the west slope of the Sierra Nevada, but even there lower-elevation populations seem to be declining, perhaps due to climate change.

Adults fly continuously along roadsides and riparian corridors and in damp meadows. They nectar at dogbanes *(Apocynum),* milkweeds *(Asclepias),* asters *(Aster),* tansy ragwort *(Senecio jacobaea),* mints *(Mentha),* and clovers *(Trifolium),* among other flowers. At rest they often hold the wings in an unusual rooflike posture, often body-basking in cool weather. The sexes are strikingly dissimilar. The male has semiopaque white wings with a reduced black pattern. His abdomen is yellowish white and densely hairy, and his genitalia are huge and heavily sclerotized. The female has much more extensive black markings and larger red spots on a semitranslucent ground. Her abdomen is black and hairless, and you normally cannot see her genitalia at all. The reason for this is the presence of a sphragis—a huge, waxy, pinkish white conelike structure attached to the lower posterior part of her abdomen. This is a vaginal plug secreted by the male during mating. It keeps her from mating again. Other males can be seen frantically trying to copulate with females so encumbered; they always fail.

The exaggerated genitalia and the sphragis are characteristic of the entire genus and appear to reflect intense male-male competition for mates. Female parnassians may attract males by releasing an airborne pheromone, a tactic not known in any other California butterflies. They lay very few, very large eggs, which overwinter. Under such circumstances, we would expect males to be very protective of paternity. The low fecundity of these butterflies has contributed to their being endangered in many parts of the world. Other factors contributing to the problem are their beauty, the historic tendency of taxonomists to oversplit and overname them, their extreme specializations, and the fact that many come from exotic or inhospitable locales, giving them an aura of mystery and, alas, a high market value. Many species are protected today under both national endangered species laws and international treaties or conventions. So far, none of our regional populations has attracted governmental attention; Strohbeen's Parnassian went to its reward long before there was a Federal Endangered Species Act.

DISTRIBUTION: In our area today, known only from Marin County.

HOST PLANTS: Bleeding heart (*Dicentra formosa,* formerly in Fumariaceae, now in Papaveraceae). The very dark larva, marked with yellow spots along the sides, is crepuscular or nocturnal and can be found in litter or under rocks by day. It appears to mimic sympatric millipedes, which secrete cyanide as a defense. The pupa is formed in a rudimentary cocoon. We do not know if this species ever uses the yellow-flowered golden ear-drops *(D. chrysantha),* a species found in chaparral, often abundant after a burn.

SEASONS: One brood, late May to July.

SIMILAR SPECIES: None.

PIPEVINE SWALLOWTAIL   *Battus philenor*
Pl. 1

This is the most characteristic, and most distinctive, of our Californian riparian butterflies. It is also one of our most specialized butterflies, feeding on only one plant species—California pipevine *(Aristolochia californica),* which can sprawl on the ground or ascend as a woody liana high into the trees. It is abundant in riparian forest in the Sacramento Valley and locally, mostly in canyons but also in coastal scrub (as in the Marin Headlands), in the Bay Area. Despite its wide distribution and seemingly robust populations, there are grounds for concern about its future. Its relatively low reproductive capacity and extreme dependency on one plant make it vulnerable to habitat loss and fragmentation, and indeed it has shown a downward trend in numbers around Sacramento for nearly 20 years. The University of California at Davis has planted California pipevine in its arboretum to encourage breeding there.

The Pipevine Swallowtail has the strangest life cycle of any California butterfly. It overwinters only as a pupa, and there is a large emergence in early spring. Eggs are laid in clusters, only on young, tender shoot tips or leaves, and the young larvae must begin their feeding on these. When the larvae pupate, some will lay over until next spring; some will hatch without diapause in a couple of weeks; and a few will hatch some time late in summer or in fall. Photoperiod seems to have nothing to do with this. Experiments have shown, not very convincingly, that water content of the host may be a factor in the decision whether to diapause. In any case, butterflies that hatch in late spring to early summer breed if they can. Pipevine stops growing most years by early summer, and after that there are no shoot tips on which to lay. Any early-summer larvae that make it to pupation perform the same way: some will hatch later the same season, while others carry over until next spring. And so on. Thus, the spring flight contains individuals from all generations of the previous year!

Despite the lack of oviposition sites, usually at least a few Pipevine Swallowtails are flying any time in summer or fall, even as late as November. Summer populations tend to have heavily male-biased sex ratios. However, females are present and are demonstrably mated and fertile. Mark-release-recapture experiments show that these butterflies regularly live for a month. If there happens to be a fire (or if the plants are cut down), pipevines regenerate quickly regardless of the time of year. Such off-season tender shoots are instantly clobbered with eggs. It would seem that the summer flight is a reproductive strategy keyed to the spotty, but not rare, production of new growth by injured plants. The long adult life spans, non-photoperiod-cued diapause, and so forth all smack of life cycles seen in butterflies of the rainfall-seasonal tropics. In fact, the genus *Battus* is a New World tropical member of a pantropical group of "*Aristolochia* swallowtails," which includes the fabulous birdwings *(Ornithoptera* and *Troides)* of the southwest Pacific. (They are arguably the biggest and showiest butterflies on the planet.)

Our populations are geographically isolated from the main body of the range, which extends from the desert Southwest to the mid-Atlantic states and south through tropical Mexico. As explained elsewhere in this book, molecular genetics has effectively disproven the hypothesis that the Pipevine Swallowtail is an ancient relict in California; instead it appears to be a recent (post-Pleistocene) arrival here. Nonetheless, our populations have diverged enough from others to warrant recognition as a subspecies, the Hairy Pipevine Swallowtail *(B. p. hirsuta),* although it is not particularly hairy. Our populations differ most conspicuously from others in having a shorter forewing—males elsewhere have the forewing somewhat prolonged apically. There are also biological differences, such as clutch size (females lay more eggs in a batch here than elsewhere, which is not surprising given how large the host plants here can be) and developmental rate, and larval morphology.

All species of pipevines produce toxic alkaloids called aristolochic acids. These are sequestered by the larva, passed on to the adult, and then transferred by the female to the eggs! The entire life history is thus protected from vertebrate predators (although invertebrate predators do eat them). The eggs (brick red), larvae, and adults display warning coloration, which as usual is enhanced by gregariousness. The pupa, however, is cryptic and may be either brown or green, in either case with a golden yellow filigree. More than half of wild-collected pupae are routinely parasitized by a small black-and-white chalcid wasp (*Brachymeria ovata,* Hymenoptera).

Because larval feeding stimulates increased levels of chemical defense, the larvae will characteristically munch a leaf for a while, then move on. No large leaf is ever entirely consumed, but by the end of the season every leaf is likely to be damaged. The larvae also eat the immature fruits, which look like small yellow bananas with fluted edges. They may reduce the seed set by 50 percent or more. However, California pipevine plants are extremely long-lived, and recruitment of seedlings is almost never observed.

Pipevine Swallowtail larvae move around a lot. Not only do they change leaves every few hours, but they often wander to seek cool retreats, and upon attaining maturity they walk for several hours before selecting a pupation site. All this mobility can be their undoing. On the American River Bikeway in Sacramento, where the species is abundant, the biggest source of larval mortality is…bicycles.

Adults nectar eagerly at California buckeye *(Aesculus californica),* yerba santas *(Eriodictyon),* dogbanes *(Apocynum),* milkweeds *(Asclepias),* thistles *(Cirsium, Carduus,* and *Silybum),* and many garden plants. In late summer the sole nectar source in many places is yellow star-thistle *(Centaurea solstitialis)*—an "upside" for California's worst weed. Males also puddle, often on sandy riverbanks. This is the only butterfly I know that visits the tubular red flowers of California fuchsia *(Epilobium canum,* formerly in *Zauschneria)* with any frequency. It also, rarely, visits vinegar weed *(Trichostema lanceolatum).*

The spring brood averages a little smaller than later ones, with slightly shorter tails and (allegedly) more "hair" on the body. Otherwise there is almost no variation. This butterfly keeps its wings in motion as it feeds; it rarely settles long but is to be seen flying through dappled light and shade and maintaining an aerial territory in sun flecks or glades. However, it wanders far from home, turning up in the city and in gardens and even occasionally colonizing an isolated, often nonnative, pipevine in someone's yard. Such events do not result in establishment of an ongoing population, however; that requires more biomass of host plant than one can probably provide.

DISTRIBUTION: All counties, but breeding only where the host occurs.

HOST PLANTS: California pipevine. Contrary to published reports, this species is not known to use our only other native member of the pipevine family, wild-ginger *(Asarum)* in California. Larvae are normally purplish black with rows of soft, orange red "horns." However, larvae that develop in full, hot sun (such as on regenerating shoots of the host after a summer fire) are usually nearly or all red, a form routinely observed in the desert Southwest as well.

SEASONS: February to November, simultaneously uni-, bi-, and multivoltine in the same location as described above. The second (late-spring) flight may be as large or larger than the first in the Sacramento Valley but is generally much scarcer in the foothills and canyons.

SIMILAR SPECIES: None.

ANISE SWALLOWTAIL
Papilio zelicaon

Pl. 1

The Anise Swallowtail is a classic example of ecotype formation—in which a taxonomic species consists of multiple "ecological races" adapted genetically to various local conditions, often presenting profound biological differences that, however, are not reflected in overall genomic differentiation or reproductive compatibility. The Anise Swallowtail belongs to a group of species found throughout the temperate Northern Hemisphere that feed on plants containing certain essential oils such as anisic aldehyde or anethole. More than 50 years ago, Vincent Dethier showed experimentally that these compounds were necessary to elicit feeding by larvae of the closely related eastern Black Swallowtail *(P. polyxenes)*. The compounds are produced by most plants in the carrot family (Apiaceae, which used to be called Umbelliferae, so the plants are often referred to simply as "umbels") and some in the rue family (Rutaceae). These families are not closely related, so the chemical similarity is probably convergent. The actual sister family to the carrot family, the ginseng family (Araliaceae), does not make them and is never used. The same is true for carrot family genera that do not make them, including the aquatic *Hydrocotyle* and, in our own area, buttoncelery *(Eryngium)* and sanicle *(Sanicula)*. On the other hand, the larvae will eat filter paper if it is coated with the oils.

The native hosts of the Anise Swallowtail are a variety of umbels found in diverse habitats, from rocky barrens (e.g., various biscuitroots *[Lomatium]*) to tule marsh *(Oenanthe* and water hemlock *[Cicuta]*). But in most of the area covered by this book, most breeding is on two species of naturalized carrot family weeds: poison hemlock *(Conium maculatum)*, which dries up by late June inland but may stay green at the coast; and sweet fennel *(Foeniculum vulgare)*—what most Californians call "anise," hence the name of the butterfly. Fennel is a perennial and offers edible material year-round. When the foliage senesces in late summer, larvae eat flowers and green fruit. Both of these abundant weeds were here by Gold Rush time and possibly before. When Lorquin was roaming California in the 1850s the Anise Swallowtail was apparently already eating fennel. This is actually not surprising: experiments have shown that it is intrinsically more attractive to ovipositing females than their own natal hosts, regardless of population.

Our weedy populations of the Anise Swallowtail, along roadsides and in vacant lots, are highly multivoltine. They have only weak, facultative pupal diapause (cued by temperature-photoperiod interaction in the third larval instar; a significant percentage of pupae may survive the winter without being in diapause). This is extremely interesting, because in our summer-arid mediterranean climate, hosts for breeding in the second half of the year would have been restricted to tule marshes inland and wet meadows in coastal prairie. Either multivoltinism evolved very rapidly de novo after fennel arrived (there was rapid selection for continuous breeding), or it was already established in the wetlands and merely spread rapidly along with fennel. Interestingly, when Anise Swallowtail larvae feed on the cauline leaves or flowering-fruiting tops of poison hemlock, a significantly higher percentage of the resulting pupae diapause than under identical conditions on either basal-rosette foliage of hemlock or foliage of fennel. Inland, poison hemlock cannot sustain breeding after late June.

Bishop's weed and toothpick weed (*Ammi* spp.), natives of the eastern Mediterranean, are naturalized spottily in the Sacramento Valley. These are summer annuals and would appear to offer an inviting target to host-starved ovipositing females, and indeed they do. However, 100 percent of the larvae die after feeding on these plants. One can predict that natural selection will either "train" these populations not to oviposit on the plants, or lead to a buildup of tolerance to their toxins. Interestingly, montane populations of Anise Swallowtail from the Trinity Alps and the Tahoe National Forest, which feed in nature on the carrot family herb angelica *(Angelica),* tolerate bishop's weed perfectly well in the lab — so we know genes exist that confer such tolerance.

Angelica is just one of several genera of native umbels used in the Coast Ranges in places where there is no, or little, fennel. Such populations are typically facultatively bivoltine. The most extreme situation is found on serpentine in the Coast Ranges, particularly in Napa County. Here the hosts—species of biscuit-root—are long-lived perennials but seasonally ephemeral; the butterflies are strictly univoltine. When reared side by side with multivoltine fennel populations on any host plant under photoperiod and temperature conditions that inhibit diapause, nearly 100 percent of the serpentine pupae diapause anyway. Of these, many will not hatch for two, three, or even up to five years! This is interpreted as a "risk-spreading" strategy in a stressful environment. We know the physiological differences are genetically controlled, but it is logistically and computationally difficult to study the genetics of threshold characters such as diapause. There is a tantalizing hint (from distorted sex ratios) of some degree of reproductive incompatibility between serpentine-univoltine and fennel-multivoltine populations; we do not know if gene flow is occurring between them.

In southern California, the Anise Swallowtail began feeding on young, tender shoots of cultivated citrus *(Citrus)* shortly before World War I. The same thing occurred, apparently independently, in Butte County near the north end of the Sacramento Valley around 1960. Lab studies have demonstrated that on most "larval performance" scales (e.g., growth rate), citrus is an inferior host compared to fennel. When potted fennel plants are set out in citrus orchards, females oviposit freely on them. It thus appears that citrus feeders are such by default—it's better to eat a poor host than to not have a host at all!

In multivoltine populations, spring-brood individuals are consistently small and pale yellow, with narrow black borders containing much blue, even in males. Summer individuals are larger, brighter yellow (sometimes slightly orange-tinted in big females), with broader black borders and little or no blue in males. Serpentine univoltines are somewhat intermediate.

Males are territorial perchers. In hilly terrain they are among the most persistent hilltoppers; in riparian forest they use sunspots. Even in the flats they will hilltop on piles of dirt or debris, but never on buildings. They are also very avid puddlers.

The distribution of Anise Swallowtail eggs and larvae shows a strong "edge effect": isolated and marginal plants get most of the ovipositions, while females rarely venture inside the huge pure stands of fennel one often sees on hillsides in the Bay Area. Thus, the abundance of the host may not accurately translate into butterfly abundance. Females flying through urban areas will leave a trail of eggs on umbels in people's gardens. Cutting or burning decadent fennel stimulates new growth and makes it much more attractive for the butterfly!

This species is easy to find and to rear and is a favorite for classroom activities. It would be even more of a favorite were it not endowed with butyric acid as its defensive secretion; one shot by an annoyed larva can fill a room with the smell of rancid butter for an hour. The Anise Swallowtail is ridiculously easy to hand pair, so you need not rely on spontaneous matings to keep a culture going.

In canyons, particularly on the often-used host California biscuitroot *(Lomatium californicum)*, up to 50 percent of larvae may be killed by the braconid (wasp) parasitoid *Cotesia lunata,* which kills in the second instar and produces only one parasitoid per host. Additional mortality is caused by a large ichneumonid wasp (in the genus *Ophion*). We do not, however, understand why populations of this butterfly (and its eastern relative the Black Swallowtail) tend to fluctuate violently from year to year.

DISTRIBUTION: All counties, in all habitats but dense coniferous forest.

HOST PLANTS: Native carrot family (Apiaceae) including biscuit-root *(Lomatium)*, angelica *(Angelica)*, *Tauschia*, yampah *(Perideridia)*, *Oenanthe*, water hemlock *(Cicuta)*, and hog fennel or cow parsnip *(Heracleum lanatum)*. Introduced carrot family including many garden species such as celery *(Apium graveolens)*, parsley *(Petroselinum crispum)*, carrot and wild carrot species *(Daucus)*, dill *(Anethum graveolens)*, true anise *(Pimpinella anisum)*, caraway *(Carum carvi)*, parsnip *(Pastinaca sativa)*, and of course sweet fennel *(Foeniculum vulgare)*. Also rue (*Ruta graveolens*, Rutaceae). No records are available of the Anise Swallowtail feeding on citrus in our area except around Corning and Orland in the northern Sacramento Valley. The form of the plant is so different that its discovery as a resource must not occur very often; females simply do not investigate plants that do not "look" right.

SEASONS: Overall, February to November. Partially bivoltine wildland populations fly mainly February to July; serpentine univoltines March to June.

SIMILAR SPECIES: This species is easily told from the three tiger-striped swallowtails by its relatively small size, black basal third of forewing, and overall checkered, rather than striped, appearance. The Indra Swallowtail *(P. indra)* is smaller still, mostly black, with yellowish white pattern and short tails.

INDRA SWALLOWTAIL
Papilio indra
Pl. 1

In our area this butterfly is essentially a serpentine endemic, found on bare, rocky ground in the Inner Coast Range, especially in Napa County. It is a small, mostly black swallowtail with very short tails and creamy white markings, flying early in the spring in the company of the Small Marble *(Euchloe hyantis)*, Spring White *(Pontia sisymbrii)*, Edith's Checkerspot *(Euphydryas editha)*, Sleepy Duskywing *(Erynnis brizo lacustra)*, and Anise Swallowtail *(Papilio zelicaon)*, which usually vastly outnumber it. Its mate-location strategy is a unique variation on hilltopping. Male Indra Swallowtails cede the summit to Anise Swallowtails. Instead, they fly back and forth parallel to and just below ridge lines, hanging suspended in space several meters beyond the reach of your net! Males also puddle. Neither sex visits flowers very often, but yellow flowers in the sunflower family (Asteraceae) and brodiaeas *(Brodiaea)* are included in its preferences. Our populations show very little variation.

DISTRIBUTION: Napa, Solano, and Sonoma Counties, mostly inland; unknown in the South Bay counties. Ranges northward through the Siskiyous, and in the Sierra. Oddly, reported in the (volcanic) Sutter Buttes; needs confirmation.

HOST PLANTS: In our area recorded only on the biscuitroot *Lomatium marginatum;* perhaps on other biscuitroots. The mature larva is ringed in pink and black and can be seen at a great distance; it usually ends up eating fruit when the foliage has all senesced. The pupa is unusually obtect and always wood brown.

SEASONS: One brood, April to May; has obligate pupal diapause, which may not break for several years.

SIMILAR SPECIES: None, except I suspect I am not the only person to have been momentarily confused by Mourning Cloaks *(Nymphalis antiopa)* in "Indra country"!

WESTERN TIGER SWALLOWTAIL   *Papilio rutulus*
Pl. 2

The Western Tiger Swallowtail is a familiar sight, coursing majestically high in the trees — be it in riparian forest, a city park, or an older residential neighborhood. It breeds on common landscaping trees. Adults visit many flowers, including in the wild California buckeye *(Aesculus californica)*, yerba santas *(Eriodictyon)*, brodiaeas *(Brodiaea)*, and related spring bulbs, vetches (annual and perennial) *(Vicia)*, milkweeds *(Asclepias)*, dogbanes *(Apocynum)*, buttonbush *(Cephalanthus occidentalis)*, and various thistles *(Cirsium, Carduus,* and *Silybum)*; in gardens a partisan of butterfly bush *(Buddleia)* and zinnias *(Zinnia)*. The spring brood is smaller, paler, with narrower black borders; otherwise there is only very minor variation. The Western Tiger is an enthusiastic puddler, and it does not hilltop. The spectacular larva, with its inflatable eyespots supposedly mimicking a snake's head, is high up in the trees and rarely seen. The pupa hibernates.

DISTRIBUTION: All counties, mostly in association with deciduous host trees.

HOST PLANTS: Sycamore *(Platanus)* and ash *(Fraxinus)* preferred—both native and exotic species. Also recorded in our area on cottonwood *(Populus),* willow *(Salix),* cherry, peach, and almond (all *Prunus*), tree privet *(Ligustrum),* lilac *(Syringa),* and (in Sacramento County) sweet gum *(Liquidambar)!*

SEASONS: Two to three broods, usually distinct; late February to November, but usually scarce after mid-August.

SIMILAR SPECIES: The Anise Swallowtail *(P. zelicaon)* is smaller and has a checkered, not a striped pattern. The Two-tailed Swallowtail *(P. multicaudatus)* is usually larger and has more yellow (the black pattern looks "thinner," with more yellow space between the elements); the second tail is longer, but this may not be evident in life. The male Pale Swallowtail *(P. eurymedon)* is white; the female is pale yellow but has more extensive black pattern and somewhat narrower wings; the contrasting orange spots near the tails are distinctive.

TWO-TAILED SWALLOWTAIL   *Papilio multicaudatus*
Pl. 2

Why is this, our largest butterfly, as scarce as it is in our area? Its habitat preference can be summed up succinctly as shaded riparian surrounded by unrelenting sun and heat. It is largely confined to canyons in the Inner Coast Range, where it soars majestically near the canopy in the company of the Western Tiger Swallowtail *(P. rutulus).* The two can be difficult to distinguish on the wing, especially in the spring when the first brood of the Two-tailed is scarcely any larger than the Western Tiger. (Some summer-brood individuals, especially females, are *huge.*) It visits tall thistles *(Cirsium, Carduus,* and *Silybum),* especially native species, eagerly. This is an urban butterfly in Mexico City, where it breeds on ash *(Fraxinus)*—but for us it is a "goodie" and mainly in wildlands.

DISTRIBUTION: Inner Coast Range counties both north and south of the Bay; south slope canyons in the Sutter Buttes; occasional along the American and Yuba Rivers on the east side of the Sacramento Valley.

HOST PLANTS: Hop tree (*Ptelea,* Rutaceae) reportedly preferred in our area; also recorded on sycamore *(Platanus),* ash *(Fraxinus),* and cherry *(Prunus)*. The distribution of the butterfly strongly argues for the hop tree association.

SEASONS: Emerges slightly later than the Western Tiger in spring; March to September (rarely early October), two to three broods.

SIMILAR SPECIES: The Two-tailed Swallowtail's wings are broader than those of the Western Tiger Swallowtail, with narrower black pattern elements (more yellow in between). The second tail is more prominent.

PALE SWALLOWTAIL   *Papilio eurymedon*
Pl. 2

Unlike the Western Tiger Swallowtail *(P. rutulus)*, the Pale Swallowtail is an infrequent visitor to "civilization." It prefers canyon riparian and chaparral habitats. Males patrol up and down canyons and streambeds and also hilltop in chaparral. The white male is very easily recognized, but the yellowish females may be difficult to tell from the Western Tiger. This species visits all the usual flowers (California buckeye *[Aesculus californica]*, yerba santas *[Eriodictyon]*, thistles *[Cirsium, Carduus,* and *Silybum]*, vetches *[Vicia]*, brodiaeas *[Brodiaea]*, etc.), but it also finds California wallflower *(Erysimum capitatum,* the orange-flowered low-altitude race) and the large lilies, especially Washington lily *(Lilium washingtonianum)*, irresistible. Both generate spectacular photo opportunities, and the latter often results in the entire lower wing surfaces being plastered with sticky orange pollen! Unlike the other striped swallowtails, the Pale Swallowtial is obligately single brooded. But it does not appear to have multiyear diapause. It is best described as a fanatical puddler.

DISTRIBUTION: General in the Bay Area. A stray downstream from the foothills into the Sacramento Valley, probably not breeding at this time.

HOST PLANTS: Coffeeberry *(Rhamnus)* and wild-lilac *(Ceanothus)* species (both in the buckthorn family [Rhamnaceae]), but recorded several times as laying eggs on white sweet clover *(Melilotus alba,* Fabaceae), which its larvae will not eat—presumably a mistake induced by an as-yet-undefined chemical similarity. White sweet clover is a naturalized alien.

SEASONS: March to August; in most years gone by late June. Never any second brood.

SIMILAR SPECIES: Yellow females resemble the Western Tiger Swallowtail, but the black pattern is more extensive (less yellow between elements), the wings narrower; typically one or two conspicuous orange lunules in the black border near the tails.

LIME SWALLOWTAIL   *Papilio demoleus*
Not illustrated

According to J.W. Tilden (*Journal of the Lepidopterists' Society* 22:187, 1968), a student in a general entomology class at San Jose State University (then still College) handed in a specimen of this common, widespread butterfly of the Old World tropics, which he supposedly caught in his back yard in Palo Alto, Santa Clara County, on April 4, 1964. Tilden believed it was the African subspecies. Whether or not the record is credible, the animal is obviously not established in the Bay Area; indeed, this remains the only record on the mainland of North America (it has recently surfaced in the Caribbean). And just as well too, since it is sometimes a serious pest of citrus *(Citrus)*.

Whites, Orange-tips, and Sulphurs (Pieridae)

The pierids, like most butterflies, are very diverse in the tropics, but this family includes the most cold-adapted of all butterflies: sulphurs fly on the shores of the Arctic Ocean, whites and sulphurs reach the subantarctic in Tierra del Fuego, both groups extend above tree line in most of the world's great mountain ranges, and a few whites reach nearly 5,500 m (18,000 ft) above sea level in the Andes and Himalayas! Our pierids are mostly medium-sized butterflies colored yellow, orange, or white with black markings. The pigments of the sulphurs belong to a unique class called pteridines, rarely found in other butterflies. Most sulphurs feed on plants in the old legume family (Leguminosae), which has now been divided into several families; the hosts of our resident species belong to the pea family (Fabaceae). Most whites and orange-tips feed on plants containing sulfur compounds called mustard oils. These are found in the mustard family (Brassicaceae), caper family (Capparidaceae), mignonette family (Resedaceae), found only in cultivation here, and nasturtium family (Tropaeolaceae), which is native to tropical America and the Andes. Many of the species are quite common and widespread, and our two most important economic pests among the butterflies belong here—the Imported Cabbageworm is the larva of the Cabbage White *(Pieris rapae)*, while the Orange Sulphur *(Colias eurytheme)* occasionally reaches high enough densities to damage alfalfa *(Medicago sativa)*. Many tropical pierids migrate in huge swarms. None of our species does this, but several are highly dispersive, and when Orange Sulphurs are leaving alfalfa fields during cutting, it certainly *looks* like a migration! The Cabbage White is not only one of our commonest and weediest butterflies, it is also the only species in our area that is naturalized from abroad.

Pierid eggs are typically spindle or milk-bottle shaped. All of our species except the Pine White *(Neophasia menapia)* lay them singly. Larvae are unspined but often downy, and usually cryptic. Pupae are attached by a button of silk at the tail and a girdle around the middle. They tend to be slender with prominent appendages in warm climates, and cobby with less conspicuous appendages where it is cool. Those of orange-tips and marbles resemble thorns. Most species overwinter as pupae in diapause.

One of our species, the California Dogface *(Zerene eurydice)*, hibernates as an adult. Several of our species have spectacular seasonal polyphenisms, and their cold- and warm-season forms are so different they were described as separate species.

PINE WHITE   *Neophasia menapia*
Pl. 3

Our Pine White is only distantly related to our other whites. Its relationships lie in the Old and New World tropics, and it is probably a really old component of our fauna (in the geologic sense!). Its life history and host plants are unique in our pierid fauna. Adults lek around the tops of pines and occasionally other trees, dropping to near the ground and then rising to near the top in a spiraling motion; they repeat the process again and again. Their flight seems effortless, with infrequent wingbeats, and they have more than once been likened to giant snowflakes wafted in the breeze. During their sporadic population outbreaks they can form a virtual snowstorm, but usually they are not very common. They do visit flowers, such as goldenrods *(Solidago)* and asters *(Aster)* but are often seen in places where nothing is in bloom. Markings and even wing shape vary quite a bit by location, and several subspecies have been described. The species is widespread in western North America, occurring from sea level to about 2,100 m (6,900 ft) in the north and 2,750 m (9,000 ft) in the south. Note the pronounced sexual dimorphism.

DISTRIBUTION: Upland forest in the Coast Ranges north of the Bay.
HOST PLANTS: Normally ponderosa pine *(Pinus ponderosa)*, also called yellow pine; reported rarely on Douglas fir *(Pseudotsuga menziesii)*. Eggs are laid in rows on the needles, up to 40 in a group; egg masses of several females may be clumped together. The larvae begin as leaf miners but soon consume entire needles. When large they feed singly, but often at high local densities. They also dangle on silk threads and blow from tree to tree. In many parts of the range there have been rare, massive outbreaks of this species leading to local defoliation of the hosts. This phenomenon seems unrecorded for the Bay Area. Pupation occurs either aloft in the trees or near the ground; pupae are rarely observed.

SEASONS: One brood, but the timing of adult flight can be highly variable from place to place. At low elevations the flight is usually in June and July, while in colder and mountainous areas it may be as late as September or even October. Overwinters in the egg; larval feeding occurs in spring.

SIMILAR SPECIES: None in our area.

SPRING WHITE or CALIFORNIA WHITE　　*Pontia sisymbrii*
Pl. 4

Characteristic of barren, rocky sites, especially on serpentine soils, this early-season species is a hilltopper but can also be found patrolling along cliffs and canyon walls. It usually co-occurs with the Small Marble *(Euchloe hyantis)*. Females are usually pale yellow rather than white. The males are always white. Our populations tend to have the ventral hindwing pattern browner and less crisply defined than those from higher elevations (e.g., in the Sierra Nevada, where this species occurs near and above tree line and on lava flows on the eastern slopes). The late-instar larva looks very un-pierid and is remarkably similar to that of the Monarch *(Danaus plexippus)*—ringed around the body in yellow, black and white and, like the Monarch, fully exposed on its host plant where it stands out like the proverbial sore thumb. This suggests either that it mimics the Monarch or that it is itself inedible and has convergently evolved warning coloration. The pupa, however, is dark blackish brown and cryptic and usually found among rocks or attached to low vegetation.

DISTRIBUTION: Bare rocky areas in most of the Bay Area counties; unrecorded from the floor of the Sacramento Valley except as a very rare stray, but may occur in the Sutter Buttes.

HOST PLANTS: Various native mustard family members (Brassicaceae) including species of jewel flower *(Streptanthus)* and rock cress *(Arabis)*. At least two of its jewel flower hosts on serpentine, *S. glandulosus* and *S. breweri,* produce orange callosities on their foliage that closely resemble the eggs of this species. It has been shown in field experiments that these actually deter oviposition by females! The real eggs are laid singly near the top of the plant, and the larva often begins by feeding on foliage but later feeds primarily or exclusively on buds, flowers, and fruit. In the eastern Sierra and western Great Basin this species is sometimes found feeding on introduced weedy mustards, especially tansy mustard *(Descurainia sophia)*.

SEASONS: Strictly single brooded, March to May (later at higher elevations). In this species pupal diapause appears obligatory, and pupae may remain dormant but viable for at least four years. This is viewed as "insurance" against bad weather during the flight or breeding period. Both development and wing pattern appear insensitive to photoperiod, unlike related species.

SIMILAR SPECIES: The heavily marked cold-season phenotypes of other *Pontia* species have the same pattern elements, but note that in the Spring White the ventral dark pattern is clearly divided into basal and submarginal portions with largely unbroken ground color between. The veins themselves are conspicuously white, edged with crisp dark lines; this is especially evident near the vein tips. Note also on the upper surface that the Spring White has a very slender black bar at the end of the discal cell, indented or notched at its center. None of the other species ever has a yellow female.

CHECKERED WHITE
Pontia protodice

Pl. 4

As recently as the 1970s this was an abundant grassland butterfly in our area and is usually treated as a "junk species" in older books. But in recent years it has become an erratic visitor, mostly in the second half of the season. It is highly dispersive, apparently with a regular seasonal rhythm of altitudinal migration in the Sierra Nevada and eastward. The late-winter brood consists of very small individuals with a complete dark vein-line pattern beneath, which appears green due to the intimate admixture of yellow and black scales; the black markings on the upper surface are reduced in extent but usually very crisp and distinct, and the wing bases are heavily melanized. The discal portion of the upper surface has a reflective sheen. This form is called *P. p. "vernalis"* and is entirely under environmental control. In the summer broods the ventral hindwing may be nearly immaculate white (males) or merely lightly veined in ochre yellow (females). Intermediates between the forms are sometimes taken in May (second brood) and more often in September and October. The markings of older females tend to fade to a light violet or brownish gray.

This species is a patroller in stands of the host plant. It may hilltop but usually is found in relatively flat or gently rolling terrain.

DISTRIBUTION: Throughout the area in short grassland, coastal prairie, shrub-steppe, and weedy habitats. In forested areas occurs only as a stray along roadsides. As noted above, this species is much less common than formerly and has not been seen in many localities for years.

HOST PLANTS: Weedy mustard family members (Brassicaceae), especially perennial mustard *(Hirschfeldia incana,* formerly *Brassica geniculata)*, less often on black mustard *(Brassica nigra)* and other species of mustards *(Brassica),* wild radish *(Raphanus),* peppergrass or pepperweed *(Lepidium),* and others. The orange eggs are laid singly, preferentially on small rosettes or plants just beginning to "bolt." The mature larva is striped lengthwise with yellow and violet gray, with many black tubercles. It is extremely similar to the larva of the Large Marble *(Euchloe ausonides)* (which has a larger head, directed slightly forward when at rest). Both species feed on buds, flowers, and fruit.

SEASONS: January to December , many broods, but often not in the same locality—this is a "fugitive species," constantly moving around. Overwintering is local and spotty; *P. p. "vernalis"* is most often seen in coastal prairie and on dredge tailings and other grasslands along the eastern side of the Sacramento Valley. Most modern records are from August through October; in many areas first seen every year in September. Diapauses as a pupa. This is one of just three species that routinely occur in hot, dry grassland with tarweeds (subtribe Madiinae, Asteraceae) in late summer in the Sacramento Valley. It is an avid flower visitor, especially to the mustard family, asters *(Aster),* goldenrods *(Solidago),* and thistles *(Cirsium, Carduus,* and *Silybum)*.

SIMILAR SPECIES: See Spring White *(P. sisymbrii)*. The Western White *(P. occidentalis)* is rare in our area but very often confused with this species, and indeed, they may occasionally hybridize. The Western White is whiter above and appears "more thickly scaled"; it has a more extensive dark pattern on the upperside, with many males reproducing most elements of the female pattern to some degree; all the dark markings are crisper and blacker, especially in females. Note the wing pattern differences identified on the plate! In the cold-season form of the Western White *(P. o. "calyce")* the veins on the underside are even darker and greener than in the Checkered White. The early stages of the two species are also extremely similar, but both the larva and pupa of the Checkered White are more slender than those of the Western White, and the lengthwise larval stripes tend to be greener or bluer in the Checkered. Be very careful before claiming to record the Western White in our area—it is a real "goodie."

WESTERN WHITE   *Pontia occidentalis*

Pl. 4

The Western White is not known to be resident in our area, but it turns up often enough to merit inclusion in this book. It normally replaces the Checkered White *(P. protodice)* northward and upslope, that is, in cooler climates. The two are broadly sympatric in the northern Great Basin and Rockies, where rare hybridizations seem to occur; where they co-occur up to about 1 percent of specimens may be indeterminable. Studies of protein polymorphisms have shown one consistent difference between the species, supporting their recognition on other grounds as distinct. They are true "sibling species," barely distinguishable but maintaining their integrity over a vast range.

The habits of the two are nearly identical, except that the Western White is a legendary hilltopper in its mountainous haunts (on the flats it patrols).

DISTRIBUTION: Three recent records on the floor of the Sacramento Valley (Yolo and Sacramento Counties; all in October) and two old but apparently valid records of the spring form in the Santa Cruz Mountains (M. Doudoroff.) These presumably represent a short-term colonization. Hilltopping individuals turn up regularly in the high North Coast Range from Colusa County northward, and the species is apparently resident in the Yolla Bollys and throughout the Klamath, Trinity, and Siskiyou Mountains. Thus strays in the North Bay counties are to be expected.

HOST PLANTS: Many mustard family plants (Brassicaceae), including weedy species; not recorded as breeding in our area.

SEASONS: Multiple brooded at its lower limits (March to November); single brooded (form *P. o. "calyce"* only) above tree line, but the cycle is often obscured by hilltoppers.

SIMILAR SPECIES: See Spring White *(P. sisymbrii)* and Checkered White. Observe your butterfly very carefully; resist the urge to call any heavily marked specimen a Western White.

GRAY-VEINED WHITE or MUSTARD WHITE *Pieris "napi complex"*

Pl. 5

The *Pieris "napi* complex" occurs completely around the Northern Hemisphere in cool, moist, usually forested environments and has been subdivided by taxonomists into a vast number of named entities. Several attempts have been made to determine how many biological (i.e., reproductively isolated) species there are in the group, at least in well-studied regions such as North America and most of Europe. The question has been addressed using experimental hybridization in the lab, micromorphology (including in the early stages), protein electrophoresis, and DNA sequencing. These efforts are still in progress at this writing. For purposes of this book it is best to avoid definitive taxonomic decisions. The true *P. napi* is European, and it is almost certain our populations are not conspecific with it. Unlike European *P. napi,* our populations have very sharp, crisp black vein lines on the hindwing below, when they have them at all. Coastal fog belt, redwoods animals south of the Bay are large, with a very strong pattern in the spring brood and at least traces of it in summer. Fog belt animals north of the Bay are somewhat smaller. The name *P. n. venosa* has been used for both. North of the Bay inland occur relatively small and more lightly marked animals, the summer brood often nearly immaculate; these have been called *P. n. microstriata.* Many inland populations produce no summer brood or do so only infrequently. However, all our populations will do so in the laboratory. Control of the seasonal phenotypes is environmental, with both photoperiod and temperature playing a role. With proper manipulation, the most extreme phenotypes of both broods can be obtained from the eggs of any given female! North of our coverage area the geography of this complex becomes very…complex. In some areas genetically different populations may come into very close proximity, reflecting ecological and climatic discontinuities.

DISTRIBUTION: Throughout the Bay Area in Douglas fir, redwood, and mixed-mesic forest, often in canyon-riparian situations, south to San Luis Obispo County. Unknown today in the Sacramento Valley but possibly collected by Lorquin near Stockton in Gold Rush days. A butterfly of dappled light and shade, never seen in wide-open spaces.

HOST PLANTS: Milkmaids, crinkleroot, toothwort (these are *Dentaria* species in older books, now usually lumped into the genus *Cardamine*); winter cress or yellow rocket (*Barbarea verna* and *B. orthoceras,* not recorded on the meadow weed *B. vulgaris* in our area); tower mustard *(Arabis glabra);* rarely hedge mustard *(Sisymbrium officinale* var. *leiocarpum)* and weedy annual mustards *(Brassica)* only on roads through woods; peppergrass (*Lepidium virginicum* and *L. campestre);* probably other mustards (Brassicaceae) in the correct habitat. Eggs yellowish white, rather large, laid singly usually on undersides of leaves, occasionally on stems; larva plain yellowish green, downy, feeding on leaves. When a second brood occurs the host it selects is almost always the naturalized watercress *(Rorippa nasturtium-aquaticum),* found restricted to streambeds. There is little risk of summer flash flooding of this plant in our area, but it is unclear what native host(s) might have been available to support summer breeding before it was introduced.

SEASONS: First brood (heavily marked) January to April; second (lightly marked), when produced, May to July; very rarely a minuscule third flight (Solano County, August). Overwinters as diapausing pupa.

SIMILAR SPECIES: The Cabbage White *(P. rapae)* has ventral hindwing usually yellower, never with black vein lines but with more or less gray suffusion, mainly in cool seasons and places. The larva of the Cabbage White is similar to that of the Gray-veined White but has a distinct bluish cast. The pupae are similar but the Gray-veined White has a longer horn on the head and when green is apple green rather than bluish green. Gray-veined White pupae typically have rather extensive dark brownish spotting, lacking or much reduced in the Cabbage White.

CABBAGE WHITE   *Pieris rapae*
or EUROPEAN CABBAGE BUTTERFLY
or IMPORTED CABBAGEWORM
Pl. 5

As far as we know, this is our only naturalized exotic butterfly. The official story is that it was introduced to North America in southeastern Canada in the mid-nineteenth century, spreading thence over the continent. The great American lepidopterist Samuel H. Scudder set up a network of observers to tell him when it reached their areas and was thus able to map its spread through time—one of the first such efforts ever. Unfortunately he did not resolve the time line in our part of the world because the West Coast lepidopterists had trouble distinguishing the Cabbage White (which they knew only from published descriptions) from the summer phenotypes of the Gray-veined White complex (*P. "napi* complex"). Scudder was not even certain the Cabbage White might not be native on the West Coast! To further confuse matters, Lorquin sent a specimen to a collector on the East Coast (Tryon Reakirt), who named it as a new species, "*Pieris yreka*," presumably after its collection locality in northwestern California. The specimen is *P. rapae*. It may have been mislabeled, but if not it might suggest a prior introduction in California, perhaps by the Spanish in the mission period. It is possible that molecular markers might help us solve this puzzle.

Meanwhile, the Cabbage White is ubiquitous in our area, occurring primarily in disturbed habitats where weedy mustards (Brassicaceae) grow or mustard family crops are grown. The garden ornamental nasturtium *(Tropaeolum)*, while not a member of the mustard family, is chemically similar and is used as a host. This plant is widely naturalized near the coast, mainly south of the Bay. The Cabbage White is primarily a butterfly of sunny, open places, but in summer it will follow hosts such as nasturtium into the woods too, where it is easily confused with the Gray-veined White.

First-brood specimens are small, with reduced or even obsolescent black wing markings and increased basal melanization. Beneath, they are usually strongly yellow and infuscated with gray on the hindwings. Summer specimens are larger, the males whiter, the females often vaguely buffy, with bold black markings above and little or no gray beneath except in the fog belt. In fall the ventral hindwing becomes infuscated again, while the dorsal pattern remains summery. The seasonal forms thus do not simply represent the relocation of a fixed amount of black pigment among wing areas, nor are they all or none. Despite the abundance of the species and its ease in rearing, the control of its seasonal phenotypes is not really understood. Presumably both photoperiod and temperature are involved.

DISTRIBUTION: All counties; everywhere but in deep shade.

HOST PLANTS: A great variety of wild mustards (*Brassica*, perennial mustards *[Hirschfeldia]*, wild radish *[Raphanus]*, *Sisymbrium*, peppergrass or pepperweed *[Lepidium]*, horseradish *[Armoracia]*, rock cress *[Arabis]*, bittercress *[Cardamine]*, watercress *[Nasturtium]*, white-top *[Cardaria]*, etc.) and cultivated mustards (*Brassica* cultivars, garden alyssum, etc.), plus nasturtiums (*Tropaeolum*, Tropaeolaceae). A number of mustards (Brassicaceae) appear immune from attack, including moneywort *(Lunaria)*, wallflowers *(Cheiranthus* and *Erysimum)*, shepherd's purse *(Capsella)*, and fanweed *(Thlaspi)*. I once found larvae abundant on stinking wart cress *(Coronopus didymus)* in a mowed lawn! In summer, hosts become scarce in unirrigated areas, and most breeding is on perennial peppergrass or tall white-top *(Lepidium latifolium)* and perennial mustard *(Hirschfeldia incana)*. This species is often heavily parasitized by the gregarious braconid (hymenopteran) parasitoid *Cotesia* (formerly *Apanteles) glomerata,* which spins masses of bright yellow cocoons on the dying or dead larva's back.

SEASONS: January to December, multiple broods, typically not flying for 4 to 8 weeks per year but occasionally seen all winter. The fall brood may straggle into January. In the Sacramento area the first sightings have varied from January 1 to February 22. The supply of hosts and the life cycle of this butterfly are typically out of phase; its populations are growing rapidly in late spring just as the huge biomass of annual mustards disappears, forcing it to contract to localized areas where summer hosts are available. In most of its global range this species overwinters as a pupa in diapause, but in our area a substantial number of larvae are still feeding into January in mild years, and these give rise to nondiapause pupae. In its original Old World range this species probably underwent a seasonal migration tracking the availability of hosts, and this is mirrored feebly in a seasonal altitudinal migration in the mountains of California.

SIMILAR SPECIES: See Gray-veined White.

LARGE MARBLE   *Euchloe ausonides*
Pl. 6

Until the late 1990s this species seemed a classic success story—a native butterfly happily adapted to and using introduced host plants in the anthropic landscape. We have a pretty good idea of what it used to do for a living. Because it always lays its eggs on the terminal flower buds of tall mustards (Brassicaceae), and there is only one native species in our area that fills the bill—California mustard *(Guillenia* [formerly *Thelypodium] lasiophylla)*—we assume that was its usual host in native pre-American grasslands. It is uncommon in our area today, and nearly all breeding is on naturalized, weedy species of mustards *(Brassica),* wild radish *(Raphanus),* and *Sisymbrium.* Or should we say "was"? This butterfly has suddenly and inexplicably disappeared from much of its recent range.

Males patrol linear habitats, along roadsides, streams, berms, and levees. They have a characteristically direct flight that, with a bit of practice, makes them easy to tell from the Cabbage White *(Pieris rapae)* on the wing. In flight, the leading edges of the two forewings appear to form a straight line nearly at right angles to the axis of the body. A peculiar characteristic of this butterfly is a row of orange scales behind the compound eyes. This same feature occurs in the South American white butterflies *Tatochila* and *Hypsochila,* which hold their wings similarly in the glide portion of a stroke. The function of these scales is unknown. Any ideas?

The decline of this species has been accompanied by the virtual disappearance of its second brood. Second-brood individuals were larger, and the ground color of the female was often partly (hindwings only) or entirely a delicate peachy orange. Montane populations are always single brooded. The adults are avid flower visitors, especially to the mustard family, thistles *(Cirsium, Carduus,* and *Silybum),* and fiddlenecks *(Amsinckia).*

DISTRIBUTION: All counties (formerly), primarily in grassland, freshwater marsh, and openings in riparian habitat. As of 2005 largely absent from the southern Sacramento Valley; still common in the East Bay in Eastshore State Park, as at the Berkeley Marina.

HOST PLANTS: Tall mustard family (Brassicaceae) plants, including mustards *(Brassica)* and wild radish *(Raphanus).* Eggs red, laid singly on terminal flower buds. Larva very similar to the Checkered White *(Pontia protodice),* the head oddly tilted forward. Pupa always light brown marked with darker brown, thorn shaped.

SEASONS: Formerly two broods, February to April and May to July, the second now rarely seen.

SIMILAR SPECIES: See Small Marble *(E. hyantis).*

SMALL MARBLE or PEARLY MARBLE   *Euchloe hyantis*
Pl. 6

In our area the Small Marble is confined to rocky areas on serpentine and other substrates, usually in chaparral. It flies with the Spring White *(Pontia sisymbrii)* and has the same host plants. Males patrol cliff edges and canyon bottoms, and both sexes visit flowers, including those of the hosts and associated species such as phacelias *(Phacelia)* and yerba santas *(Eriodictyon)*. Fairly common in its specific localities, but rarely straying elsewhere. It almost never co-occurs with the Large Marble *(E. ausonides)*.

DISTRIBUTION: Napa, Solano, Sonoma Counties, rocky areas in the Coast Ranges.

HOST PLANTS: Species of jewel flower *(Streptanthus)* and rock cress *(Arabis)*. Eggs red, laid singly near top of plant. Larva plain green with a white stripe on each side, feeding on buds, flowers, and fruit. Infested plants are usually easy to spot due to "stemming." One larva can consume the entire seed output of a small host. When about to pupate the larva turns a livid reddish purple; the pupa, which is thorn shaped, is initially the same color but turns gray brown over several days. As in its associates the Spring White and the Indra Swallowtail *(Papilio indra)*, this species often demonstrates extended pupal diapause—up to 5 years.

SEASONS: Obligately single brooded, March to May.

SIMILAR SPECIES: The Large Marble is usually larger, the white ground color duller and often slightly yellowish, the black discal spot tending to be more slender, and the green marbling beneath less intense and dark. The Small Marble has a distinctly pearly quality to the white ground color; its black markings are crisper, its green marbling more intense and usually more extensive. It is never yellowish or peachy. The habitat and strict univoltinism are usually enough to establish its identity! Some older works doubted that these two were really separate species; their authors clearly never saw the larvae. Each is more closely related to relatives in the Old World than to the other Californian species.

SARA ORANGE-TIP   *Anthocharis sara*
Pl. 6

The beautiful Sara Orange-tip is a characteristic member of the spring foothill fauna. It only rarely occurs on the floor of the

Sacramento Valley, however. A beast of riparian-canyon habitat in foothill woodland and chaparral, it has adapted itself well to breeding on introduced mustards (Brassicaceae) but still uses native species when it can. In cooler and moister sites it is usually partially double brooded; second-brood adults are large and bright and have less extensive green marbling beneath. Bright yellow females are taken throughout the range, but are rare.

The orange tip of the male is UV reflective, and this looks "bee purple" to the butterfly, while the female's tip absorbs UV and looks orange. Presumably, then, Sara Orange-tips are better at sexing one another in flight than we can be. An avid flower visitor, with a special attraction to the flowers of fiddlenecks (*Amsinckia*, Boraginaceae) and blue dicks (*Dichelostemma capitatum*, Liliaceae). Males patrol streamsides and roadsides in canyon bottoms.

DISTRIBUTION: Resident throughout the Bay Area. Infrequent in the Sacramento Valley, rarely recorded as breeding.

HOST PLANTS: Mustard family (Brassicaceae), in our area especially on the native milkmaids, crinkleroot, and toothwort *(Dentaria)* and yellow rocket *(Barbarea)*, and the naturalized hedge mustard *(Sisymbrium officinale* var. *leiocarpum)*. Eggs blue green turning to bright red, laid on stems and leaves; larvae eat buds, flowers and fruits. It is not uncommon to find this species and the Gray-veined White *(Pieris "napi")* using the same individual host plant but keeping out of each other's way—Sara Orange-tip on the reproductive structures and Gray-veined White on the leaves. The larva is plain green and sits lengthwise among the fruits. The light tan brown, thorn-shaped pupa overwinters and may remain in diapause as long as 3 years.

SEASONS: First brood late January to April; second May to early July; occasionally the two generations may overlap.

SIMILAR SPECIES: None in our area.

BOISDUVAL'S MARBLE or GRAY MARBLE   *Anthocharis lanceolata*

Pl. 6

"Gray Marble" is a recent and rather unfortunate coinage, since the marbling is not gray but claret red. Although locally common elsewhere in its range, this species is infrequently encountered in our area—north of the Bay in mixed-mesic, riparian canyon, and rocky cliffside situations. It is easily mistaken for the Gray-veined White *(Pieris napi)* in flight and is thus probably underreported. Females are larger than males but do not differ from them in color or pattern. Both sexes are eager visitors to flowers, especially dogbanes *(Apocynum)*, wallflowers *(Erysimum)*, and—in some places but not others—woolly sunflower *(Eriophyllum lanatum)* and yerba santas *(Eriodictyon)*.

DISTRIBUTION: Napa and Sonoma Counties, generally well inland.

HOST PLANTS: Tower mustard *(Arabis glabra)* preferred; other mustard family (Brassicaceae) plants including hedge mustard *(Sisymbrium officinale* var. *leiocarpum)*, rapeseed *(Brassica napus)*, and peppergrass *(Lepidium virginicum)* occasionally used. Yellow eggs are laid singly on the upper part of the plant. The downy green larva is mostly a foliage feeder but sometimes consumes fruits as well. Pupae may remain dormant up to 3 years.

SEASONS: One brood, April to June.

SIMILAR SPECIES: None. Extremely rare hybrids between this species and Sara Orange-tip *(A. sara)* have been reported; watch for them!

ORANGE SULPHUR or ALFALFA BUTTERFLY   *Colias eurytheme*

Pl. 7

One of the most abundant and generally distributed butterflies in California and indeed in most of North America, the Orange Sulphur occurs in all open habitats but almost never in forest. It ranges from sea level to tree line and above, and at low elevations flies nearly all year. In the Central Valley, where alfalfa *(Medicago sativa)* is grown, it sometimes swarms by the millions in the second half of the season. An alfalfa field covered with thousands of moving flecks of color is an astonishing sight. The flecks come in various colors because this is one of the most variable butterflies in our fauna. The variation is partly genetic and partly environmental in nature. Up to half the females are white, due to an autosomal dominant gene called "alba," which is expressed only in that sex. "Alba" acts by interrupting the series of chemical reactions leading to the production of the usual orange pigment—but it does not affect the orange spot in the middle of the hindwing upperside. This gene is found in nearly all members of the genus *Colias* worldwide, but only one species (from Chile and

Argentina) produces *only* white females. Several rare mutations (called "whitish" and "blonde") occur that produce pearly white or yellowish white males; examples have been collected in our area, mainly in the North Coast Range. Mostly in late summer and fall, females occur that are intermediate between orange and "alba"; these seem to be produced by gene-environment interaction and are poorly understood. Very rarely, bright yellow specimens of either sex may turn up. These are superficially extremely similar to the closely related Yellow Sulphur *(C. philodice,* also called *C. eriphyle),* which occurs only east of the Sierra Nevada, but appear to represent mutations of local origin. Where the Orange Sulphur and *C. eriphyle* co-occur they hybridize, producing evenly colored orange yellow hybrids never seen in our area. A weak salient of *C. eriphyle* once crept through the Inyo-Kern area to near Visalia but died out.

The Orange Sulphur shows extreme environmentally controlled variation. Cold-season animals are very small; they have relatively narrow wings with less pronounced apices, narrow black borders above—which may disappear altogether!—and very heavy gray scaling on the ventral hindwing. This last trait has been shown to be adaptive for thermoregulation; *Colias* species are lateral baskers. The summer broods have broader wings with pronounced apices in the males, broad black borders, and little or no gray beneath. The shape differences reflect different patterns of flight. Cold-season animals typically fly low, near the ground, keeping out of the wind (often on the lee side of hills, facing south or southwest); summer animals often exploit rising currents of warm air to give them a passive boost aloft. This was the first American butterfly to be studied rigorously from an ecophysiological standpoint to document the significance of seasonal variation. As usual for midlatitude species, photoperiod is paramount in the control of variation, but there are thermal inputs too. One can induce whatever phenotype(s) one wants from the eggs of a given female by rearing under appropriate regimes. There is much that we still do not know about the system, however. Are all individuals in all broods equally plastic? Are there intergenerational physiological effects (or "maternal effects") on phenotypic plasticity? Are the animals sensitive to the direction or rate of change in photoperiod? The larvae are also highly variable and in cool seasons may be heavily marked with black. This too appears to function in thermoregulation.

For reasons that are not understood, the Orange Sulphur overwinters poorly in alfalfa fields in the Sacramento Valley. Most breeding in late winter and early spring is on annual vetches *(Vicia)*, particularly on warm south- or southwest-facing highway embankments. This is not, however, simply a matter of the adults being drawn to these habitats by thermal needs. Alfalfa tends to be colonized as the vetch begins to dry up in May and June, but where vetch can stay green in summer the butterfly continues to breed on it. Once alfalfa has been colonized, numbers of butterflies rise rapidly, occasionally peaking as early as July but more often in September and October. When the alfalfa is cut the butterflies disperse. There are no organized migrations, but large numbers may move through urban areas near alfalfa, and there is a generalized outflow into wild habitats and upslope. When the cut alfalfa regenerates, females flow back into the fields to oviposit on the young growth. Alfalfa is an introduction from the Old World. The seasonal dynamics of the Orange Sulphur in California probably reflect an ancient rhythm of up- and downslope colonization to track the availability of native host plants with the march of the seasons. Most of the host plants used by this species today are in fact exotics.

Despite certain problems (vulnerability to disease, difficulty in securing matings), the Orange Sulphur has proven to be a very useful model system for physiological, ecological, and molecular studies. Alluding to a famous bacterial lab system, Ward B. Watt of Stanford University once said he hoped to make the Orange Sulphur "the *E. coli* of population biology."

DISTRIBUTION: Everywhere except in deep shade.

HOST PLANTS: Papilionaceous legumes (Fabaceae), both native and introduced, including alfalfa *(Medicago sativa)*, other *Medicago* species, vetches (species of *Vicia* and *Lathyrus*, but not *L. latifolius*, the naturalized perennial sweet pea), clovers *(Trifolium)*, sweet clovers *(Melilotus)*, *Lotus* species (annual and perennial, including deer weed *[L. scoparius]* and birdfoot trefoil *[L. corniculatus]*); locoweeds *(Astragalus)*; occasionally on lupines *(Lupinus)*, especially the common annual arroyo lupine *(L. succulentus)*.

SEASONS: January to December, usually with a 4 to 8 week winter hiatus, but in severe drought years flying all winter; local early in the season in warm pockets with vetch, spreading out over the landscape in March and April; many broods. The larvae do not enter true diapause but are merely torpid in winter.

SIMILAR SPECIES: In our area, only the Western Sulphur *(C. occidentalis chrysomelas).*

WESTERN SULPHUR   *Colias occidentalis chrysomelas*
or GOLD-AND-BLACK SULPHUR

Pl. 7

This magnificent insect may be the only forest butterfly in the genus *Colias* in the world. The genus is strongly grassland and steppe associated, with its highest diversity in central Asia. But this butterfly of the North Coast Range occurs in Douglas fir, mixed-mesic, and Oregon White Oak forest—mainly along roadsides—and is local and generally uncommon. Males "fly a beat," patrolling along a roadside. Both sexes are strong fliers, difficult to approach or to net. They do visit flowers, especially yerba santas *(Eriodictyon)*, milkweeds *(Asclepias)*, and dogbanes *(Apocynum)*, but tend to be wary. In selectively logged areas they visit the introduced bull thistle *(Cirsium vulgare)* freely. Pale females occur but are extremely rare.

DISTRIBUTION: North Bay counties, mainly inland; commoner northward, for example, in the Mendocino National Forest. Intensely local. Records from Solano County, as well as all records from the Sierra Nevada, are almost certainly based on misidentifications.

HOST PLANTS: Perennial lupines *(Lupinus)* and *Lotus crassifolius.*

SEASONS: One brood only, May to early July, varying by locality. The third-instar larva diapauses.

SIMILAR SPECIES: Males should be unmistakeable. They are bright yellow with heavy, wide black borders; there is no trace of orange above, although there may be an orange flush on the forewing below; the hindwing below is warm buff, the discal spot with a single, narrow purple rim; larger than the Orange Sulphur *(C. eurytheme).* Females in flight are readily confused with the female California Dogface *(Zerene eurydice)* and the Cloudless Sulphur *(Phoebis sennae marcellina).* The female Dogface differs in the shape of the forewing apex, as well as in its usual lack of a well-defined dark border on the forewing above. Unfortunately, these are not conspicuous differences in flight, and female Dogfaces routinely fly 5 m (16 ft) or more above ground level. The two species may co-occur, but the Dogface has more than one brood a year. The Cloudless Sulphur has a different wing shape but again tends to fly high, making identification difficult. Unlike the others, it is a rare immigrant in our area. All three species are "goodies" in any case!

CALIFORNIA DOGFACE or DOGFACED SULPHUR   *Zerene eurydice*

Pl. 8

This is California's state insect, but relatively few people have knowingly seen it. It occurs in the North and East Bay and in a few relict populations in the Sacramento Valley, where it and its host plant were probably much more widespread in riparian habitat in the past. A very strong flier, it strays widely (even, once, to my driveway in Davis!). The male cannot be confused with anything else in our area, but the all-yellow female is readily confused with both the Western Sulphur *(Colias occidentalis chrysomelas)* and the Cloudless Sulphur *(Phoebis sennae marcellina)*, leading to many questionable records of all three. The Dogface (and the Cloudless Sulphur) typically flies high up in the trees, 5 to 10 m (15 to 30 ft) above the ground. It is often spotted from a moving vehicle but seldom netted except at flowers—it particularly visits thistles *(Cirsium, Carduus,* and *Silybum)* and tall blue verbena *(Verbena bonariense)*. Males patrol and characteristically fly down side canyons, reach the mouth, turn around, and fly back up. These canyons contain the host plant, typically growing mixed with poison oak *(Toxicodendron diversilobum)* in or along the boulder-strewn creekbed in places almost impossible to reach. There is almost no phenotypic variation.

DISTRIBUTION: North and East Bay counties, mainly inland; Sacramento Valley. Riparian forest only.

HOST PLANTS: False indigo *(Amorpha)*, a pea family (Fabaceae) shrub with dull purple flowers. The Silver-spotted Skipper *(Epargyreus clarus californicus)*, which also breeds on this plant, has switched in many areas onto the introduced tree black locust *(Robinia pseudoacacia)*, but the Dogface has not. Perhaps it ultimately will, and become commoner as a result. Another candidate host plant is the Argentine flame pea *(Sesbania punicea)*, which is spreading aggressively in the Sacramento Valley and Delta.

SEASONS: Adults apparently hibernate and have been seen flying as late as Thanksgiving; records from February to November, resolving into two broods per year, in early summer and in fall (reappearing in spring). Although adults begin flying in March, the host rarely leafs out before April.

SIMILAR SPECIES: See Western Sulphur. Keep in mind that the Cloudless Sulphur is a rare immigrant and lacks the large round, black discal spot of the Dogface, and that the Western Sulphur is very local in our area. Context is often helpful in discriminating among these three on the wing.

CLOUDLESS SULPHUR *Phoebis sennae marcellina*
Pl. 8

This butterfly is a stray from the subtropics, recorded sporadically in our area but occasionally immigrating in numbers, as in 1992. In its normal range it is well known as a seasonal mass migrant. No native host plants grow in our area, so that breeding, if it occurred, would be restricted to cultivated hosts. Typically the Cloudless Sulphur flies high off the ground. It is an avid flower visitor, especially attracted to hibiscus *(Hibiscus)* blooms. Females are very variable.

DISTRIBUTION: Can turn up anywhere in our area, usually as a singleton. I have seen it flying across West Texas Street in Fairfield, Solano County, and in an alfalfa *(Medicago sativa)* field in eastern Lassen County, to name two very different venues.

HOST PLANTS: Senna *(Cassia)*, occasionally cultivated in our area.

SEASONS: Local records from April through October, with most records from August and later.

SIMILAR SPECIES: See Western Sulphur *(Colias occidentalis chrysomelas)* and California Dogface *(Zerene eurydice)*. Other tropical and subtropical species of *Phoebis* occasionally stray north and might eventually turn up here.

MEXICAN YELLOW *Eurema mexicana*
Not illustrated

The Mexican Yellow is an extremely infrequent stray ranging northward from the southwestern deserts. It has been recorded once each in Sonoma and Yolo Counties and is capable of turning up anywhere, though perhaps not in one's lifetime!

DAINTY SULPHUR   *Nathalis iole*

Pl. 8

Another stray from the southwestern deserts, the Dainty Sulphur is rare in our area but could turn up anywhere. In 1992 it was found in several counties, but only in small numbers. It shows strong seasonal polyphenism—winter broods are very heavily infuscated beneath—but this is not seen in our area, where breeding has not been observed. This is a very peculiar species whose roots lie in South America. In addition to its distribution in North America, it occurs in the high-altitude *páramos* ("alpine grassland-steppe" above tree line) in northeastern Colombia and adjacent Venezuela; its sister species, *N. plauta,* looks like a miniature orange *Colias* and is from the high Andes. We believe the Dainty Sulphur entered the Northern Hemisphere relatively recently, probably as part of the great American interchange across the Isthmus of Panama about two million years ago.

DISTRIBUTION: Can turn up anywhere, normally in open country.

HOST PLANTS: Not known to breed locally. Many of the recorded hosts are common weeds, but most of these are almost certainly in error. All authentic records known to me are on genera of the sunflower family (Asteraceae) containing photosensitizing compounds, including marigolds *(Tagetes)* and beggar's ticks or sticktights *(Bidens).* Records on other plants need confirmation.

SEASONS: Records here from late April to early October; could theoretically occur at any time. Breeds all winter in Mexico.

SIMILAR SPECIES: None in our area.

Coppers, Hairstreaks, Blues, and Metalmarks — The Gossamer-winged Butterflies (Lycaenidae)

This is the largest family of butterflies, with over 6,000 species worldwide — and counting. New species are being discovered all the time, and not just in the tropics: a new blue was found in California and named as recently as 1998!

Lycaenids are small to (at most) medium-sized butterflies. Many have relatively small bodies and fragile-looking wings (hence "gossamer-winged"). Both the adults and the early stages have many distinctive anatomical traits that set them off from all other butterflies. Males have the forelegs moderately reduced and clawless; females have six fully developed legs. The base of the antenna is adjacent to and usually indents the compound eye. The anatomy of the genitalia is so unusual that it has its own distinctive vocabulary. Many species have iridescent colors on the wings in at least one sex. These are produced by extraordinarily beautiful and intricate physical structures on the scales that act as diffraction gratings, manipulating the reflected light. There is recent evidence that this feature may be influenced by environmental factors during development. Many species have a "false head" at or near the anal angle of the hindwing, sometimes equipped with hairlike tails that seem to simulate antennae. Tropical hairstreaks may have patterns that focus the viewer's eye on this "false head." It appears to be a device to entice predators into striking at an expendable part of the insect, sparing the body.

Lycaenid eggs are turban shaped, relatively large, and in our species laid singly. Under a scanning electron microscope they display beautiful lacelike textures, which may be of taxonomic value. Some are amply endowed with air-retaining structures, allowing them to be submerged in water for weeks or months without harm. Many of our species overwinter as eggs, which may be placed in litter or soil near the host rather than on the host itself.

The larvae are downy and usually described as sluglike. The head is small and tucked under the thorax. Most secrete honeydew to attract ants, and some communicate with the ants by stridulating (making rasping noises). The ants protect them from predators and parasites and sometimes even take them into their nests, where the caterpillars may feed on ant larvae! Some species pupate in the ant nest; the adult has to find its way to the surface before expanding its wings. We have much more to learn about ant-lycaenid relationships in our own fauna. How obligatory are they? How specific is the relationship—are some lycaenids associated with only one kind of ant?

Lycaenid larvae tend to be rather nondescript, either plain green or reddish, or marked dorsally with chevronlike markings; they are very difficult to identify and are best reared out. The most commonly seen lycaenid larva in our fauna is that of the Common Hairstreak *(Strymon melinus)*, which breeds on many common weeds and some garden plants. Lycaenid larvae should be reared individually; they are notoriously cannibalistic.

The pupae are dark brown, obtect (without prominences), and resemble pine nuts. They may be formed loose in litter or at the base of the host (or in an ant nest!), or anchored by a silken girdle. Like the larvae, they all look very much alike.

A few of our lycaenids, despite their fragile form, are highly dispersive and excellent colonizers (Western Pygmy Blue *[Brephidium exile]*, Marine Blue *[Leptotes marina]*, Eastern Tailed Blue *[Everes comyntas]*, Common Hairstreak). Most, however, are very sedentary and prone to form discrete colonies that may not move significantly for decades. A given species may feed on various plants across its geographic range, but local populations often specialize on only one of them. Because they routinely lek and mate on the host plant, this is a situation tailor-made for the formation of "host races" or ecotypes, which under the right circumstances could evolve into full species. This sort of thing is happening in many lycaenid lineages right now (since the end of the Pleistocene) in the western United States, providing us with many of our thorniest and most exciting evolutionary and taxonomic problems. When you see something referred to as a "complex," it means we are not sure what the limits of biological species in that group are.

The metalmarks are overwhelmingly a tropical group. In tropical America they display a dazzling array of shapes and colors and often enter into mimicry relationships. They are sometimes considered a separate family, the riodinids (Riodinidae).

TAILED COPPER   *Lycaena (Tharsalea) arota*

Pl. 9

This butterfly is a typical species of mixed-mesic, canyon riparian habitats in the Coast Ranges, extending into the redwoods, but absent in the Sacramento Valley. Males perch on leaves in sun-flecks, often 3 m (12 ft) above the ground, and engage in swift, aerobatic chases. Fortunately, both sexes visit flowers freely, especially those in the sunflower family (Asteraceae) and white buck-wheat *(Eriogonum nudum)*. The sexual dimorphism is striking. It is seldom considered common in our area, but its rather secretive habits in shaded, rocky stream bottoms make it easy to overlook.

DISTRIBUTION: All Bay Area counties (though only one known colony in Solano County). Unknown in the Sacramento Valley today, but perhaps once present in riparian woodland.

HOST PLANTS: *Ribes* species (currants and gooseberries, perhaps any species will do).

SEASONS: Single brooded; records from late April to early October, but the vast majority are from May to early August.

SIMILAR SPECIES: None.

GREAT COPPER   *Lycaena xanthoides*

Pl. 9

The Great Copper is widespread, sometimes locally common in grassland, marsh, and riparian habitats, but overall it is a rather mysterious beast. Most populations today breed on the natural-ized weed curly dock *(Rumex crispus)*. This weed is ubiquitous, but the butterfly tends to form discrete, localized, highly persis-tent colonies and is absent from large areas of seemingly ideal habitat. Some colonies have not moved 30 meters in 30 years! The Great Copper is also rather picky about nectar sources; the avail-ability of these may be just as important to it as the presence of the host plant. The favored nectar plants are gum plant *(Grindelia camporum)*, dogbanes *(Apocynum)*, especially the tall Indian hemp *(A. cannabinum)*, horehound *(Marrubium vul-gare)*, the marsh umbel *Oenanthe*, tamarisk *(Tamarix)*, and per-ennial peppergrass or tall white-top *(Lepidium latifolium)*. Re-markably, it does not seem to visit other species in the gum plant genus! The Great Copper occurs from sea level in the Bay Area and Sacramento Valley to at least 1,700 m (5,600 ft) in the high

North Coast Range, but in the Sierra Nevada it is replaced above about 1,000 m (3,200 ft) by the smaller Edith's Copper *(L. editha)*. In recent years numbers of this species have declined dramatically in much of the Bay Area and Suisun Marsh, though most Sacramento Valley and upland populations are doing well.

Both sexes vary considerably in the intensity of the ventral spot-pattern. Males are uniformly pearl gray above, but females may have little or much orange and more or less of a spot-pattern.

Males perch on low vegetation and are territorial from midday through midafternoon. They are extremely strong, swift fliers.

DISTRIBUTION: All counties, but local.

HOST PLANTS: Docks (larger *Rumex* species), both native and introduced, elsewhere including sheep sorrel *(R. acetosella)*. Overwinters as an egg, usually laid on dry, dead leaves around the base of the host and capable of surviving prolonged bouts of flooding; eggs can tolerate 3 months underwater, but late-spring flooding can be devastating to the larvae.

SEASONS: One brood, with records from April to late August, but mostly late spring to early summer (May to early July in Sacramento Valley, as usual more diffuse near the coast). At higher altitudes flies into October and visits rabbitbrush *(Chrysothamnus)*!

SIMILAR SPECIES: In our area the male is distinctive. The female may be confused with the female Gorgon Copper *(L. gorgon)* (paler above, typically straw yellow rather than orange, the overall effect more "checkered"; detail differences on ventral hindwing, see pl. 9). Dark females may be confused with dark females of the Blue Copper *(L. heteronea)*; pattern more obscure in that species, ventral hindwing very lightly if at all spotted, without orange; rare and local with us.

GORGON COPPER   *Lycaena gorgon*

Pl. 9

This sexually dimorphic copper is locally common in rocky foothill sites, in chaparral and oak woodland, and occasionally in grassland, often on serpentine. Males perch, often on the host plant, and are territorial. Interspecific interactions with the Great Copper *(L. xanthoides)* are very frequent where they co-occur. However, they never hybridize.

In some areas this species shows an extreme preference for the nectar of woolly sunflower *(Eriophyllum lanatum)*, while in others it ignores this plant, instead visiting dogbanes *(Apocynum)* and the hosts (wild buckwheats *[Eriogonum]*, California buckeye *[Aesculus californica]*, etc.).

DISTRIBUTION: All Bay Area counties except San Francisco; rare stray in the Sacramento Valley, but breeding at the edges in the far north.

HOST PLANTS: White buckwheat *(Eriogonum nudum)* (sensu lato) and *E. latifolium,* perhaps other perennial wild buckwheats. The white buckwheat group includes a number of genetic entities, which seem to be used differentially as hosts by lycaenid butterflies. There is a good study for someone in this.

SEASONS: One brood, records from late March through July, but the vast majority in May and June; overwinters as egg.

SIMILAR SPECIES: See Great Copper, female.

BLUE COPPER or "VARIED BLUE"   *Lycaena heteronea*
Pl. 16

The brilliant sky blue color of the male Blue Copper is unique, unlike any true blue in our fauna. The pointed forewing apex and very plain ventral hindwing are also unique. To see this insect in our area you must go to cold, foggy places north of the Golden Gate. Our handful of populations are certainly Ice Age relicts, persisting in unusual climates. The species is common and widespread in the northern California mountains and in the Sierra Nevada even to tree line; an isolated population existed in the high country of Colusa County but has not been seen recently.

In typical copper style, males perch on vegetation, often the host plant, usually at about waist height. Both sexes visit wild buckwheat *(Eriogonum)* flowers but are infrequently seen nectaring elsewhere.

DISTRIBUTION: Coastal Marin and Sonoma Counties; very local.

HOST PLANTS: In our area, usually *Eriogonum latifolium;* elsewhere on the white buckwheat *(E. nudum)* complex and the sulphur flower *(E. umbellatum)* complex. It is possible that two genetic species with host-plant differences are concealed under this name; further molecular studies are warranted. In any case, only one of them occurs in our area!

SEASONS: One brood. Records from mid-May through August, perhaps on into September; overwinters as egg.

SIMILAR SPECIES: Males are easily told from all of our blues by color, wing shape, and ventral hindwing pattern. (Or lack thereof!) The female may be confused with dark specimens of the Great Copper *(L. xanthoides)* or the Gorgon Copper *(L. gorgon),* but it has a very plain ventral hindwing. Note that on the upper hindwing, the marginal lunules of the female Blue Copper do not have dark "pupils" like the others.

PURPLISH COPPER   *Lycaena helloides*
Pl. 9

Since the mid- to late 1990s this usually abundant and even weedy species has declined precipitously in abundance and distribution throughout our area. Many populations have simply disappeared, while others are seldom observed except in fall. In days of yore, the Purplish Copper was equally at home in a variety of open habitats: grassland, sloughs, tule marshes, riparian forest openings, urban vacant lots, and gardens. Now it seems increasingly confined to marshlands. Its increasingly dominant pattern of greatest abundance in September and October seems to be converging to what has long been its seasonal rhythm in the Great Basin—but why?

In the 1970s a large and brightly colored "race" of the Purplish Copper was discovered in the Suisun Marsh, where it fed on *Potentilla anserina* subsp. *pacifica* (also known as *P. egedei*) (in the rose family [Rosaceae]) rather than the usual knotweed family (Polygonaceae) hosts. Other populations that feed on the rose family are known elsewhere in the range, and the underlying secondary chemistry seems to be similar. Although the Suisun animals were formerly abundant, they appear to have gone extinct in the 1990s. Cinquefoil-associated populations may still occur in the Delta or North Coast and should be looked for.

The Purplish Copper has considerable minor variation in wing pattern, and females may be either bright orange or strongly clouded with brown above. Major aberrations with variably fused black pattern, as well as bilateral gynandromorphs, have been recorded in our area.

This species visits low flowers with shallow corollas, especially heliotropes *(Heliotropium)* and lippias *(Lippia),* but in fall it often visits tall asters *(Aster)* and coyote brush *(Baccharis pilularis)*. It displays neither territoriality nor frequent puddling.

DISTRIBUTION: All counties, but increasingly local.

HOST PLANTS: Knotweeds and smartweeds *(Polygonum),* including the common prostrate knotweeds of the *P. aviculare* complex; docks *(Rumex);* in the Suisun Marsh (formerly) *Potentilla anserina* subsp. *pacifica.* Before the decline, fugitive populations were commonly observed on waste ground, breeding on short-lived patches of prostrate knotweed and then moving on.

SEASONS: Multiple broods, March to December, but in recent years in most areas seldom seen before July or August, and common (if at all) only in fall. Overwinters as egg.

SIMILAR SPECIES: All our other coppers are larger, except perhaps some Tailed Coppers *(L. arota),* with different wing shape and ventral hindwing pattern.

GOLDEN HAIRSTREAK   *Habrodais grunus*
or GOLDEN OAK HAIRSTREAK
Pl. 10

Who will figure out what this butterfly does for a living? There is only one brood of Golden Hairstreak per year, the adults emerging in late spring or early summer, estivating, and re-emerging in fall to lay eggs. Marked individuals (in the Sierra Nevada) have been recaptured after three and a half months. During the heat of summer these butterflies seem to disappear but can be found sitting quietly inside bramble thickets, in leaf litter, or on cool, moist forest floors, and even inside old buildings. I have found them packed densely together on the shaded cement foundation of a long-gone structure now deep in the woods. They come out in late afternoon to fly around the host plant and are often still active in twilight. When perched on the host they are highly cryptic, resembling dead leaves; sometimes amazing numbers can be roused by jostling the branches. But what are they living on all summer? I have never met anyone who claimed to have seen them visit flowers. They visit wet sand and mud, but butterflies cannot live on water alone—at least not for long. I have seen them do unusual things with their extremely short proboscis, including probing unopened buds of California buckeye *(Aesculus californica)* flowers and the golden fuzz covering the acorn cup of the host tree. I once had a student sit in the midst of a large colony of this insect for several weeks, trying to find out what the adults eat. She never saw one feed at all.

This very plain, almost unpatterned butterfly and its sister species from Baja California, Poody's Golden Hairstreak *(H. poodyi),* are phylogenetically isolated in North America. Although their lifestyle might suggest an origin in the desert or seasonal tropics, their relatives are all in the temperate Old World, and they are probably Tertiary relicts here. There is very little variation (but then, there is little pattern to vary).

This butterfly occurs virtually everywhere its principal host occurs, and rarely anywhere else. It is our most shade-tolerant butterfly and, as noted above, performs its odd dance—presumably a lek—around the host deep into the gathering evening gloom. In over 30 years afield, I have never seen either the act of mating or a copulating pair.

DISTRIBUTION: Bay Area counties (except San Francisco) in rocky canyon situations with the host plant; absent from the Sacramento valley.

HOST PLANTS: Golden oak *(Quercus chrysolepis),* known also as maul oak or canyon live oak, but never other species of true oaks; also recorded from tanoak *(Lithocarpus densiflorus)* and chinquapin *(Chrysolepis).*

SEASONS: One very extended brood, April (rare) through October, creating the illusion of two broods since seldom seen in hot weather; males uncommon late in the season, and both sexes worn. Overwinters as egg.

SIMILAR SPECIES: None.

GREAT PURPLE HAIRSTREAK   *Atlides halesus*
Pl. 10

A shimmering bit of the tropics, this may be a great hairstreak but it is not a purple one—there's no purple on it anywhere. Go figure! It is a fairly common species in most of our area, but infrequently seen because it lives in the canopy and seldom comes down near eye level. Most often you will see it as an instantaneous flash of blue high in the trees, caught by your peripheral vision. To understand its distribution you must understand the ecology of its hosts, mistletoes of the genus *Phoradendron: P. villosum* (also known as *P. flavescens* var. *villosum),* which are restricted to oaks and occur primarily in oak woodland; *P. macrophyllum* var.

macrophyllum and *P. tomentosum*, which occur on a wide variety of broadleaf trees *other than* oaks; and *P. juniperinum*, on junipers (apparently not in our area). *Phoradendron macrophyllum* originally infested native riparian trees such as Fremont cottonwood *(Populus fremontii)*, white alder *(Alnus rhombifolia)*, and California sycamore *(Platanus racemosa)*. It has now spread to many exotic trees planted in cities and suburbs, especially Modesto ash *(Fraxinus velutina)*, which was very widely used as a street tree some 40 years ago. The Great Purple Hairstreak thus occurs in oak woodland, riparian forest, and older urban and suburban residential neighborhoods and parks, though the odd adult may turn up almost anywhere.

The butterfly nectars (seasonally) on milkweeds *(Asclepias)*, dogbanes *(Apocynum)*, members of the carrot family (Apiaceae), California buckeye *(Aesculus californica)*, perennial peppergrass or tall white-top *(Lepidium latifolium)*, and goldenrods *(Solidago)*, among other plants. It is hard to miss. Male-male chases occur high in the trees, but while nectaring a male pays no attention to other individuals. It is often found perched on hilltops, especially in chaparral or chaparral-oak woodland mixes.

The apple green larva drops to the ground when mature and pupates in litter or under loose bark. Overwintering is apparently in the pupa, but most pupae found in the field prove to be parasitized. Numbers of this species fluctuate, and it has occasional population outbreaks—but usually you cannot count on seeing more than a handful a day.

DISTRIBUTION: General inland, including the Sacramento Valley; seemingly very rare or absent in the fog belt.

HOST PLANTS: Mistletoes (*Phoradendron,* Viscaceae) as explained above.

SEASONS: January to November, multiple broods typically unsynchronized from locality to locality; usually commonest in late June to early July and September to October.

SIMILAR SPECIES: None!

CALIFORNIA HAIRSTREAK *Satyrium californicum*
Pl. 11

This is a widespread species in oak woodland in the Coast Ranges but now very rare and regionally threatened with extinction in the southern Sacramento Valley, where its riparian habitat has been severely fragmented. In its foothill strongholds its numbers fluctuate wildly from year to year. There is some local phenotypic variation. Sacramento Valley specimens have a more lustrous gray ground color beneath and (usually) sharper black markings. Coast Range specimens from the hot interior are duller, matte gray, and less crisply marked beneath. Males rarely have significant orange on the disc of the wings above, but females are very variable and some are mostly orange.

This species, like many hairstreaks, has a very strict daily routine. These butterflies nectar from about 10:30 A.M. to 2:00 P.M.; in our area they especially visit California buckeye *(Aesculus californica)*, milkweeds *(Asclepias)*, dogbanes *(Apocynum)*, yerba santas *(Eriodictyon)*, horehound *(Marrubium vulgare)*, and coffeeberries *(Rhamnus)*. Males perch on or near the host and are territorial in late afternoon, typically peaking around 4:00 to 4:30 P.M. Note that I do not specify any early-morning activity—we don't know where they are then!

DISTRIBUTION: Nearly all counties, but now very rare and local in the southern Sacramento Valley.

HOST PLANTS: Oaks *(Quercus)*; in chaparral sometimes on wild-lilacs *(Ceanothus);* on the east slope of the Sierra and perhaps in places in the Coast Ranges, on mountain-mahoganies *(Cercocarpus)*. The Sacramento Valley ecotype feeds on valley oak *(Q. lobata)* in rich bottomlands.

SEASONS: One flight, April to July, lasting 3 to 5 weeks in any given location. Overwinters as egg; larva feeds on tender foliage shortly after bud burst. This seasonality is characteristic of feeders on tannin-rich plants, which become less and less edible as the season goes on.

SIMILAR SPECIES: The Sylvan Hairstreak *(S. sylvinum)* and Tailless Sylvan Hairstreak *(S. dryope)* have essentially the same color and pattern but are washed-out in appearance, the black dots below usually indistinct (sometimes obsolescent); the ground color below, as in the California Hairstreak, varies from dull/ashen to rather lustrous. The lunules at the base of the tails (or, in the Tailless Sylvan, where the tails would be) below are less well defined than in the California Hairstreak; often only one has any red, and the large blue spot has little or no red above it. Both the Sylvan and the Tailless Sylvan Hairstreaks are strict riparian specialists, associated with willows *(Salix)*—not oaks. The Common Hairstreak *(Strymon melinus)* has ventral pattern of lines, not dots, and very prominent red eyespot on the hindwing above and below. The male's abdomen is orange above.

SYLVAN HAIRSTREAK
or WILLOW HAIRSTREAK
Satyrium sylvinum

TAILLESS SYLVAN HAIRSTREAK
or WILLOW HAIRSTREAK or DRYOPE
Satyrium dryope

Pl. 11

The true relationship between these two entities is unclear. Are they species, subspecies, or…? They are very similar except for the absence of tails in the Tailless Sylvan Hairstreak. Statistically, the Tailless Sylvan tends to be paler, ash gray beneath with much-reduced or even obsolescent black spots. Females are usually strongly orange washed above. The Sylvan Hairstreak tends to be more lustrous beneath, with more distinct spots; the females are very variable. The problem is that some tailed populations, especially in the Sacramento Valley, look very like the tailless otherwise. Tailed populations occur in cooler and moister habitats, while tailless ones are in hotter and drier habitats. The Tailless Sylvan predominates in Santa Clara County, the Livermore Valley, and along the eastern edge of the Coast Range south of the Delta. All other populations in our area are tailed. Where streams come down into the Sacramento Valley from the North Coast Range, foothill populations tend to be more lustrous and strongly marked than valley floor populations in the same drainage, though all are tailed. All in all, a complicated situation. To make it more complicated, butterflies that look just like our Tailless Sylvan also occur in the western Great Basin, even in Carson Valley just east of Lake Tahoe. Are they the same entity distributed disjunctly (compare the Yuma Skipper *[Ochlodes yuma]* and Field Crescent *[Phyciodes campestris]*), or are they convergent? Perhaps molecular genetics will tell us.

Males of both are highly territorial, perching on twig tips of the host plant in mid- to late afternoon. They nectar around midday, especially on dogbanes *(Apocynum)* and milkweeds *(Asclepias)*, whose saddlebag pollen sacs they often carry. In the 1970s the Sylvan Hairstreak was immensely abundant in the Sacramento Valley, but by the late 1990s it was in serious decline and is now absent from many of its old haunts. As usual, there is no obvious reason. It is curiously absent from large stretches of seemingly ideal habitat in the Coast Ranges, as well.

DISTRIBUTION: All counties, always with willows *(Salix)*, but not uniformly distributed and becoming more and more local.

HOST PLANTS: Native willows *(Salix)*, including both glabrous and pubescent species; the Tailless Sylvan usually on sandbar willow *(Salix exigua,* also called *S. hindsiana).* Local populations often seem associated with particular species and sometimes even certain individual trees.

SEASONS: One brood, May to July, generally starting and ending 1 to 3 weeks later than the sympatric California Hairstreak *(S. californicum).* Overwinters as egg, which can survive immersion in flood waters for several weeks.

SIMILAR SPECIES: See California Hairstreak.

GOLD-HUNTER'S HAIRSTREAK   *Satyrium auretorum*
Pl. 11

This species has a reputation for scarcity, but it is sometimes quite common in warmer, drier parts of the Coast Ranges where it joins the California Hairstreak *(S. californicum)* on buckeye flowers *(Aesculus californica).* It is restricted to oak woodland and nearly extinct in the lower Sacramento Valley. Its daily regimen is like that of the California Hairstreak, and the two routinely co-occur in the hills. Males are very triangular in profile when backlit sitting high in a buckeye tree. The only significant variation is in the amount of orange on the female upperside. Females are noticeably larger than males and have more squarish forewings. An unnamed, larger subspecies occurs at midelevations on the Sierran west slope.

DISTRIBUTION: General in Coast Ranges, except apparently absent in the fog belt. Rare in the Sacramento Valley, and perhaps now extinct near Sacramento where it once occurred at least in Rancho Cordova and North Sacramento.

HOST PLANTS: Oaks *(Quercus)* including interior live oak *(Q. wislizenii),* blue oak *(Q. douglasii),* and scrub oak *(Q. berberidifolia),* among others.

SEASONS: One brood, April to July, commonly first emerging in a given place on the same day as the California Hairstreak, or else a day or two earlier, and ending its flight a little earlier too. It usually has little temporal overlap with the Mountain-mahogany Hairstreak *(S. tetra),* which it resembles in silhouette.

SIMILAR SPECIES: The Mountain-mahogany Hairstreak, with similar shapes in both sexes, is dark charcoal gray, not brown.

MOUNTAIN-MAHOGANY HAIRSTREAK   *Satyrium tetra*
or GRAY HAIRSTREAK
Pl. 11

A chaparral species par excellence, it is also found in rocky canyons where the host plant grows. Males are very active territorially in late afternoon, perching on the host plant. This butterfly often visits flowers of toyon *(Heteromeles arbutifolia)*, not usually thought of as a butterfly flower. Its numbers vary greatly from year to year. There is no phenotypic variation in our area. The name Gray Hairstreak appears in the literature but should be reserved for *Strymon melinus*.

DISTRIBUTION: Chaparral and rocky canyons except in the fog belt. Unrecorded in the Sacramento Valley except near Red Bluff, but the host occurs in the American River drainage below Folsom, and probably elsewhere in the lower foothills.

HOST PLANTS: Mountain-mahogany *(Cercocarpus betuloides)*.

SEASONS: One flight, May to July, later than the Gold-hunter's Hairstreak *(S. auretorum)* and California Hairstreak *(S. californicum)*, on average, where they co-occur.

SIMILAR SPECIES: The Gold-hunter's Hairstreak has a similar silhouette but a wood brown ground color; the Mountain-mahogany Hairstreak is charcoal gray, frosted on ventral hindwing, especially in the female.

HEDGEROW HAIRSTREAK   *Satyrium saepium*
or SEPIA HAIRSTREAK
Pl. 11

Often abundant in chaparral, this species is less common in canyons. The Hedgerow Hairstreak is typically the last *Satyrium* to emerge and to fly in any given locality, often extending its season into midsummer. In canyons it often sits in dappled light and shade or in moist streambeds. Coastal specimens tend to have a brassier upperside color and a slightly darker ground color below. Visits yerba santas *(Eriodictyon)*, California buckeye *(Aesculus californica)*, dogbanes *(Apocynum)*, thistles *(Cirsium, Carduus,* and *Silybum)*, white buckwheat *(Eriogonum nudum)*, and the sunflower family (Asteraceae).

DISTRIBUTIONS: All Bay Area counties; absent from Sacramento Valley except in extreme north.

HOST PLANTS: Various species of wild-lilac *(Ceanothus,* Rhamnaceae).

SEASONS: One flight, mid-April to September (!), usually May to August; the flight especially protracted in the fog belt.

SIMILAR SPECIES: The underside reminiscent of the Gold-hunter's Hairstreak *(S. auretorum),* but the male silhouette and upperside color are quite different. In riparian canyons it can be briefly confused with the Tailed Copper *(Lycaena arota)* in flight.

Bramble Hairstreak
(Callophrys "dumetorum complex")

This is one of our worst taxonomic nightmares. Scarcely any two authorities agree on the limits of species, or to what named entity various populations should be assigned, or even on the correct biological entity to which some of the names refer. Fortunately— whatever one chooses to call them—there appear to be two biological entities in our area that can be appreciated and studied if names are either avoided or used with a healthy dose of both skepticism and tact. At this writing, some authorities are calling these things *C. affinis!*

"INLAND" BRAMBLE HAIRSTREAK

Pl. 10 *Callophrys dumetorum* of older books; *C. perplexa perplexa* of Brock and Kaufman

Emmel, Emmel, and Mattoon (1998) concluded in the absence of direct evidence that the type *C. dumetorum* was probably collected in or near San Francisco, and that name, which had been used for inland populations since at least 1944, actually applied to the coastal entity. A popular guide such as this is not the place to rehash all the relevant issues, especially since the basic *biological* evidence remains to be explored. Whatever you call it, this is a common butterfly of chaparral and foothill woodland, occasional in riparian canyon habitats. Males are territorial perchers. The male is slate gray above with a prominent stigma; the female, orange brown. The green underneath is "warmer" (less blue) than in the coastal entity.

DISTRIBUTION: Bay Area, probably all counties inland from the fog belt; not recorded in San Francisco. Local and spotty in the Sacramento Valley, mainly on dredge tailings on the east side of the valley; absent from wetlands and agricultural land.

HOST PLANTS: Deer weed *(Lotus scoparius)* and perennial wild buckwheats *(Eriogonum)*. The interchangeability of these hosts for several lycaenids strongly implies a chemical commonality, not yet identified.

SEASONS: Supposedly one brood, with records from February to mid-June, but in the Inner Coast Range there are hints of two flights (March to April, May to June), which could reflect a second brood or differing phenology on north- and south-facing slopes; the matter invites investigation.

SIMILAR SPECIES: The "Coastal" Bramble Hairstreak is colder, bluer green below with (usually) a more complete white pattern; the male is darker slate gray above; confined to fog belt.

"COASTAL" BRAMBLE HAIRSTREAK

Pl. 10   *Callophrys viridis* of older books;
C. perplexa viridis of Brock and Kaufman;
true *C. dumetorum* per Emmel, Emmel, and Mattoon

Confined to the coastal fog belt in scrub and dune communities, this animal—one of our most beautiful and photogenic butterflies—may or may not be conspecific with the "Inland" entity, but the two seemingly never co-occur. Habits similar to the "Inland" insect.

HOST PLANTS: Deer weed *(Lotus scoparius)* and perennial wild buckwheats *(Eriogonum)*.

SEASONS: One very extended flight, February to June.

SIMILAR SPECIES: The "Inland" Bramble Hairstreak is duller, yellower green below, usually with little if any white pattern.

THICKET HAIRSTREAK   *Mitoura (Callophrys) spinetorum*

Pl. 12

Generally regarded as uncommon and sporadic, this lovely butterfly spends much of its time high in the trees and may be more widespread than we think. It visits flowers, including yerba santas *(Eriodictyon)*, California buckeye *(Aesculus californica)*, milkweeds *(Asclepias)*, dogbanes *(Apocynum)*, wild-lilacs *(Ceano-thus)*, and yellow-flowered members of the sunflower family (Asteraceae), but most captures seem to be of single females (!) on mud puddles or wet rocks. Its host is a parasite on pines *(Pinus)*, in our area usually on gray or foothill pine *(P. sabiniana)*. Although this pine grows on many kinds of soils, pine mistletoe *(Arceuthobium campylopodium)* is much commoner on trees growing on serpentine, and the butterfly tends to be commoner there as well. During a warm, dry period a few thousand years ago, vast quantities of pine mistletoe pollen were deposited— suggesting that both the plant and its butterfly may have been much commoner and more widespread then, and the current distribution is relictual. Males are territorial and lek on and around the host tree. There is virtually no variation.

Many authorities combine the genera *Mitoura* and *Incisalia* as subgenera of *Callophrys*. Others divide them even more finely, for example, placing the Thicket Hairstreak and Johnson's Hair-streak *(M. johnsoni)* in a separate genus, *Loranthomitoura*. You have been warned.

DISTRIBUTION: Scattered localities in the Coast Ranges, mostly in-land and on serpentine. Absent from the Sacramento Valley ex-cept perhaps in the far north.

HOST PLANTS: Pine mistletoe *(Arceuthobium campylopodium)* on two- and three-needle pines, in our area usually on Gray Pine.

SEASONS: Apparently multiple brooded, records from February to October with a concentration in March through June; the later broods often fly when very little is in flower and seemingly keep to the treetops. Where rabbitbrush *(Chrysothamnus)* occurs, it is an irresistible lure in fall.

SIMILAR SPECIES: Johnson's Hairstreak is somewhat similar, but brown above with no hint of blue.

JOHNSON'S HAIRSTREAK   *Mitoura (Callophrys) johnsoni*
Pl. 12

Generally considered rare, the Johnson's Hairstreak has been recorded sporadically in the Inner Coast Range, mostly in spring. It is especially partial to flowers of skunkbush *(Rhus trilobata)*, also called lemonade bush or squawbush, but also visits wild-lilacs *(Ceanothus)*, dogbanes *(Apocynum)*, milkweeds *(Ascle-pias)*, yerba santas *(Eriodictyon)*, and so forth. As in the Thicket

Hairstreak *(M. spinetorum)*, most collections are of single females, often at mud. This species does not seem particularly associated with serpentine; it is recorded variously in foothill woodland, canyon riparian, and Douglas fir–dominated habitats, and its ecology is poorly understood. It may be briefly common in a locality, then vanish altogether.

DISTRIBUTION: Counties of the Inner Coast Range.

HOST PLANTS: Conifer mistletoes (*Arceuthobium campylopodium* on pines, *A. douglasi* on Douglas fir).

SEASONS: Spotty, from late February to late September; probably multiple brooded, but most records from March through June.

SIMILAR SPECIES: See Johnson's Hairstreak.

Juniper Hairstreak
(Mitoura [Callophrys] "gryneus complex")

This group consists of numerous named entities across temperate North America, feeding on different species of the juniper-cypress family (Cupressaceae). Usually the populations feeding on a particular host are distinctive in color, pattern, and sometimes phenology vis-à-vis populations feeding on different hosts in the same region, but they do not differ significantly in genital morphology or (as a rule) at the level of molecular genetics. When the proper experiments are done, we find that host preference (by ovipositing females) and larval performance indeed differ and are under genetic control. It appears that the various entities in northern California, at least, are in the process of differentiating into different species. But despite behavioral isolation in the field, they are fully capable of hybridizing in the laboratory. Two entities in this dynamic complex occur in our area.

JOHN MUIR'S HAIRSTREAK　　　*Mitoura gryneus muiri*
Pl. 12

Generally very dark on the ventral hindwing and flying very early in the year, John Muir's Hairstreak is nearly restricted to serpentine sites where its two edaphic-endemic hosts grow. At Mount Diablo State Park and a few other sites south of the Bay, what appears to be this entity occurs with California juniper (*Juniperus californica*) on nonserpentine soils. These populations deserve further study. Males lek on the hostplant in the afternoon, and nearly all the males in the population may be clustered on one or a handful among many trees on any given day.

DISTRIBUTION: Found with the host cypresses on serpentine soils, in chaparral and foothill woodland in several Bay Area counties. Very localized with juniper in grassland-foothill woodland in Contra Costa and Alameda Counties, but also with Sargent cypress *(Cupressus sargentii)* on serpentine at Cedar Mountain Ridge, Alameda County.

HOST PLANTS: Sargent cypress *(Cupressus sargentii),* a tree growing mostly along streamsides on serpentine; MacNab cypress *(C. macnabiana),* a usually multistemmed, large shrub on serpentine "cedar roughs" mostly inland; and California juniper *(Juniperus californica),* a large shrub found on various soils in hot, dry, sunny situations, for us mainly in the Inner Coast Range south and east of the Bay. Visits yellow-flowered members of the sunflower family (Asteraceae) and other flowers, and mud.

SEASONS: One brood, March to early July, mostly early. As with other entities in this complex, very rarely a fresh individual will turn up in late summer or fall, suggesting at least the potential for a second brood.

SIMILAR SPECIES: See Nelson's Hairstreak.

NELSON'S HAIRSTREAK   *Mitoura gryneus nelsoni*
Pl. 12

This is a very abundant species in mixed-mesic forest statewide, wherever its sole host plant, incense cedar *(Calocedrus decurrens),* grows. In our area it is only reported in Napa County but is likely to be more widespread. It visits wild-lilac *(Ceanothus),* yerba santas *(Eriodictyon),* woolly sunflower *(Eriophyllum lanatum),* and other spring flowers, as well as mud. It leks on and near small specimens of the host tree. There is considerable variation in color below and in the extent of orange in the female above, but it is always much lighter on the ventral hindwing than in John Muir's Hairstreak *(M. g. muiri).* Any Juniper Hairstreak *(M. gryneus)* found away from the two cypresses and on nonserpentine soils may be this entity.

DISTRIBUTION: Napa County, probably elsewhere inland in the North Bay.

HOST PLANT: Incense cedar *(Calocedrus decurrens).*

SEASONS: One brood, April to July, with the occasional fresh individual August to October elsewhere.

SIMILAR SPECIES: See John Muir's Hairstreak.

WESTERN BROWN ELFIN
Pl. 12

Incisalia (Callophrys)
augustinus iroides

A common species in the Bay Area, especially in chaparral, it is infrequent to rare in the Sacramento Valley, where it is restricted to wildlands. The Western Brown Elfin is territorial beginning in the late morning, but mating pairs are seldom seen before late afternoon. There is considerable variation in color and pattern, mainly on the ventral surface. This, combined with highly variable seasonality and an astonishing range of host associations, has led to speculation that more than one genetic species may be involved. The conspecificity of our animals (or some of them) with the Eastern Brown Elfin *(I. augustinus)* is quite uncertain as well.

Adults visit yerba santas *(Eriodictyon)*, yarrow *(Achillea mille-folium* and *A. borealis)*, the sunflower family (Asteraceae), and a variety of other flowers; this is one of the few butterflies that nectar at redbuds *(Cercis)* early in the season.

DISTRIBUTION: All Bay Area counties, in a great variety of habitats and plant communities, scarcest in forest and commonest in chaparral and in rocky canyon or cliff situations. In the Sacramento Valley, occasional in oak woodland and especially on dredge tailings, where soap plant occurs. Astonishingly, has bred on dodder *(Cuscuta)* in the Suisun Marsh!

HOST PLANTS: Larva feeds on buds, flowers, and developing seeds. The usual foothill and grassland host is soap plant *(Chlorogalum pomeridianum,* Liliaceae); in chaparral also recorded on buck-brush *(Ceanothus cuneatus;* probably other species) and manzanitas *(Arctostaphylos)*; in mesic habitats near the coast recorded on madrone *(Arbutus menziesii)* and salal *(Gaultheria shallon)*; in coastal dune and brush habitats and in the Suisun Marsh on dodder *(Cuscuta)*—overall the most bizarre host catalogue of any of our butterflies. (Two blues in Chile and Argentina also feed on dodder; otherwise this association is unique.)

SEASONS: February (sometimes one of the first nonhibernators out) to early July, seemingly univoltine in some places and bivoltine in others, or perhaps more than one species is involved.

SIMILAR SPECIES: All of our *Incisalia* species except the Western Pine Elfin *(I. eryphon)* are confusing, though nearly always your specimen will turn out to be the Western Brown Eflin. The various subspecies of Moss's Elfin *(I. mossii)* look quite different from one another, but all have a white or whitish marginal line (absent in the Western Brown) and at least some grayish on the hindwing below, though it may be confined to a patch near the anal angle. Moss's Elfin has distinctly scalloped and frosted hindwing margins and is overall much more subtly colored than the Western Brown, usually with little or no brick red tinge. Male Moss's Elfins are gray above, females orange brown; the Western Brown Elfin is usually brown in both sexes. The Western Pine Elfin looks like an Oriental rug beneath.

MOSS'S ELFIN   *Incisalia (Callophrys) mossii*
Pl. 12

Several named subspecies of Moss's Elfin with mostly very restricted distributions are found in northern California. One, the San Bruno Elfin *(I. m. bayensis)* is federally listed as endangered and is "off limits" to collecting or disturbance. All are strikingly colored, with subtle underside patterns. All are single brooded and typically fly very early in the year. All feed on the stonecrop family (Crassulaceae) growing on rocks—typically on cool, damp, north-facing cliffs, which may be in redwood forest, coastal scrub, or even oak woodland contexts. Our populations feed on sedum *(Sedum),* and the larvae, which often feed on the flowers, are usually easier to find than the adults. In fact, the Marin County subspecies, the Marin Elfin *(I. m. marinensis),* was discovered by rearing from wild larvae collected near Lagunitas Creek.

Because the host plants are usually north facing and in shade most of the day, adults are only infrequently observed on or even very near them. Instead, males are territorial in bright sun, often on the other side of the canyon, perching on shrubs—coyote brush *(Baccharis pilularis)* is a favorite. Both sexes nectar at various early flowers, including the carrot, mustard, borage, and phlox families (Apiaceae, Brassicaceae, Boraginaceae, and Polemoniaceae), and both sexes visit puddles.

DISTRIBUTION: The San Bruno Elfin occurs in San Mateo County only, in coastal scrub. The Marin Elfin is found in Marin County.

Populations in Sonoma and Napa Counties become gradually more like the "inland" subspecies, Wind's Elfin *(I. m. windi)*, which as currently understood occurs in the Coast Ranges north of our area and on the west slope of the Sierra Nevada, where it reaches the subalpine zone. Doudoroff's Elfin *(I. m. doudoroffi)*, (type-locality Big Sur) occurs on the central coast just south of our area. It may extend into Santa Cruz County. Finally, an as yet unnamed subspecies occurs around Mount Diablo in Contra Costa County!

HOST PLANTS: In our area, Pacific stonecrop *(Sedum spathulifolium)*; recorded elsewhere on other sedums and on dudleya *(Dudleya)*.

SEASONS: One flight, February to April, depending on locality and year, often flying so early as to be missed by people looking for it.

SIMILAR SPECIES: See Western Brown Elfin.

WESTERN PINE ELFIN   *Incisalia (Callophrys) eryphon* or WESTERN BANDED ELFIN

Pl. 12

The underside pattern of this exquisite butterfly is reminiscent of a fine Oriental rug. The pattern is quite visible when the territorial male perches on a tall plant in a sunfleck. The male is dark brown above; the female, suffused with orange brown. This butterfly is fairly common along roads or paths through either natural or planted stands of pine *(Pinus)*.

The history of this species in the Bay Area has been carefully documented by Jerry A. Powell of the University of California at Berkeley ("*Callophrys eryphon* colonizes urban and suburban San Francisco Bay area," *Journal of the Lepidopterists' Society,* 51:176–179, 1997). There are two recorded hosts in our area, Bishop pine *(P. muricata)* and Monterey pine *(P. radiata)*. The former is widespread near the coast but the latter is native in only a handful of sites on the Monterey Peninsula. Apparently San Francisco had native pines in Gold Rush days, but we do not know what species. Monterey pine was very widely planted in the late nineteenth and twentieth centuries as an ornamental and to stabilize what had been migratory dunes, as in the Presidio and Golden Gate Park. The Museum of Comparative Zoology at Harvard has three nineteenth-century specimens of the Western Pine

Elfin labeled "San Mateo, Cal.," apparently collected by Alexander Agassiz. There do not seem to be any modern records from San Mateo County. We have no reason to doubt that the Western Pine Elfin was widespread in stands of Bishop pine from Marin County north- and northeastward. The first San Francisco records were by W.D. Field and R.G. Wind in 1929. From then until at least 1950 it appears to have been resident in the city, apparently breeding on Monterey pine. It may now be extinct there. However, since the 1970s it has appeared in a variety of new locations around the Bay, from Strawberry Point and Belevedere, Marin County, to sites in western Alameda and Contra Costa Counties, where it seems to be established, again, on planted Monterey pines. Ironically, the one known population in Monterey County uses Bishop pine—the butterfly does not occur in the native groves of Monterey pine! In the uplands of the North Coast Range this species uses ponderosa or yellow pine *(P. ponderosa)*. All of these are two- or three-needle "hard" pines. Any new localities should be reported to local butterfly authorities. The spread of the exotic pine disease pitch canker in our area may adversely impact this butterfly.

DISTRIBUTION: North Bay counties plus western Alameda and Contra Costa Counties, including Berkeley. Apparently absent from San Francisco and southward at this time, but "should be" in the Santa Cruz Mountains. Unknown in the Sacramento Valley.

HOST PLANTS: Bishop, Monterey, and ponderosa pines *(Pinus muricata, P. radiata,* and *P. ponderosa);* probably other "hard" pines.

SEASONS: March to June, as usual flying later in the fog belt.

SIMILAR SPECIES: The complex hindwing underside pattern is unique.

COMMON HAIRSTREAK or GRAY HAIRSTREAK
Strymon melinus

Pl. 10

Found almost everywhere, this very common hairstreak is if anything more likely to turn up in your garden than in wildlands; it is rare only in deep woods. It shows pronounced seasonal variation. Late winter to early spring (and some late fall) individuals are small, very dark above and slaty gray below, usually with small red markings. Summer ones are bigger, lighter both above and below (usually pearly beneath), with large red markings. There is plenty of minor variation in the pattern and in the amount of red below. "Albinos," with the normally red markings pale whitish or yellowish, are not rare, especially in the cold-season broods.

Males are avid hilltoppers but in flat terrain will perch on tall plants in late afternoon. This species is capable of extremely swift flight and dazzling aerobatics. It has a distinctive habit of flying in tightening circles around a flower or perch, often popping up onto it from below. It is the only hairstreak you are likely to see visiting clover *(Trifolium)* or lippia *(Lippia)* flowers in mowed lawns in the city.

Its diet, while not so bizarrely eclectic as that of the Western Brown Elfin *(Incisalia augustinus iroides)*, is second in breadth only to that of the Painted Lady *(Vanessa cardui)* in our fauna. In its overall range it is recorded from dozens of plants scattered over many families, quite a few with no natives in our flora. In most places it is at least partially (often seasonally) dependent on exotic hosts; in the Central Valley it often has only one or two native hosts available (turkey mullein *[Eremocarpus setigerus]* or Spanish lotus *[Lotus purshianus]*). The larvae feed on buds, flowers, fruit, and even leaves, perhaps largely by default. Any lycaenid larva found on a weed or garden plant is very likely to be this species.

Adults visit many flowers, including those in the wild carrot family (Apiaceae) (wild carrots *[Daucus]* and bishop's weed *[Ammi]* are favorites), heliotropes *(Heliotropium)*, lippias, clovers, yarrow, butterfly bush *(Buddleia)*, wild buckwheats *(Eriogonum)*, perennial peppergrass or tall white-top *(Lepidium latifolium)*, dogbanes *(Apocynum)*, milkweeds *(Asclepias)*, coyote brush *(Baccharis pilularis)*, asters *(Aster)*, goldenrods *(Solidago)*, …and on and on.

DISTRIBUTION: All counties, scarcest in fog belt and in dense forest, commonest in urban, suburban, and agricultural contexts.

HOST PLANTS: Mallow family (Malvaceae): cheeseweeds *(Malva neglecta, M. nicaeensis,* and *M. parviflora)*, hollyhock *(Alcea)*, alkali mallow *(Malvella leprosa)*; pea family (Fabaceae): alfalfa *(Medicago sativa)*, *Lotus* species (both native and introduced, including Spanish lotus *[L. purshianus]* and birdfoot trefoil *[L. corniculatus]* and probably many others); spurge family (Euphorbiaceae): turkey mullein *(Eremocarpus [Croton] setigerus)*; recorded ovipositing on bottlebrush *(Callistemon,* Proteaceae!)

and crape myrtle (*Lagerstroemia indica*, Lythraceae!) in Davis. Probably many more. In many places consecutive generations must feed on different plants, due to their quick seasonal turnover. Occasionally reported as a minor pest of cultivated peas and beans.

SEASONS: January to December, with a flight season of 10 to 11 months; multiple broods; overwinters as pupa. In the Sacramento Valley the broods tend to form distinct pulses, while near the coast the species flies more continuously but is less common.

SIMILAR SPECIES: The Common Hairstreak is distinctive for its large, black-pupilled red spot on both surfaces of the hindwing; the pattern of white- (or white-and-red) edged lines rather than spots beneath; the male's orange red abdomen; and no distinct stigma. The black-pupilled red spot and the highly mobile tails are believed to constitute a "false head" that diverts predators away from the real body. Similar "false heads" are common in tropical hairstreaks and in some blues.

EASTERN TAILED BLUE   *Everes comyntas*
Pl. 13

The Eastern Tailed Blue is "weedy," occurring in disturbed habitats, vacant lots, and annual grassland. It is also frequently common in freshwater marshes, along swales and sloughs, and in Central Valley riparian forest. It is the only tailed blue in the Central Valley and Delta but is absent or rare in the coastal fog belt. It has not been found co-occurring with the Western Tailed Blue (*E. amyntula*) in our area.

Most "populations" of the Eastern Tailed Blue are obliged to move around seasonally, using a succession of host plants. In spring this species is often found on roadsides and highway embankments, breeding on introduced annual vetches *(Vicia)*. After midsummer it can be found (usually with the Acmon Blue [*Plebejus acmon*]) in stands of Spanish lotus *(Lotus purshianus)*, a warm-season annual often found in and along the dry beds of seasonal streams. Two perennial hosts that permit repeated breeding cycles in the same locality are native sweet peas *(Lathyrus)*, and the naturalized alien birdfoot trefoil *(Lotus corniculatus)*.

Flight is low and near the ground. Males patrol patches of

host plant to find females. Occasional individuals disperse widely, turning up in urban and suburban lawns and gardens. Adults often rub the hindwings together when at rest with the wings over the back. This causes the "tails" to move up and down. It has been suggested that the tails simulate antennae and the adjacent orange lunules form a false head—the combination serving to distract predators away from the body. The Eastern Tailed Blue visits many low flowers, but especially heliotropes *(Heliotropium)*, clovers *(Trifolium)*, and lippias *(Lippia)*. It supposedly overwinters as a larva but is still one of the first species to emerge in spring.

March to April individuals are very richly colored, with a deeper blue ground color in males and usually quite a bit of blue in females. Summer males are duller, and females are usually entirely slate gray to dull black above. It has been suggested that this species was naturalized from eastern North America, primarily because it is associated with disturbed habitats and because the earliest Bay Area records date from only 1949. Molecular genetics has failed to resolve this question. However, our specimens are subtly different from eastern ones in a variety of ways. The orange lunules on the hindwing above tend to be reduced or obsolescent in California; in females they are usually whitish or very pale yellow. The ground color of summer females is much duller than in the east. These differences could, of course be induced by environmental factors rather than being genetically determined

DISTRIBUTION: Throughout the Sacramento Valley and Delta including the Suisun Marsh. Rarely found on the coast, and generally rarer in cool-summer than in hot-summer localities. Entirely absent in conifer and oak forests and in the foothills above about 1,000 m (3,000 ft).

HOST PLANTS: All pea family (Fabaceae): both introduced annual vetches *(Vicia)* and native perennial sweet peas *(Lathyrus* spp., but never on the widely naturalized alien *L. latifolius)*; both annual and perennial species of *Lotus,* especially Spanish lotus *(L. purshianus)* and the naturalized alien birdfoot trefoil *(L. corniculatus)*. Larvae eat fruits and seeds.

SEASONS: March to November, breeding continuously but tracking a seasonal succession of host plants.

SIMILAR SPECIES: The Western Tailed Blue averages larger, with broader forewing; has the under surface paler and more ashen, with generally smaller, less-contrasting black dots; and occurs only in or near forest in the coastal counties and foothills, associated with native perennial hosts. Globally the two species are not each other's closest relatives, but in at least two locations outside our area they co-occur and apparently hybridize.

WESTERN TAILED BLUE   *Everes amyntula*
Pl. 13

The ranges of this species and the Eastern Tailed Blue *(E. comyntas)* are complementary in our area, and they seem never to co-occur here. The western species is found primarily near the coast, commonly in forest openings, along streams and roadsides through redwood, Douglas fir, and oak forests where its host plants grow, and often in clear-cuts. It flies low, near the ground, and rarely strays far from its host plant. Males puddle. There are generally two broods a year in the Bay Area (but only one above about 1,200 m [3,900 ft]; the species extends above tree line in the Sierra and northern mountains). Flights are in midspring and early to midsummer. Overwinters as a larva, supposedly completing feeding in very early spring. However, the larvae eat seeds inside the pods, and these are not available in spring! When feeding inside a pod the larva closes off its entrance hole with silk, apparently excluding parasitoids. Like the Eastern Tailed Blue, this species rubs its hindwings together and may thereby simulate a "head" at the anal angle. There is less seasonal variation in the Western Tailed Blue than in the Eastern Tailed Blue, though some summer females are quite dark.

DISTRIBUTION: Coastal counties in cool, mostly forested areas; partly shaded often north-facing canyons and canyon bottoms inland. Absent from the Sacramento Valley, the Delta, and both the Inner Coast Range and Sierra foothills below about 1,000 m (3,200 ft).

HOST PLANTS: Native perennial vetches *(Vicia)* and sweet peas *(Lathyrus);* elsewhere and possibly in our area, locoweeds or milkvetches *(Astragalus)*. Unrecorded on alien hosts in our area, perhaps in part due to adult habitat selection.

SEASONS: Usually two broods, March to May and June to August, with stragglers.

SIMILAR SPECIES: The Eastern Tailed Blue averages smaller, with narrower wings; more lustrous ground color beneath with usually more contrasting dot pattern. Western Tailed Blue females are more likely to have significant blue scaling on the upper surface in summer. Male genitalic differences are reliable. Locality and habitat are usually sufficient to differentiate these two species!

SILVERY BLUE   *Glaucopsyche lygdamus*
Pl. 16

This common and widespread spring species fed originally on native perennial legumes, but sometime before the 1970s it "discovered" the abundant annual species of exotic vetches *(Vicia)* naturalized in California and rapidly developed a race or ecotype adapted specifically to their use. This ecotype spread mainly along freeway embankments and is now found throughout the Sacramento Valley. It has also moved up into the hills, where it sometimes contacts populations still using their traditional native hosts. Oddly, a similar annual-vetch ecotype emerged in the Northeastern states at about the same time, leading to a significant range expansion for the species (and a risk of genetic swamping for the rather rare, endemic subspecies found mainly on shale barrens in the northern Appalachians).

Populations of this species fluctuate in numbers from year to year. This is particularly true in the Sacramento Valley, where "boom and bust" cycles appear routine, although one little population somehow persisted near Davis for over 20 years.

Individual variation is substantial. The ventral spots may be small or large, strongly or weakly contrasting; most females are mostly blackish above with only a few blue scales, but a few are very rich blue. One extraordinary male, collected April 8, 1982, in Sacramento County (illustrated in *Journal of Research on the Lepidoptera* 20:240–241, 1983), showed a complete pattern of black spots on the upper surface matching those of the related Old World genus *Maculinea*. Such cases are generally interpreted as expressions of an ancestral genetic program, normally suppressed in the descendant species; to put it another way, it suggests that the ancestor of our Silvery Blue had spots like those of *Maculinea*. Such cases are called "reversions" or "atavism."

This is a devout puddler. In big Silvery Blue years, aggregations of a hundred males or more can sometimes be seen. There seems to be no territoriality. Males patrol patches of the host plant. Courtship, mating, and oviposition—which takes place on the inflorescences—are all easily observed in this species.

The larvae, which are attended by ants, feed on flowers and green fruit. They develop quickly and then remain quiescent as the pupa for some 9 months! The pupae may sometimes overwinter in ant nests.

Strays of this species (both sexes) tend to follow linear habitats—roadsides and watercourses—so that new colonizations are frequent. A population in Suisun Marsh on the scarce native vetch the Delta tule pea *(Lathyrus jepsonii jepsonii)* is now apparently extinct (though the plant is fine), but the marsh was recently colonized by the annual-vetch ecotype (on landfill!).

DISTRIBUTION: Throughout the area in a variety of habitats: dunes, coastal prairie, coastal scrub, rocky canyons, roads, or trails through forest; annual-vetch ecotype on freeway embankments and weedy roadsides, dredge tailings, and in riparian forest and tule marsh where the hosts occur.

HOST PLANTS: Native perennial pea family plants (Fabaceae) including a variety of lupines (the annual *Lupinus succulentus* is also occasionally used); golden pea or false-lupine *(Thermopsis);* vetches *(Vicia),* and sweet peas *(Lathyrus).* In the Sacramento Valley and adjacent foothills and spreading along highways, on naturalized annual vetches (several species). Recorded in central and southern California on deer weed *(Lotus scoparius)* but not seen using it in our area.

SEASONS: March to May inland, February to June in the fog belt; one brood with a rare, rudimentary second flight in the Inner Coast Range.

SIMILAR SPECIES: Three similar-sized blues routinely co-occur in our area, feeding on perennial lupines: the Silvery Blue, Arrowhead Blue *(G. piasus),* and Boisduval's Blue *(Plebejus icarioides).* All have blackish females that occasionally have plenty of blue above. The male Silvery Blue is, well, silvery—a bright sky blue. The male Arrowhead Blue is deep violet blue and has checkered fringes. The male Boisduval's is less violet than the Arrowhead but more violet than the Silvery. If you are a color connoisseur you would call it ultramarine. On the underside of both sexes, the complex dark pattern of the Arrowhead Blue is unmistakeable. The Silvery Blue has an area of bright blue green scales at the base of the hindwing; the ground color is usually fawn gray; and there is never any pattern in the submarginal area of the wing. The Boisduval's Blue, in most of its incarnations, lacks the blue green patch and always has more or less of a submarginal spot-band;

the ground color is usually light ashen gray. You know you are a "pro" when you can pick out the Arrowhead Blues on the wing!

XERCES BLUE   *Glaucopsyche xerces*
Pl. 14

This species is globally extinct.

While a few authorities still argue that the Xerces Blue was not a "good species" distinct from the Silvery Blue *(G. lygdamus)*, most of the evidence says it was. It was confined to dune habitats on the San Francisco Peninsula, and its extinction can probably be attributed to habitat loss. It had once been common, but by the 1930s it was clearly in trouble, and R.G. Wind tried repeatedly—and unsuccessfully—to establish new colonies at seemingly suitable sites. It was last seen alive near Lobos Creek in the Presidio in May 1941. Now it gives its name to the Xerces Society, founded to promote invertebrate conservation.

There were two "forms." One (typical *G. xerces*) had large white spots on the ventral hindwing, contrasting with the ashy ground color. The other (*G. x.* "*polyphemus,*" after the one-eyed giant in the *Odyssey*) had black "pupils" in the spots, giving a more normal appearance for a blue. Specimens of both remain in numerous old collections.

DISTRIBUTION: Northern half of the San Francisco Peninsula only.
HOST PLANTS: Recorded on yellow bush lupine *(Lupinus arboreus), Astragalus nuttallii,* deer weed *(Lotus scoparius,* as *L. glaber),* and possibly other perennials of the pea family (Fabaceae).
SEASONS: One brood, March to April with stragglers to early July.
SIMILAR SPECIES: Males were slightly more lavender tinged than the other large blues. Otherwise this species was similar to the Silvery Blue, slightly larger, the large round white spots below being the most distinctive field mark. After more than 60 years you are not likely to need to sight identify this species.

ARROWHEAD BLUE   *Glaucopsyche piasus*
Pl. 16

Generally considered a rarity in our area, the Arrowhead Blue occurs in mesic forest habitats, commonly along road- or stream-sides, invariably flying with the commoner Silvery and Bois-duval's Blues *(G. lygdamus* and *Plebejus icarioides).* It also occurs in coastal scrub. Its range in the Bay Area seems to have diminished, and it is currently unrecorded in several counties where it had a historic presence. It should be kept in mind during blue season! It is an indifferent puddler, often observed perching on or near the host plant. In some inland localities it seems to show a preference for lupines *(Lupinus)* with silvery-pubescent foliage. Adults nectar on a variety of flowers, including yerba santas *(Eriodictyon)*.

DISTRIBUTION: Bay Area counties, but spotty and local. Unknown in the Sacramento Valley.

HOST PLANTS: Perennial lupines *(Lupinus)*.

SEASONS: One brood, March to June.

SIMILAR SPECIES: See Silvery Blue.

MARINE BLUE   *Leptotes marina*
Pl. 13

This is a subtropical species, frequently abundant in the desert Southwest, where it breeds on mesquites *(Prosopis)*, acacias *(Acacia)*, and other shrubby members of the pea family (Fabaceae). When the South African ornamental shrub leadwort *(Plumbago)* was introduced into the southern California landscape, this butterfly began breeding on it, despite its being quite unrelated to the native hosts. Presumably this reflects a similarity in secondary chemistry. The Marine Blue is now abundant in residential neighborhoods and office parks in central and southern California. Its appearance in our area is irregular and unpredictable, and it may invade from both the south and the east (in 2003, for example, a wave of Marine Blues moved across the Sierra along the Interstate 80 corridor, appearing consecutively in Sacramento, Davis, Vacaville, and Suisun City). Most years it is unrecorded. When present, it may occur as sporadic individuals or as dense local breeding colonies. At any rate, it seems unable to overwinter here, with only a handful of records before midsummer.

The flight is rapid and more hairstreaklike than that of most blues. It specifically resembles the Common Hairstreak *(Strymon melinus)* in its habit of approaching in a series of narrower circles. Adults are most likely to be seen on or around host plants but visit a great variety of flowers, showing a special predilection for the perennial mustard *Hirschfeldia incana*. There is almost no variation in color or pattern in this species. When at rest the hindwings are rubbed together, as in many other blues.

DISTRIBUTION: Can turn up anywhere; usually in highly disturbed habitats, ranging from urban gardens to vacant lots to alfalfa fields; has been found on the grounds of the State Capitol in Sacramento!

HOST PLANTS: Leadworts (*Plumbago*, Plumbaginaceae) and the following pea family plants (Fabaceae): alfalfa *(Medicago sativa)*, wild licorices *(Glycyrrhiza)*, and white sweet clover *(Melitotus alba)*; probably others.

SEASONS: April to November; most records after July. Multiple broods.

SIMILAR SPECIES: None in our area. The "wavy" underside pattern is unique.

REAKIRT'S BLUE   *Hemiargus isola*
Not illustrated

This blue is an infrequent to rare stray that could turn up anywhere, though most likely southward. It is common in the southern California deserts, where it breeds on woody legumes. I have seen it three times in northern California in 33 field seasons.

DISTRIBUTION: Recorded individually in sporadic locations; not known to breed anywhere in our area.

SEASONS: Multiple brooded; probably most likely to stray northward in late summer.

SIMILAR SPECIES: The Ceraunus Blue *(H. ceraunus)*, which is an even less likely stray, has washed-out pattern on underside. The "cut-off" apex of the forewing is very distinctive, unlike any of our resident blues.

CERAUNUS BLUE   *Hemiargus ceraunus*
Not illustrated

A very rare stray recorded from Alameda and Contra Costa Counties, this is a subtropical, mostly desert butterfly associated with mesquites *(Prosopis)* and acacias *(Acacia)*. It could turn up anywhere at any time as a singleton, but do not hold your breath looking for it. It is distinguishable from Reakirt's Blue by the washed-out pattern beneath, much less contrasty, except for the one large eyespot near the anal angle of the hindwing.

WESTERN PYGMY BLUE   *Brephidium exile*
Pl. 13

Our smallest butterfly, this species is instantly recognizable but often overlooked. The Western Pygmy Blue prefers habitats rarely frequented by butterfly enthusiasts: salt and alkali marshes, industrial neighborhoods, railroad rights-of-way, and the like. Its host plants are mostly succulent or semisucculent members of the goosefoot family (Chenopodiaceae).

The Western Pygmy Blue is a textbook example of what population biologists call "source-sink dynamics." Overwintering occurs only very locally in places with evergreen perennial host plants that are not regularly submerged in water. Despite statements to the contrary, this species does not appear to diapause and will continue to breed all winter in suitable sites, not only on the central and south coasts but in the Central Valley as well! With the advent of warm weather, generation follows generation every 3 weeks, and as numbers of individuals increase they spread out over the landscape and colonize new localities and different host plants—especially Russian thistles or "tumbleweed" (*Salsola*), a late-summer annual. Away from the perennial ("source") populations, adults may be encountered, rarely, as early as April, more commonly in June, but routinely by August or September. Numbers peak in October, when these tiny butterflies can be found dancing around Russian thistles, like swarms of gnats. The plants die by early December, and their associated populations die out as well, for lack of hosts—hence the term "sink."

Adults often perch on the host plant, and in cool weather bask with the wings spread. There is virtually no variation in color and pattern, but the largest females may be twice the size of the smallest males. The larvae, which frequently feed on flowers and fruits, are usually attended by ants.

In the Suisun Marsh and near the coast, peak Western Pygmy Blue season usually coincides with the flowering time of coyote brush *(Baccharis pilularis)*, and the numbers of these insects visiting this common shrub can defy the imagination.

DISTRIBUTION: All counties in alkaline and saline situations; in the second half of the season spreading to vacant lots, waste ground, transportation corridors, and so forth, where Russian thistle grows. Rarely seen in forest or in upland areas.

HOST PLANTS: Native perennial, succulent or semisucculent goose-foot family plants (Chenopodiaceae), especially *Suaeda;* rather infrequently on orachs *(Atriplex patula* and *A. hastata)* and pickleweeds *(Salicornia);* sea purslanes (*Sesuvium,* Aizoaceae). In disturbed habitats on Russian thistles or "tumbleweed" *(Salsola)*. Two or three species of this genus are in our area, and they do not appear to be used equally by this butterfly, but the taxonomy is confused. In older books they tend to be lumped under the name "*S. kali,*" no longer in use for our species. In the 1970s this species bred extensively in our area on Australian saltbush *(Atriplex semibaccata)*, naturalized for erosion control on highway embankments. This plant has become rare. The Western Pygmy Blue does not appear to breed on the common weedy annual species of *Atriplex,* such as redscale *(A. rosea)*, nor is it recorded accurately on any species of the genus *Chenopodium* in our area. It is found on big saltbush *(A. lentiformis)* around the Bay.

SEASONS: All year in permanent ("source") localities. In transient ("sink") areas, usually mid- to late summer to early December, breeding continuously until hard frost or all the Russian thistles are dead.

SIMILAR SPECIES: None.

SONORAN BLUE   *Philotes sonorensis*
Pl. 15

Thought by many to be our most beautiful butterfly, the Sonoran Blue—which does not occur in the Mexican state of Sonora—is often the first nonhibernating butterfly to emerge in this area (in January!). Its early flight season reflects the phenology of its succulent host plant, dudleya *(Dudleya)*, which "fattens up" in the rainy season, blooms in spring, and dries up and goes completely dormant by early summer. Eggs are laid singly and are easy to find. The larva bores into a leaf and lives inside it. As it eats it creates a translucent window; its droppings can often be seen through it. The larvae may be attended by the ant *Tapinoma sessile*. When full-grown the larva tucks itself in among the leaf bases and pupates. The plant shrivels, and there the pupa sits in diapause for 9 months.

Adults fly along the bases of cliffs and rocky outcrops and also move vertically up and down them. They visit very early flowers such as milkmaids *(Dentaria)* and fiddlenecks *(Amsinckia)* but are often out and about before there are any flowers to visit. The brilliant sky blue color with orange patches on the forewing (both sexes, including the underside) and hindwing (female upperside) is a unique pattern in our fauna.

Dudleya is widespread, but the butterfly is intensely colonial and recorded only in the south and southeast part of the Bay area; it has never been taken north of the Bay, though it ranges north in the Sierra Nevada to near Downieville. Some colonies are on serpentine, but it is by no means a serpentine associate or endemic.

DISTRIBUTION: Santa Clara, Alameda, and Contra Costa Counties.

HOST PLANTS: Dudleyas *(Dudleya),* our native hen-and-chickens (in the stonecrop family [Crassulaceae]); in our area on *D. cymosa* var. *setchellii.* The plant is much more widespread than the butterfly.

SEASONS: One brood, January to April, varying by locality and weather of the year.

SIMILAR SPECIES: None.

SAN BERNARDINO BLUE or SQUARE-SPOTTED BLUE   *Euphilotes bernardino*

Pl. 15

This was formerly considered a subspecies of *E. battoides* (also called Square-spotted Blue) and listed for the Bay Area under that name. It is now considered a full species with subspecies of its own. It is included here on the basis of a questionable record from Santa Clara County. Its host plant occurs in chaparral and coastal scrub from San Benito County southward, and the butterfly very rarely wanders far from it. The flowers of this plant serve as both the food of the larva and the principal nectar source of the adult. The species ranges widely in the southern half of California.

DISTRIBUTION: Dubiously present. Recorded from Santa Clara County.

HOST PLANTS: California buckwheat *(Eriogonum fasciculatum* and *E. f.* var. *foliolosum).* These plants have been used for erosion control and naturalized along highways in the Bay area and northward at least to Glenn and Tehama Counties, but there is no evidence that the butterfly has followed them (yet?).

SEASONS: One brood. April to July in the South Coast Ranges, depending on locality.

SIMILAR SPECIES: All *Euphilotes* species are maddeningly difficult to identify without recourse to the male genitalia. Compared to our subspecies of the Dotted Blue *(E. enoptes)*, this species has a better defined dark marginal line below, a broader black border in the male above, and a ventral hindwing submarginal orange band that is brighter and more continuous; on average all the black dots below are larger, blacker, and squarer. As a general rule, any small blue that is not the Acmon Blue *(Plebejus acmon)* and is collected in association with California buckwheat in the southern end of our area is likely to be this species and would be significant as a confirmation of its northern limits.

DOTTED BLUE *Euphilotes enoptes*
Pl. 15

The Dotted Blue has several named subspecies, most of which are ecotypes associated with specific host plants. Like all *Euphilotes*, these butterflies stick very close to their host plants, forming discrete colonies. The adults nectar almost entirely at the flowers of the host, mate at the plants, and roost on the flower heads overnight. Their flight is weak and dancing. There is substantial variation in size and markings among individuals. Males puddle.

Two named subspecies, Tilden's Blue *(E. e. tildeni)* and the Bay Region Blue *(E. e. bayensis)* supposedly occur in our area. They are exceedingly similar and can be told apart only statistically. Moreover, they feed on the same plants—members of the very common white buckwheat *(Eriogonum nudum)* group. Historically, upland populations from the Mount Hamilton Range (Santa Clara County) were regarded as Tilden's Blue (type-locality Del Puerto Canyon, Stanislaus County), while coastal fog belt populations from all around the Bay were called Bay Region Blue. Southward, some Tilden's Blues feed on annual species of wild buckwheat *(Eriogonum)*. There is a gap in known Dotted Blue populations in the East Bay south of Point Richmond; that area might yet be found to harbor such Tilden's Blue populations. Claims of the subspecies Smith's Blue or "Santa Cruz Blue" *(E. e. smithi)* in our area are not justified. Its type-locality is Burns Creek, Monterey County, and its host is *Eriogonum parvifolium*. It is more distinct than the other subspecies in our region, as both adult and larva. There is an enigmatic population at the Marina Dunes, Monterey County, that feeds on *E. latifolium* and seems closer to the Tilden's Blue and Bay Region Blue than to Smith's Blue, all of which is beyond this book's limits!

DISTRIBUTION: Most counties around the Bay. Absent from the Sacramento Valley, but the nominate subspecies, *E. e. enoptes* (host white buckwheat *[Eriogonum nudum]*), might sidle in along the eastern foothills!

HOST PLANTS: White buckwheat *(Eriogonum nudum)* and *E. n.* var. *auriculatum,* in our area.

SEASONS: Early May to mid-July, one brood flying about 3 weeks, seasonality varying by locality. Diapauses as pupa for as long as 9 months.

SIMILAR SPECIES: The Acmon Blue is larger, lighter violet blue with more extensive orange on upper surface. The San Bernardino Blue *(E. bernardino)* may barely enter our area.

ECHO BLUE or SPRING AZURE   *Celastrina ladon echo*
Pl. 14

You will find this insect under a bewildering array of generic names and specific epithets, depending on where you look in the literature. The Spring Azure complex includes a number of apparent cryptic species in North America (and more in Eurasia), and great uncertainty remains concerning their biological status, relationships, and geographic ranges. Amazingly, none of that impinges directly on our area, where no one has raised any suspicion that more than one entity occurs. That entity is known as *echo,* and whatever species it ends up in (or even if it becomes its own species), at least we can call it the "Echo Blue" and know what we are talking about.

The Echo Blue is common to abundant in much of the Bay Area but almost absent in the Sacramento Valley. It flies very early in spring, then has a second brood in late spring when California buckeye *(Aesculus californica)* blooms. It is one of our most enthusiastic puddlers, and "mud puddle clubs" of dozens of males are often seen (and photographed). Adults visit a great variety of mostly small flowers but spend much of their time high in the treetops. They will visit redbud *(Cercis)* flowers very early in the season. Second-brood individuals are larger and a little brighter than firsts, as a rule, but there is little variation.

An amazing array of host plants has been recorded for the Spring Azure complex. The larvae feed on buds, flowers, and young fruit. In much of our area the usual host of first-brood larvae is wild-lilacs *(Ceanothus)* and of second-brood ones California buckeye. Adults occur in foothill woodland, canyon riparian, chaparral, coastal scrub, and occasionally urban situations.

DISTRIBUTION: All Bay Area counties; only sporadic records in Sacramento Valley, with breeding colonies apparently few and far between (recently extinct in metropolitan Sacramento, for example).

HOST PLANTS: Wild-lilacs *(Ceanothus)* and dogwoods *(Cornus)* most frequently early; California buckeye *(Aesculus californica)* late. Others recorded in our area are huckleberries *(Vaccinium)*, sumacs *(Rhus)*, chamise *(Adenostoma fasciculatum)* (!), blackberries *(Rubus)*, English ivies *(Hedera)*, and, questionably, birdfoot trefoil *(Lotus corniculatus)*.

SEASONS: Mostly two broods, February to April and May to early July, but scattered records as early as mid-January and as late as mid-October. Records after July are mostly near the coast and could represent a partial third or even fourth brood. There are also very infrequent singletons recorded in the Inner Coast Range in September and October; all I have seen were males.

SIMILAR SPECIES: The Echo Blue is distinctive in lacking orange completely in both sexes. The wings look thin and papery; the underside is chalky white with a very delicate black spot-pattern. The broad black borders of the females above are also distinctive.

ACMON BLUE　　　　　　*Plebejus (Icaricia) acmon*
Pl. 15

This species is common and generally distributed in open country and even along roadsides through woods. Like the Bramble Hairstreak (*Callophrys "dumetorum* complex"), the Acmon Blue feeds interchangeably on various papilionaceous legumes (mostly in the genus *Lotus*) and on wild buckwheats (*Eriogonum,* Polygonaceae). Such a repeating pattern strongly suggests a chemical commonality between them, which remains to be investigated. This set of hosts, coupled with the multivoltinism displayed by this butterfly everywhere, leads to it showing two different lifestyles. Where perennial buckwheats (usually the white buckwheat *[Eriogonum nudum]* complex) are available, Acmon can produce multiple generations in the same place and does so. The larvae feed on both leaves and flowers, depending on the time of year. When it is using legumes, these are often strongly seasonal and are available for only one or two generations, forcing it to emigrate in search of new hosts. It eats many small, herbaceous, annual *Lotus* but is especially frequent on Spanish lotus *(L. purshianus)*, which is a summer to late-summer annual. I recently found it breeding on the very different meadow lotus *(L. oblongifolius)* in cool, semishaded swales. On the east side of the Sacramento Valley it uses the shrubby deer weed *(L. scoparius)*. In some localities it breeds on weedy populations of prostrate knotweeds or yard grass (*Polygonum aviculare* complex), but it will not oviposit or feed on other populations of the same complex. When it is using deer weed it usually co-occurs with the Eastern Tailed Blue *(Everes comyntas)*. When using knotweed in urban vacant lots, it formerly occurred with the now-scarce Purplish Copper *(Lycaena helloides)*. On wild buckwheat it co-occurs with the various races of the Dotted Blue *(Euphilotes enoptes)* and with the Mormon Metalmark *(Apodemia mormo mormo)*!— and it shares many of its hosts with the Common Hairstreak *(Strymon melinus)*, too.

This species is tended by ants.

The phenology of the Acmon Blue has been a source of confusion. In upland sites it generally flies nearly all year. In the Sacramento Valley it often breeds in late summer on Spanish lotus in dry creek beds. When these flood over winter, the diapausing larvae are carried away, and the species is not seen until late spring or early summer, when it manages to colonize again from elsewhere. In any case, it has a succession of broods into late fall or early winter.

Early-spring males are more brilliantly colored than summer ones; they are more blue and less violet, and the lunules on the hindwing above are more orange than lilac. Spring females may be mostly brilliant blue above, with very strong orange lunules. The underside pattern is more distinct and on a more ashen ground. The spring form is informally called *P. a.* "*cottlei.*" Summer females are all black on top, with reduced orange lunules. Summer individuals, especially females, may be very tiny—as small as Western Pygmy Blues *(Brephidium exile)*. In fall the orange band on the male's upperside may be broken into separate spots, or the orange color may even disappear.

The Acmon Blue visits many small, low flowers such as heliotropes *(Heliotropium)* and lippias *(Lippia)*. It is usually most abundant in fall and swarms over coyote brush *(Baccharis pilu-*

laris) flowers, especially male plants. It is nonterritorial and is not much of a puddler, unlike the Lupine Blue *(P. lupini).*

DISTRIBUTION: All counties.

HOST PLANTS: Knotweed family (Polygonaceae): various perennial wild buckwheats *(Eriogonum),* especially the white buckwheat *(E. nudum)* group, and some forms of the prostrate knotweeds *(Polygonum aviculare* complex); pea family (Fabaceae): many species of *Lotus,* including (rather infrequently) birdfoot trefoil *(L. corniculatus);* white sweet clover *(Melilotus alba).*

SEASONS: Multivoltine, all year in some localities, March to November more usual; in areas where extirpated in winter, shows up by June. Overwinters as larva in litter.

SIMILAR SPECIES: See Lupine Blue. The Melissa Blue and Lotis Blue *(Lycaeides melissa* and *L. idas)* are similar but not in our area today; males have no orange above, females have at least traces on forewings above; larger.

LUPINE BLUE   *Plebejus (Icaricia) lupini*

Not illustrated

Of all the species listed in this book, this is the one whose status in our area is most uncertain. It has been recorded from Santa Cruz and Santa Clara Counties, but whether the specimens are correctly identified and whether there are breeding populations (and in what ecological context) remain to be determined. The ill-named Lupine Blue never feeds on lupines *(Lupinus).* Its hosts are always perennial wild buckwheats *(Eriogonum).* As generally understood, its populations fall into two broad groups: the nominate subspecies *P. l. lupini* from montane northwestern California and the Sierra-Cascade axis, and the southern California entity *P. l. monticola* found in the South Coast Ranges, the Transverse Ranges, and on into Kern County and southward into Baja California. *Monticola*-type animals begin in San Benito County, along with their usual host, California buckwheat *(Eriogonum fasciculatum).* Our records thus fall squarely in the gap between the two, though geographically more associated with *P. l. monticola.* Variation within this complex is actually much more intricate than this summary suggests. Furthermore, some authorities recognize a third entity, named *P. l. lutzi* or *P. l. spangelatus,* as occurring in the northern part of the state. The situation is so fluid that we had best wait for Paul Opler to complete his revision of the group, now in progress, and hope that it clarifies these questions (including whether any *P. "lupini"* actually occur in the area covered by this book).

The Lupine Blue in the broad sense is single brooded. It averages larger than the Acmon Blue *(P. acmon)* and is similarly marked. The hindwing lunules in the Lupine are more orange, not pinkish as in (summer) the Acmon, and there is a fine black line along the basal edge of the row of lunules (rarely seen in the Acmon). The upperside black borders are wider and more contrasting than in male Acmons, and the blue color is less violet tinged. The female is apt to have considerable grayish blue above. On the underside, the spot-pattern tends to be very crisp, the spots small, with white haloes against a more ashy ground than in the Acmon. Note that the Lupine, which generally flies in late spring, resembles the early-spring form *(P. a. "cottlei")* of the Acmon in most traits. Note also that the phenotypic differences between these two sibling species parallel those between the Eastern and Western Tailed Blues *(Everes comyntas* and *E. amyntula)*. Have fun!

GREENISH BLUE   *Plebejus saepiolus*
Pl. 14

The Greenish Blue is a characteristic species of cool, moist, grassy meadows and of streamsides and hillside seeps where the host plants, clovers *(Trifolium)*, grow. It is abundant in the higher mountains of northern and parts of southern California and on the north coast, but in our area is at the edge of its tolerances in the North Bay. Adults fly just above the ground and nectar at low flowers, especially the host plant. In a few localities outside our area, local ecotypes have been found breeding on dryland clovers and on *Lotus* species. Sexual dimorphism is extreme in our populations, but on the far north coast some females have a fair bit of blue above, and in the White Mountains (Inyo County) they all do. Males puddle.

DISTRIBUTION: North Bay counties and northward; perhaps now extinct in Marin County.

HOST PLANTS: Clovers *(Trifolium,* generally native).

SEASONS: One brood, April to June, males emerging a week or so before females as a rule. Part-grown larvae overwinter and appear to be tolerant of flooding.

SIMILAR SPECIES: In our area, no other blues are easily confused with this one except an occasional, unusual male Boisduval's Blue *(P. icarioides)*. The greenish blue color of the upperside, chalky blue white underside with sharp black markings, strong submarginal spot row, and pointed forewing apex should allow easy recognition of male Greenish Blue. The Boisduval's Blues that resemble them are invariably more violet blue above and usually differ in most or all of the other characters listed, as well.

BOISDUVAL'S BLUE or ICARIOIDES BLUE
Plebejus (Icaricia) icarioides

Pl. 16

The suffix -*oides* means "similar to." This widespread, large blue was thought to resemble the common European species named *P. icarus*. It actually doesn't. It does, however, display extraordinary variability both within and among populations, and Bay Area ones have been divided among five subspecies—one of which, the Mission Blue *(P. i. missionensis)*, has become something of a poster child for federally endangered species. The species is very widely distributed in California in a great variety of habitats, from coastal strand to tree line, feeding on a great variety of lupines *(Lupinus)* as hosts. The butterflies can be found primarily in and around stands of hosts. Males puddle vigorously. The densest populations of Boisduval's Blue often occur around road or power-line cuts, where lupines have come in during the course of succession. Females may have much or little blue above. Their wide dark borders are sometimes quite brownish, and there may be a little orange around the anal angle above.

Subspecies Pheres Blue *(P. i. pheres)* was endemic to San Francisco and, not surprisingly, is extinct. Somewhat similar-looking populations occur at Point Reyes, however. These have recently been named Point Reyes Blue *(P. i. parapheres)*. The Pheres Blue, which resembled the typical form of the also-extinct Xerces Blue *(Glaucopsyche xerces)* in having the black dots obsolete below, leaving large "blind" white spots in their stead, occurred on sand dunes in the western part of the city. According to J. W. Tilden, it was very common at 14th Avenue and Taraval Street in April until 1926, when it was wiped out by construction there. In 1930 it was rediscovered on dunes west of 20th Avenue, but was wiped out there by 1940. Tilden wrote in 1956 that it was definitively gone by 1950.

The usual East Bay subspecies, Pardalis Blue *(P. i. pardalis)*, is well marked beneath. Its type-locality was the former San Antonio, now paved over in the sprawl between Alameda and Oakland. A neotype (new type-specimen to replace the apparently lost original one) had to be designated from Oakland (from Lake Chabot, collected 1946!). The Pardalis Blue tends to have the outermost (submarginal) row of dark spots on the ventral hindwing somewhat arrowhead shaped, pointed toward the base.

The federally endangered Mission Blue is also strongly marked beneath, though the submarginal spots are less prominent than in the Pardalis Blue, and usually much smaller. The largest remaining population is on the east, northeast, and south slopes of San Bruno Mountain, where a Habitat Conservation Plan is in place for it and other rare and endangered species (including the San Bruno Elfin *[Incisalia mossii bayensis]*), and San Mateo County administers an 810 ha (2,000 acre) preserve for it. Other known populations are in McLaren Park and Twin Peaks, San Francisco, at and near Fort Baker, Marin County, and at Milagra Ridge, San Mateo County. In all locations they are fully protected by law.

The nominate subspecies, *P. i. icarioides*, is something of a wastebasket into which more inland populations, as well as those in much of Santa Cruz County, have been dumped. In some cases there is a smooth gradation from one population to another; in other cases radically different-looking individuals may occur within the same population. There is an oft-voiced suspicion that many of the traits used to distinguish subspecies in Boisduval's Blue may be directly influenced by environmental conditions during development, but this remains uninvestigated.

DISTRIBUTION: All Bay Area counties, divided among subspecies as follows: Pheres Blue *(P. i. pheres)*, west side of San Francisco, extinct; Point Reyes Blue *(P. i. parapheres)*, Point Reyes area, Marin County; Mission Blue *(P. i. missionensis)*, scattered localities in San Mateo, Marin, and San Francisco Counties; Pardalis Blue *(P. i. pardalis)*, East Bay, grading into the nominate *P. i. icarioides* inland, south, and north (Napa, Sonoma, Solano Counties). Perhaps still exists in the Sacramento Valley, but not recorded recently, even in the Sutter Buttes.

HOST PLANTS: Perennial lupines including bush lupine *(Lupinus* *albifrons), L. formosus, L. variicolor,* and sometimes others. Of these, *L. formosus* still exists in the Sacramento Valley.

SEASONS: One brood, the flight season varying with location, March to July (late March to mid-June on San Bruno Mountain). Males generally emerge a little before the first females.

SIMILAR SPECIES: See Silvery and Arrowhead Blues *(Glaucopsyche lygdamus* and *G. piasus);* the three routinely co-occur in the Coast Ranges. Some individuals, mostly the Pardalis Blue, may be small enough to be confused with the Greenish Blue *(P. saepiolus).*

MELISSA BLUE   *Lycaeides melissa*
Not illustrated

The Melissa Blue supposedly occurred in Monterey and Santa Cruz Counties in the past. If it truly did, it was an occurrence disjunct from other known California populations far to the south and east. The Melissa Blue is another "complex" that probably conceals more than one genetic species. Molecular genetic studies suggest rapid evolution in progress in the group. This is a butterfly of sunny, open habitats, originally bunchgrass-shrub steppe, high desert, and the alpine zone of the Sierra Nevada. In much of western North America it has adapted to cultivated alfalfa *(Medicago sativa)* as a host and is found mostly in irrigated agricultural land, but "ag *melissa*" is notably absent from the California Central Valley. The Melissa somewhat resembles the Acmon Blue *(Plebejus acmon)* below. Above, the male is rich ultramarine blue with no orange at all, while the female is mostly or all black and has bright orange lunules on both fore and hindwings, which may fuse into a band. Lowland populations are two to four brooded.

LOTIS BLUE   *Lycaeides idas lotis*
Not illustrated

The Lotis Blue is still officially protected under the Endangered Species Act, but in fact it has not been seen for at least 20 years and can safely be presumed extinct. It enters our area only by virtue of some old, vague records for Sonoma County. The last two known occurrences were at Point Arena and near Russian Gulch State Park, Mendocino County. Unfortunately, little was known of the biology of the beast at the time it was given federal protection. There is reason to suspect that excessive prudence in managing the last known colony led to its extirpation: the culprit was vegetational succession. (That is, fear of disturbing the insect led those in authority to allow the shrub layer to shade out the probable host plant!) In any case, the Lotis Blue *(L. idas)* occurs in moist and cool habitats, usually along streamsides where it breeds on perennial legumes (usually species of *Lotus,* sometimes large-leaved lupine *[Lupinus polyphyllus]*). Populations of other subspecies often occupy successionally transitory situations and are constantly colonizing up- and downstream. The Lotis Blue was probably a victim of widespread agriculturization of land in its range in the nineteenth century, making it increasingly difficult for it to find patches of host plant along boggy streams or swales. It was quite similar to the Melissa Blue *(L. melissa),* but slightly larger, and has been characterized as an "*idas* with a *melissa* wing pattern." It would be wonderful if someone discovered a colony alive somewhere (and kept quiet about it until it could be safeguarded). Remember the Ivory-billed Woodpecker!

Metalmarks (Riodininae)

The metalmarks are an enormous group in the New World tropics and subtropics. Our two representatives are very distinctive in our fauna.

MORMON METALMARK   *Apodemia mormo mormo*
and LANGE'S METALMARK   **and** *A. mormo langei*

Pl. 19

The Mormon Metalmark occurs in widely scattered colonies in the Coast Ranges, typically in hot dry places—on cliffs or in rocky canyons, sometimes in chaparral or grassland, and often along roadsides. Lange's Metalmark is known only from the relict sand dunes on the south shore of the San Joaquin River at Antioch, Contra Costa County. Its habitat is enclosed in a fence, and the subspecies is federally listed as endangered, due to its extreme localization. Its wing pattern is somewhat reduced, especially basally. Somewhat similar specimens occur in several colonies in the inner North Coast Range, which display a wide range of individual variation.

This butterfly has a very characteristic erratic flight, often making aerial figure-eights before alighting. Once on a flower it remains active, keeping its wings open but waving them constantly. It is almost impossible to follow in flight. It emerges late (July at the earliest; the Antioch population is unusually early, and in general, the hotter the site, the later the animal emerges!) and flies for 2 to 4 weeks. The eggs hatch during the winter but the larva does not begin feeding until the host begins putting on new growth in late winter or early spring. It starts out eating foliage but eventually switches to the inflorescences. It is mostly nocturnal and very odd looking, lilac with numerous hair-bearing tubercles. Because of its unusual seasonality and behavior, this butterfly is often overlooked.

Elsewhere in its range, different supposed subspecies or "races" of *A. mormo* feed on different hosts and may occur sympatrically (in the same place) but allochronically (at different times of year), behaving as full species—which they may in fact be, or be in the process of becoming.

DISTRIBUTION: Scattered colonies in several Bay Area counties, others probably undetected. Unknown from Sacramento Valley except perhaps at the edges in the far north. Lange's Metalmark only at Antioch Dunes, within the confines of a federal wildlife refuge established to protect it. The site is now closed to the public except for guided tours. As of the 2006 season, Lange's Metalmark seems to be declining and in serious trouble despite the attention lavished on it.

HOST PLANTS: Various wild buckwheats *(Eriogonum)* in different parts of the range; sometimes wildly different species in the same place. Among recorded hosts in our area are *E. latifolium, E. parvifolium,* and Wright's buckwheat *(E. wrighti).* At Antioch only on *E. nudum* var. *auriculatum,* and on other white buckwheat–group entities elsewhere in the northern part of the state. Adults visit many flowers but are especially likely to visit wild buckwheats.

SEASONS: July (Antioch) to October; one isolated late March record. One brood, often flying in truly parched late-season landscapes.

SIMILAR SPECIES: The upper surface has a very vague resemblance to the crescents; another member of the genus *Apodemia* is actually named *A. phyciodoides.* The wings are short and broad with a silvery gray underside having a unique pattern, the antennae are very long, and the compound eyes are apple green, all of which makes misidentification hard indeed.

Brushfoots (Nymphalidae)

The nymphalids comprise some 350 genera and nearly 6,000 species, rivaling the lycaenids as the largest butterfly family. But they are much more diverse in appearance and biology than the lycaenids. Currently 10 subfamilies are recognized, most of which have been (and by some authorities still are) considered families in their own right. The subfamilies represented in our fauna are the Heliconiinae, which are now construed to include the fritillaries; the Nymphalinae, which include the crescents and checkerspots as well as the tortoiseshells, ladies, and their relatives; the Limenitinae, represented in our fauna by Lorquin's Admiral *(Limenitis lorquini)* and the California Sister *(Adelpha bredowii californica)*; the Satyrinae, the meadow browns and satyrs; and the Danainae, the milkweed butterflies (including our Monarch *[Danaus plexippus]*). Even this arrangement differs from that in Brock and Kaufman's book (2003), which does not separate the Limenitinae from the Nymphalinae. Our concepts of relationships within this family continue to evolve rapidly under the combined influence of molecular genetics and cladistic methodology. The last word has certainly not been written.

All these subfamilies are united in having the forelegs reduced and brushlike—hence the name "brushfoots." The male forelegs do not have any known function, but those of the female are sensory organs used to recognize host plants. They are "drummed" against the plant in characteristic fashion.

Because the subfamilies have gone off in different eco-evolutionary directions, it is difficult to generalize about the family as a whole. Many of the nymphalids (except the meadow browns and satyrs) are rather large, tough bodied, and long-lived butterflies. Most of ours have the sexes very similar in appearance. The Monarch, at least some of the checkerspots, and the California Sister employ chemical defenses against predators. Nymphalid eggs are usually more or less globular and ribbed. Some groups (crescents, checkerspots, and tortoiseshells) lay them in masses; most lay them singly. There are four basic larval types in our fauna. Heliconiine and nymphaline larvae bear branching spines. Limenitine larvae have large dorsal spines behind the head. Satyrine larvae are smooth, with bifid tails. Danaine larvae are smooth, with filamentous appendages on the body. Nearly all nymphalid pupae hang vertically, head down, attached to a button of silk by the cremaster. Nearly all are cryptic.

The fritillaries feed on the rather closely related violet and passionflower families (Violaceae and Passifloraceae) and only rarely on other plants. Crescents are mostly associated with the sunflower family (Asteraceae), at least in our fauna, and checkerspots with the figwort and plantain families (Scrophulariaceae [snapdragons, monkey flowers] and Plantaginaceae). The tortoiseshells and their relatives have a diverse diet, with hints that the nettle and elm families (Urticaceae and Ulmaceae) may be ancient associations. The meadow browns and satyrs feed on grasses and sedges, and the milkweed butterflies on the milkweed family (Asclepiadaceae).

Overwintering strategies also differ among groups. The true fritillaries diapause as neonate larvae; crescents, checkerspots, meadow browns and satyrs, and our limenitines as larvae; tortoiseshells and their relatives as adults; and the Monarch as an adult in specific wintering grounds. Our most migratory butterflies—the Monarch, California Tortoiseshell *(Nymphalis californica)*, Painted Lady *(Vanessa cardui)*, and Buckeye *(Junonia coenia)*—all belong to the Nymphalidae.

There are brief overviews before the species accounts for the true fritillaries and the anglewings because they are difficult to identify, and for the meadow browns and satyrs generally, because they are so different from our other nymphalids.

Fritillaries (Heliconiinae)

GULF FRITILLARY   *Agraulis vanillae*
Pl. 17

The Gulf Fritillary is our only butterfly that has no native host plant in California. It is thus entirely dependent on introduced species. Clearly it cannot be native here, but colonized after hosts became available. There seem to be only a couple of Bay Area records before 1955. The oldest, however, is before 1908! It had been established several decades earlier in coastal southern California. A tropical and subtropical weedy species, it cannot persist in areas where killing frosts are frequent. It has never established resident breeding populations in the Sacramento Valley, and even strays are rare there despite a seeming "critical mass" of host plants in some neighborhoods. This extremely showy, conspicuous insect visits many garden flowers, especially lantanas *(Lan-*

tana), butterfly bush *(Buddleia)*, and tropical milkweed *(Asclepias curassavica)*, and is often common in the East and South Bay.

DISTRIBUTION: Bay Area, mainly in older residential neighborhoods; rarely seen outside of cities; especially common in Berkeley. Rare stray in the Sacramento Valley, recorded breeding sporadically as far east as Fairfield. This species is often bred for release at social functions but does not seem to persist.

HOST PLANTS: Passionflowers *(Passiflora)*; perhaps many species will do. Most records are on the commonest species, maypop *(P. incarnata)* and *P. caerulea* and its hybrid *P. alatocaerulea.* Larvae are loosely gregarious, orange with black spines, and very conspicuous; pupa very slender, oddly shaped, extremely "wriggly" when disturbed.

SEASONS: Nearly all year; inland strays July to November, mostly in fall.

SIMILAR SPECIES: True fritillaries do not have the long, pointed forewing apex, and if silvered beneath, the spots are rounded, not long and narrow as in this species. None of our true fritillaries is urban, and all are single brooded.

VARIEGATED FRITILLARY   *Euptoieta claudia*
Not illustrated

This is a very rare stray in the Sacramento Valley, recorded at Davis, Yolo County, on October 7, 1988, and in the same year in the western Sierran foothills in Plumas and Mariposa Counties. It is a common multivoltine species in the southwestern deserts but seldom recorded even in southern California and never yet in the Bay Area.

WESTERN MEADOW FRITILLARY   *Boloria epithore*
or PACIFIC FRITILLARY
Pl. 17

If Mark Twain never actually said "The coldest winter I ever spent was summer in San Francisco," he should have—and it would explain the distribution of this butterfly, which belongs to a very cold-adapted group that reaches the shores of the Arctic Ocean. The Western Meadow Fritillary is widespread in the mountains of California. In northwest California it reaches the vicinity of Russian Gulch State Park, Mendocino County; inland, the high

North Coast Range near Mendocino Pass in Glenn County. Then it jumps to the Santa Cruz Mountains, where it occurs in splendid isolation along roadsides and streams in redwood, Douglas fir, and mixed-mesic forest, flying in late spring and visiting wild-lilacs *(Ceanothus)*, tansy ragwort *(Senecio jacobaea)*, thistles *(Cirsium, Carduus,* and *Silybum)*, and other flowers. But it apparently once occurred in the city of San Francisco—before 1910, when Francis X. Williams reported that Cottle and Mueller (two pioneering collectors) had gotten it there "a good many years ago."

DISTRIBUTION: Santa Cruz Mountains. Unrecorded in the North Bay counties.

HOST PLANTS: Western heartsease *(Viola ocellata)* and possibly other violets.

SEASONS: One brood, mid-April into July, mostly in May and June—approximately the same time it flies in the midelevation northern Sierra Nevada.

SIMILAR SPECIES: This species' wing shape is different from that of all the larger fritillaries. The violet sheen on underside of the hindwing is distinctive.

True Fritillaries *(Speyeria)*

These beautiful insects all look very much alike, and their identification is a real challenge. In many cases sympatric populations of several species look more like one another than they resemble allopatric populations of their own species. Sometimes habitat is a better indicator of species identity than color and pattern. All the species are single brooded, and all feed only on violet species *(Viola)*. Little is known of their larval biology because the larvae typically feed only at night. Eggs are laid singly, in some cases on or near the completely dried-out remnants of the year's violet foliage. The larva hatches quickly, eats its eggshell, and then becomes dormant until the following spring—sometimes for 7 months or more! The Callippe Fritillary *(S. callippe),* is usually the first species to emerge (in late spring). The Crown Fritillary *(S. coronis),* and Zerene Fritillary *(S. zerene)* are often seen flying into October. Some populations of these two species estivate as adults and do not begin laying eggs until the short days of fall. Research by S.R. Sims showed that adult seasonality is correlated with larval desiccation tolerance; larva of species that emerge or oviposit later are less able to deal with the hot, dry summer conditions than can the Callippe Fritillary. Despite their extreme conservatism in color, pattern, and overall life history, genetic studies have failed to turn up any evidence that these butterflies ever hybridize.

A warning: diagnostic characters given in this book are for *our area only.* By and large they will not work, and may be horribly misleading, for material collected anywhere outside the geographic limits of the book. And even here, there may be a small residue of specimens you simply cannot name.

CROWN FRITILLARY
or **CORONIS FRITILLARY**

Speyeria coronis

Pl. 18

If you catch a large, gloriously silvered female fritillary in the middle of the Suisun Marsh in late September or October, it's a Crown Fritillary.

Not that the species breeds there!

This is one of our two largest true fritillaries and certainly our most dispersive. It breeds primarily in open grassland. Both sexes emerge relatively early (May to June) and can often be seen nectaring at California buckeye *(Aesculus californica)* blossoms on isolated trees at the woodland-grassland transition, or sometimes in canyons. They then disappear. After Labor Day the females reappear, generally looking none the worse for wear, and begin laying eggs. This pattern, reflecting adult reproductive diapause/estivation, has given rise to a myth of two-brooded-ness. The fall females have even been seen in metropolitan Sacramento, though at present no breeding is known on the Sacramento Valley floor. They show up regularly in the Delta and Suisun, nectaring on asters *(Aster)* and goldenrods *(Solidago)* far from their natal colonies. The Sheep Moth, a large diurnal moth (*Hemileuca* or *Pseudohazis eglanterina*, Saturniidae), flies at the same time and looks similar in the air but does not visit flowers.

There is a breeding colony of this species in the Vaca Hills near Vacaville (Solano County), where no species of violet *(Viola)* has been collected in many years. This merely shows that we are not done inventorying our flora! (The old records are of V. *quercetorum* and pine violet *[V. lobata]*, from single sites on the highest peak in the range!)

DISTRIBUTION: Bay Area counties, straying widely.

HOST PLANTS: Violets *(Viola)*.

SEASONS: May to October, seldom seen in midsummer; population mostly or entirely female in fall.

SIMILAR SPECIES: The Crown Fritillary is about 10 percent larger on average than the Callippe Fritillary *(S. callippe)* where they occur together. These two are often very similar. Both have strong show-through of the ventral silver spots on the dorsal hindwing surface—even stronger in the Callippe than in the Crown Fritillary. Both have the black markings generally rather finely formed. The upperside of the Callippe, however, has a more "checkered" look than the Crown. The early flight season of the Callippe (as early as late April) is a valuable indicator. Even in moist areas the Callippe is normally finished by August, while the Crown hasn't yet got its second wind. A few female Crowns are much larger than any Callippe. The Zerene Fritillary *(S. zerene)* is as big as the Crown Fritillary but usually decidedly redder. Its black markings are coarser and on the underside of the forewing may be large and squarish to arrowhead shaped, as compared to more linear in the Crown.

ZERENE FRITILLARY
Speyeria zerene

Pl. 18

Another large "frit," the Zerene in our area is at the southern edge of its range. Northward it is a very common species of upland forest and roadsides. With us it is primarily a species of coastal prairie and scrub. Two subspecies occurring in our area are listed as federally endangered, mainly due to habitat loss. They are Myrtle's Fritillary *(S. z. myrtleae)*, and Behrens' Fritillary *(S. z. behrensii)*. Behrens' Fritillary barely enters our area in northwestern Sonoma County, extending from Stewart's Point to about Mendocino. Myrtle's Fritillary was originally described from San Mateo County. At the time it was declared endangered its range was understood to include populations in Marin County and much of coastal Sonoma County as well. Subsequent study by John and Thomas Emmel and Sterling Mattoon has led to a potentially bizarre legal conundrum. According to them, the original concept of *S. z. myrtleae* actually embraced three phenotypically different entities—one from each county. They thus named two new subspecies, *S. z. puntareyes* from (yup!) Point Reyes and

S. z. sonomensis from (yup! again) Sonoma County. Because the Zerene Fritillary is apparently extinct in San Mateo County, subspecies Myrtle's Fritillary in the restricted sense is also extinct. Are the two new subspecies automatically protected under the Endangered Species Act because they were included as part of Myrtle's Fritillary? Or did federal protection end when they were severed from Myrtle's? Who knows? Anyway, it is suggested you not bother them. The new subspecies *S. z. puntareyes* occurs in supposedly well-known areas, including not only Point Reyes but Bodega Bay and Jenner. The type-locality of *S. z. sonomensis*, however, is restricted to a small area near Sears Point. Phenotypically they seem to form a gradient, with *S. z. sonomensis* intermediate between *S. z. puntareyes* and the (lost) Myrtle's Fritillary. Some males from near the coast are uncharacteristically buffy and may even have a vaguely greenish sheen, quite unlike the ruddy inland and upland populations. We clearly have a lot to learn.

A strong flower visitor, often seen at white buckwheat *(Eriogonum nudum)* and thistles *(Cirsium, Carduus,* and *Silybum).* Both sexes fly early, but females are long-lived and occasionally seen well into fall. In the Pacific Northwest some subspecies of the Zerene Frittilary have summer adult diapause and may undergo a seasonal migration away from and back to their breeding grounds. These phenomena have not been specifically documented in our area.

DISTRIBUTION: San Mateo County (Myrtle's Fritillary; extinct); Marin County (*S. z. puntareyes*); Sonoma County (coastal, *S. z. puntareyes;* far northwest, Behrens' Fritillary; far south, *S. z. sonomensis*); Napa County (undefined, upland, for the moment falling "near *S. z. conchyliatus*").

HOST PLANTS: Violet *(Viola)* species, undetermined; both western dog violet *(V. adunca)* and johnny-jump-up *(V. pedunculata)* present in habitat.

SEASONS: Myrtle's Fritillary, late April to August; *S. z. sonomensis,* mid-May to early July; *S. z. puntareyes,* early July to September. Upland populations to the north fly June to October. Scattered coastal records of females as late as December (!).

SIMILAR SPECIES: The Zerene Fritillary is larger than the Callippe Fritillary *(S. callippe),* with coarser black markings above and below and little *(S. z. puntareyes)* to moderate show-through of the silver spots on upper hindwing surface; it is redder than the Crown Fritillary *(S. coronis),* the silver spots smaller and less brilliant. The large, squarish median black spots on the forewing (especially ventrally) are the best character for the Zerene Fritillary, at least where there are no Hydaspe Fritillaries *(S. hydaspe)!*

CALLIPPE FRITILLARY

Speyeria callippe

Pl. 18

The Callippe Fritillary is our smallest and earliest-emerging true fritillary. It previously occurred in all the counties of the Bay Area but has been extirpated in many areas by loss of habitat. The nominate subspecies, *S. c. callippe,* is listed as federally endangered. It previously occurred from La Honda in the south to Twin Peaks (San Francisco) and from Richmond to Castro Valley. Of seven original populations, only the one on San Bruno Mountain and a very small one in Alameda County are left. The San Bruno site also houses the endangered San Bruno Elfin *(Incisalia mossii bayensis),* Bay Checkerspot *(Euphydryas editha bayensis),* and Mission Blue *(Plebejus icarioides missionensis)*—all on one mountain!

The other subspecies in our area are the Liliana Fritillary *(S. c. liliana),* from the North Bay and North Coast Range, and—by some interpretations—Comstock's Fritillary *(S. c. comstocki),* the southern California subspecies, extending into the South Bay. To add confusion, one text (Scott) treats nominate *S. c. callippe* as synonymous with *S. c. comstocki* ("San Francisco to Baja California")! At any rate, coastal populations are typically heavily silvered, with a strongly two-toned effect above, especially in females; inland populations are ruddier and less two-toned, and those from relatively arid sites may have reduced silvering, especially in males; and some serpentine-grassland and serpentine-chaparral populations have a relatively pale, buffy ground color. About 20 named subspecies are described from California, and the situation is likely to get worse. No one has done the obvious experiments to determine how much of this complicated pattern of variation is genetic and how much is under environmental control.

The butterfly is an eager visitor to thistles *(Cirsium, Carduus,* and *Silybum),* mints *(Mentha),* and especially coyote mint *(Monardella),* as well as a variety of other flowers, and males are avid puddlers. Different populations occur in a variety of habitats, from woodland and canyon riparian to open grassland.

DISTRIBUTION: The nominate *S. c. callippe* is today restricted to San Mateo and Alameda Counties, as noted. Other subspecies are spottily distributed in other Bay Area counties (the type-locality of the Liliana is Mount St. Helena, Napa County). There has been a population at Hunter Hill, Solano County, and individuals are taken sporadically in Inner Coast Range localities where no breeding colonies are known. The species (subspecies undefined) has been taken once in Davis (Yolo County) on the floor of the Sacramento Valley, May 9, 1972, raising the possibility that breeding colonies may yet exist there. Sightings are recorded along the eastern edge of the valley, which would presumably involve the Sierran foothill and midelevation subspecies *S. c. juba*.

HOST PLANTS: Reportedly only on johnny-jump-up *(Viola pedunculata).*

SEASONS: Late April through July, invariably earlier than other sympatric true fritillaries but usually overlapping them in late spring.

SIMILAR SPECIES: The Crown Fritillary *(S. coronis)* may be extremely similar to the Callippe Fritillary in the area south of the Bay, but the Callippe is almost always smaller (males about 5.1 cm [2 in.], females 6.3 cm [2.5 in.] in wingspan). Also the Crown is significantly smaller than the Zerene Fritillary *(S. zerene)* and without the terra-cotta cast on the ventral hindwing. Note the collection dates!

HYDASPE FRITILLARY   *Speyeria hydaspe*
Not illustrated

Recorded once, questionably, from Napa County, this is a common species at midelevation in the North Coast Range, co-occurring with the Zerene and Callippe Fritillaries *(S. zerene* and *S. callippe)* in many areas. It is somewhat smaller than the Zerene, very bright orange above with a heavy black pattern, and the hindwing below with the pattern unsilvered and the ground color markedly purplish. Like all the others, it has one brood in summer and feeds on violets *(Viola).*

UNSILVERED FRITILLARY   *Speyeria adiaste*
Pl. 17

Restricted to the southern Bay Area counties and southward, this species is distinctive in its washed-out, unsilvered "ghost" pattern beneath. Although it is not federally listed, its range has contracted and it appears to be in global decline; no one knows why. It is found primarily in openings in redwood forest near the coast. Adults visit a variety of flowers, including thistles *(Cirsium, Carduus,* and *Silybum)*, tansy ragwort *(Senecio jacobaea),* and late-blooming brodiaeas *(Brodiaea).* Populations farther south have paler females than ours. The southern California subspecies *S. a. atossa* went extinct in 1959. Ours seems determined to follow it to that big flowery meadow in the sky.

The Unsilvered Fritillary was formerly lumped in with the Egleis Fritillary *(S. egleis),* and you will find it so listed in older sources. The Egleis occurs north of our area (the southernmost Coast Range population seems to be on Anthony Peak near Mendocino Pass) and in the Sierra Nevada.

DISTRIBUTION: Currently known only from San Mateo and Santa Cruz Counties in our area. Formerly recorded in Santa Clara County. The species ranges south in Monterey and San Luis Obispo Counties.

HOST PLANTS: Presumably violets *(Viola);* several species occur in its habitats.

SEASONS: One brood, mid-May to mid-August.

SIMILAR SPECIES: The hindwing underside is unique.

Crescents, Checkerspots, Tortoiseshells, Ladies, and Their Relatives (Nymphalinae)

FIELD CRESCENT   *Phyciodes campestris*
Pl. 19

You may find this species referred to as *P. pratensis* and *P. pulchellus* in different works. Whatever you call it, it is locally common with the host plant in moist, springy places near the coast and in the Delta, becoming increasingly colonial inland. One of its largest colonies in the Sacramento Valley, at Willow Slough north of Davis (Yolo County), appears to be extinct after some 30 years of monitoring. Many of its populations are transient on much shorter time scales.

Sacramento Valley populations display pronounced seasonal

polyphenism: summer individuals are small and have considerably more black above than those from overwintered larvae. Suisun Marsh individuals are similar but larger, and the median band above, which is usually pale, is especially so there. All of these populations are at the mercy of winter Xooding. Although diapausing larvae can clearly survive several weeks of immersion, presumably in air pockets in the litter, mortality in wet years is very heavy, and the survivors may metamorphose so late that an entire generation is lost. The overall pattern of geographic variation in this complex is extremely interesting, though most of the exciting things are happening to our east.

The Field Crescent visits dogbanes *(Apocynum)*, milkweeds *(Asclepias)*, thistles *(Cirsium, Carduus,* and *Silybum)*, mints *(Mentha)*, small species of purple loosestrife *(Lythrum)*, and so forth. Not territorial, though desultory male-male chases are observed. In fall very large numbers of this species can be found on aster flowers *(Aster)*, usually mixed with Buckeyes *(Junonia coenia)*, Sandhill Skippers *(Polites sabuleti)*, Monarchs *(Danaus plexippus)*, and whatever else is around.

DISTRIBUTION: All Bay Area and Delta counties; scattered colonies in Sacramento Valley, many of them ephemeral.

HOST PLANTS: The very common *Aster chilensis/lentus* complex, in tule marsh, riparian habitat, coastal prairie, and so forth. Perhaps occasionally other asters.

SEASONS: Three broods inland (March to October), occasionally only two (May or June to October) in very wet years; up to four coastwise (February to December). In some areas only the fall (September to October) flight is usually observed (why?). Young larvae feed gregariously and hibernate half-grown.

SIMILAR SPECIES: The Field Crescent has a heavy, complete black pattern above with a lighter median shade of orange; the antennal club is brownish. The Mylitta Crescent *(P. mylitta)* is mostly orange above, the pattern (often obsolescent) of narrow black lines only; the medial area is paler only in females and if so, it is whitish, not yellowish. The California Crescent *(P. orseis)* looks basically like a very dark, rather large Mylitta. Both Mylitta and California Crescents have orange antennal clubs, and the apex of the forewing is more truncate than in the Field Crescent (especially in males).

CALIFORNIA CRESCENT   *Phycodes orseis*
Pl. 19

This species is extinct in our area—we have no idea why—but it is given room on pl. 19 because it is not extinct globally and it might return some day. As of now, we have to go to the Klamath region to see it alive, and even in that region it is local and uncommon.

This is a very enigmatic species. It has two subspecies, ours (nominate *P. o. orseis*) and Herlan's Crescent *(P. o. herlani)* from the Sierra Nevada. Ours looks like a Mylitta Crescent *(P. mylitta)* with a very heavy black pattern. In fact, it looks like what you might imagine a Mylitta Crescent–Field Crescent hybrid to look like. Subspecies *P. o. herlani* looks like a hybrid between the Mylitta and the endemic Sierran member of the Field Crescent complex (*P. campestris* complex, *P. c. montana*). Is this just a coincidence? Parallel evolution? Or might it actually *be* a stabilized hybrid entity? The two "parental" species co-occur in many places and do not hybridize today. To make matters more interesting, both subspecies of the Field Crescent are extraordinarily variable at the individual level, something we might expect of hybrids.

At any rate, in my experience, most *P. o.* "*orseis*" in collections are *not P. o. orseis,* and many real *P. o. orseis* are misidentified as one of the other species. I wish there were a single infallible *orseis* character one could use to spot these things. If there is, no one has identified it yet.

Males perch and puddle along streamsides (in the Klamath region, generally in coniferous forest or on meadows). This species, if it is a species, is always univoltine and never weedy.

DISTRIBUTION: Formerly, San Francisco and the North Bay. Regionally extinct.

HOST PLANTS: Native thistles *(Cirsium).*

SEASONS: One brood, May to early July.

SIMILAR SPECIES: Good luck! *Phyciodes o. orseis* averages slightly larger than the Mylitta Crescent. Forewing shape and pattern details match the Mylitta rather than the Field Crescent, but the pattern is as dark as the Field Crescent's, and the medial shade may be a little lightened. The antennal club is orange, as in Mylitta; the ventral hind wing is similar to that of cold-weather Mylittas of the same sex.

MYLITTA CRESCENT
Phyciodes mylitta
Pl. 19

To find this species in true wildlands you must go to marshes or
bogs where native thistles *(Cirsium)* grow. To find it anywhere
else you need only go outdoors. One of the great beneficiaries of
European colonization, the Mylitta Crescent is just about as
ubiquitous today as its weedy thistle hosts of Old World origin.
There is a catch, though: most of those weedy hosts are winter annuals and dry up and die in early summer. As a result, breeding in
the second half of the season is restricted to the common bull
thistle *(Cirsium vulgare)*. By the time the last brood of adults is
flying in fall, seeds of the annuals are germinating, and the young
plants are available for oviposition.

Males are territorial, often sitting in paths. Cold-season
broods have more complete ventral hindwing patterns; females
may be largely whitish, even vaguely silvery, with heavy red
brown markings on this wing. There is much individual variation. This would be no problem if the California Crescent *(P. orseis)* did not exist!

DISTRIBUTION: All counties, anywhere there are weedy thistles
(Cirsium, Carduus, and *Silybum)*. Otherwise, marshes and wetlands with native species of thistles *(Cirsium)*.

HOST PLANTS: Any species of thistles *(Cirsium)* and plumeless
thistles *(Carduus),* and milk thistle *(Silybum marianum)*. Not
convincingly recorded from star-thistles *(Centaurea)*. Larvae
feed gregariously when young and hibernate half-grown. They
can often be found sunbathing on thistle plants on sunny late-
winter days.

SEASONS: Three to five broods, January to November.

SIMILAR SPECIES: See Field Crescent *(P. campestris)* and California
Crescent.

NORTHERN CHECKERSPOT
Chlosyne palla
Pl. 20

Common throughout the Bay Area, this species is only a rare
stray from the foothills in the Sacramento Valley. It is found in
chaparral, oak woodland, canyon riparian, and mixed-mesic forest environments, most commonly along roadsides or streams.
Males often sit in paths and are territorial perchers.

In our area there are three "forms" of the female Northern Checkerspot. One is mostly orange and resembles the male. Another is black with yellow and red spots and is apparently a mimic of the Variable or Chalcedon Checkerspot (Euphydryas chalcedona). The third is variable and intermediate between these. The mimicry hypothesis has not been tested experimentally. Strikingly, at high elevations the Variable Checkerspot is mostly red, and the dark forms of the Northern Checkerspot do not occur there. In general, if only one sex is mimetic, it is the female (why?) — so this presumed case is consistent with a pattern. Much minor variation is superimposed on the sexual polymorphism.

The Northern Checkerspot is "northern" in that it is replaced from the Central Coast Ranges southward by the quite similar Gabb's Checkerspot (C. gabbii), and in the southern California deserts by the Sagebrush Checkerspot (C. acastus).

An eager flower visitor (California buckeye [Aesculus californica], yerba santas [Eriodictyon], woolly sunflower [Eriophyllum lanatum], mule's ear [Wyethia], coyote mint [Monardella], dogbanes [Apocynum], milkweeds [Asclepias], etc.) and puddler.

DISTRIBUTION: All Bay Area counties. Spotty records on Sacramento Valley floor as a stray; apparently not resident except perhaps in extreme north.

HOST PLANTS: Recorded on a variety of sunflower family members (Asteraceae), including asters (Aster) (in our area mainly on broad-leafed aster [A. radulinus]) and goldenrods (Solidago); elsewhere on rabbitbrush (Chrysothamnus) and ragwort or butterweed (Senecio). Not known to breed on the asters found in tule marsh, or to occur in that habitat.

SEASONS: One brood, March to July; half-grown larvae overwinter.

SIMILAR SPECIES: The Northern Checkerspot is larger than the crescents, with a much more precise, well-defined checkered ventral pattern. The dark female can be told from the Variable Checkerspot by wing shape, size (slightly smaller), and the absence of any yellow or white spots on the hindwing upperside basad of the single one in the cell. The Leanira Checkerspot (Thessalia leanira) is similar in size or slightly smaller; on the ventral hindwing it has a black postmedian band containing white spots, quite unlike any other checkerspot in our area.

BORDERED PATCH   *Chlosyne lacinia*

Not illustrated

This common and very variable butterfly of the desert Southwest
was discovered breeding in the vicinity of Camp Pollock, a Boy
Scout camp in North Sacramento, and in a nearby location in
Yolo County in the early 1970s. It died out soon thereafter. It almost certainly represented an introduction, as all other California records are from the southeastern corner of the state. The
host plant in Sacramento was the common annual sunflower
(*Helianthus annuus*).

LEANIRA CHECKERSPOT   *Thessalia leanira*

Pl. 20

This is usually a rather uncommon and local species found in
open, rocky areas, including but by no means limited to serpentine. Although it is more or less the same size as the Northern
Checkerspot (*Chlosyne palla*) and Edith's Checkerspot (*Euphydryas editha*), this species has a distinctive "look" that is unmistakable, even in flight. Males are usually smaller than females.
It visits woolly sunflower (*Eriophyllum lanatum*) and coyote
mint (*Monardella*) avidly, as well as other flowers. Males patrol
along rocky embankments. This species seems to be a mimic of
the Variable Checkerspot (*Euphydryas chalcedona*), but we know
nothing of its own palatability. Populations in desert regions are
mostly orange. Unlike the Northern Checkerspot, Leanira does
not climb to high altitudes, where the Variable Checkerspot is
mostly red.

DISTRIBUTION: All Bay Area counties, but local. Not in Sacramento
Valley.

HOST PLANTS: Indian paintbrush (*Castilleja*, Scrophulariaceae),
especially tomentose species. Larvae feed primarily on the flowering shoot tips (but keep in mind that the "flowers" are mostly
colored bracts). Hibernates as half-grown larva.

SEASONS: One brood, March to July, mostly in May and early
June.

SIMILAR SPECIES: See our other checkerspots. Among the unique
attributes of Leanira are that the abdomen is black, ringed with
white; the palpi (thus, "face") are bright orange; and it has a simple underside pattern on the hindwing, with a black postmedian
band containing round white spots.

VARIABLE CHECKERSPOT or CHALCEDON CHECKERSPOT

Euphydryas chalcedona

Pl. 20

This is our commonest checkerspot. In other parts of its range it presents difficult taxonomic and biological issues, but in the area covered by this book it is apparently straightforward. Just why the name "Variable Checkerspot" has been judged superior to "Chalcedon Checkerspot" is unclear. The mineral chalcedony is a quite different color, and there is no obvious connection of the butterfly to the city of Chalcedon (now Kadikoy) in Asia Minor, where a council of Church fathers was held in A.D. 451. On the other hand, all of the checkerspots are "variable," both individually and geographically. Our populations are less variable than most; aberrations are rare. Females are substantially larger than males and fly much less (and less well). Males are territorial perchers, often in sunflecks in oak woodland or riparian forest, or in chaparral clearings. This species is a very avid flower visitor, especially at skunkbush *(Rhus trilobata),* also called lemonade bush or squawbush*;* yerba santas *(Eriodictyon),* California buckeye *(Aesculus californica),* coyote mint *(Monardella),* dogbanes *(Apocynum),* milkweed *(Asclepias),* mule's ears *(Wyethia),* large clovers *(Trifolium),* and numerous others. Males are occasional puddlers. Despite a wide range of acceptable habitats, this species seldom strays into built-up areas and is a rare sight in gardens unless they are very close to canyons.

This species breeds on plants containing iridoid glycosides, bitter-tasting compounds, which it apparently utilizes for its own defense. It is warningly colored and appears to be a model for various mimics, of which the dark form of the female Northern Checkerspot *(Chlosyne palla)* is the best. Eggs are laid in masses and the young larvae feed gregariously in a web. They then diapause over winter. In late winter they can be found sunbathing on warm days, and in March they typically resume feeding, but singly. At this time they will feed on many plants normally not selected by females to receive eggs. Coast Range and Bay Area larvae are black and orange, with only traces of a white pattern. Larvae in the Sierra Nevada are boldly striped lengthwise in white; this is the form found in populations along the eastern edge of the Sacramento Valley. No one has yet found this butterfly in the Sutter Buttes. If someone does, will the larvae be "eastern" or "western"?

We do not know if there were any populations on the valley floor; there certainly aren't now.

Where this species and Edith's Checkerspot *(E. editha)* co-occur, the latter usually begins emerging significantly earlier.

DISTRIBUTION: All counties, but not resident on the floor of the Sacramento Valley, though strays are recorded as far in as Davis and Woodland. Breeds on dredge tailings and similar locations along the eastern edge of the valley (e.g., Fair Oaks, Sacramento County). Yes, the larvae are "eastern."

HOST PLANTS: Various figworts (Scrophulariaceae) and, after diapause, plantains (Plantaginaceae) as well. Females oviposit only on large hosts capable of sustaining all the larvae from one egg mass. This usually means bush monkeyflower *(Mimulus* or *Diplacus aurantiacus)* or one of the shrubby penstemons *(Keckiella)* or, infrequently, bee plants *(Scrophularia)*. After diapause, will feed on these, plus Chinese houses *(Collinsia)* and various others, plus species of plantain *(Plantago)*. Captive females will oviposit on *Plantago,* and in the eastern United States the related Baltimore Checkerspot *(E. phaeton)* is now doing so in some areas. This species never uses lippias (*Lippia,* Verbenaceae) unlike the Buckeye *(Junonia coenia),* another iridoid specialist.

SEASONS: One brood, March to July. A spotty second brood has been claimed in some South Bay locations (August to October). The circumstances responsible for this are unclear, and it is highly atypical of the group.

SIMILAR SPECIES: This is our largest and blackest checkerspot. The forewing apex of the male is pronounced. The various subspecies of Edith's Checkerspot all have rounder forewings in the male. The Variable Checkerspot in our area never has a postmedian band of large red spots on the hindwing above. But it can have such a band in other places! The dark females of the Northern Checkerspot are smaller, with different wing shape and only one yellow spot above on the basal half of the hindwing. The antennal club of the Variable is all orange, and the abdomen is white spotted. Neither is true of Edith's Checkerspot.

EDITH'S CHECKERSPOT *Euphydryas editha*
Pl. 20
This is one of the most-studied organisms on Earth from the standpoint of population biology, thanks to the work of Paul

Ehrlich and his students and associates over more than 40 years. (See the book *On the Wings of Checkerspots,* edited by Ehrlich and Ilkka Hanski, Oxford University Press, New York, 2004, for a scientific retrospective.) The focal population for this work, at Jasper Ridge on the Stanford University campus, is now extinct. That extinction highlighted the metapopulation structure of the species. Within our area there are at least three such metapopulations, differing in both biology and phenotype and recognized as taxonomic subspecies. One of these, the Bay Checkerspot *(E. e. bayensis),* was federally listed as threatened in 1987. It is now extinct in San Francisco, where it had once occurred on Twin Peaks and possibly Mount Davidson. The type-locality of the subspecies is Hillsborough, San Mateo County. The largest remaining population is at Morgan Hill, just south of San Jose and near the limit of the range.

Globally, Edith's Checkerspot is a local, colonial animal. Some of this may be due to habitat fragmentation, but some is certainly due to host plant preferences and the history of California vegetation. Most populations at low elevation in northern California are on serpentine grassland, where invasive Mediterranean annuals are largely excluded by soil chemistry, and the native plants needed by Edith's Checkerspot remain abundant. However, we do not know how widely these plants occurred on nonserpentine soils before the weedy invasions. Although local populations of the butterfly may be abundant, their long-term survival is never assured.

Edith's Checkerspot emerges quite early in the year. Herman Behr wrote in 1863 that it is much rarer than the Variable Checkerspot *(E. chalcedona)* and "of a more restless disposition. It makes its appearance before [the Variable Checkerspot] and is one of our first vernal butterflies." It is under considerable pressure to emerge early. Its larvae have to reach the third instar to be able to diapause successfully, and their host plants have an unfortunate propensity to dry up early in hot, dry years. Prediapause larval mortality can be immense, so even a few days' head start in March may be the difference between life and death. On the other hand, March weather in northern California can be foul; adults can be kept grounded and fail to reproduce; eggs can be washed away and larvae can drown. It may even snow every decade or two! The serpentine races of the Anise Swallowtail *(Papilio zelicaon)* have reacted to a similar problem by developing multiyear pupal diapause as a buffer against short-term catastrophe. Edith's

Checkerspot can spend 2 years as a larva if need be. But first it has to get to the third instar.

Males generally emerge several days before females. They are perchers and puddlers and eager visitors to flowers, including yellow-flowered members of the sunflower family (tidy-tips and goldfields [*Layia* and *Lasthenia*, Asteraceae]), bulbs (wild onions [*Allium*], brodiaeas [*Brodiaea*]), biscuitroots *(Lomatium)*, yerba santas *(Eriodictyon)*, and others. In places where they emerge late enough, they often visit flowers of skunkbush *(Rhus trilobata)* and coffeeberries *(Rhamnus)*.

DISTRIBUTION: Three subspecies: Bay Checkerspot in San Francisco (extinct), San Mateo, and Santa Clara Counties; Luesther's Checkerspot *(E. e. luestherae)* in Contra Costa, Alameda, and Santa Clara Counties; and Baron's Checkerspot *(E. e. baroni)* in North Bay counties. Not in the Sacramento Valley or the Delta.

HOST PLANTS: Bay Checkerspot: *Plantago erecta* (Plantaginaceae) late in spring and after diapause, *Castilleja densiflora* (also called *Orthocarpus densiflorus*) (Scrophulariaceae); Luesther's Checkerspot: Indian warrior *(Pedicularis densiflora*, Scrophulariaceae), a showy perennial hemiparasite usually found in chaparral. Local populations in the North Bay may feed on different figworts as well as (postdiapause) on plantain *(Plantago)*.

SEASONS: March to June, one brood flying 3 to 4 weeks in a given locality, and varying with the weather of the year.

SIMILAR SPECIES: See Northern Checkerspot. Edith's Checkerspot is smaller than the Variable Checkerspot, and the forewing much rounder at apex; the abdomen is without white spotting; the hindwing is almost always with a complete postmedial row of large red spots (this character will not work outside our area in places where the Variable Checkerspot is mostly red); and the antennal club has the basal part usually blackish. For the Northern Checkerspot see under that species.

Anglewings or Commas *(Polygonia)*

This is a very confusing group, and you will find inconsistencies in classification as you go from one reference book to another. Except for the Satyr Anglewing *(P. satyrus)*, which is distinctively colored beneath, accurate sight identification of these animals is a skill that must be acquired through experience. They are very much like the *Empidonax* flycatchers, which can be used to judge a birder's degree of skill. But the anglewings don't sing, alas.

In general, keep in mind that the Satyr Anglewing is the most widely distributed (and the only one in the Sacramento Valley and Delta); the Rustic *(P. faunus rusticus)* is the next most widely distributed, but rare; the Oreas *(P. oreas)* is the commonest in the redwoods, and the Zephyr *(P. zephyrus)* is found in many cool inland canyons. Got that? Good luck.

SATYR ANGLEWING   *Polygonia satyrus*
Pl. 21

The Satyr Anglewing is never common but occurs throughout our area in riparian habitat where the native stinging nettle *(Urtica holosericea)* grows. Males are territorial in sunflecks, often sitting in paths through the woods. Like most anglewings, this species is seldom seen at flowers, preferring sap fluxes, rotting fruit, moist fungi, and such. There are pronounced seasonal forms, but their control has not been investigated. This is our most heat-tolerant anglewing.

DISTRIBUTION: All counties in riparian and tule marsh habitats, seldom seen elsewhere.

HOST PLANTS: The native stinging nettle *(Urtica holosericea)*. Eggs are laid in short "stacks" of three or four; the larvae live singly in rolled-leaf nests, much like the Red Admiral *(Vanessa atalanta)*, with which this species usually co-occurs. Not recorded on weedy nettles.

SEASONS: Two to three generations. Because adults are active on warm winter days there are records for all months but December, but inland the broods tend to be well defined (hibernators in March, adults in June and again August to September).

SIMILAR SPECIES: This is our only anglewing with a reddish brown underside. It also has a less extensive black pattern and a brighter orange ground color above than any of the others.

RUSTIC ANGLEWING   *Polygonia faunus rusticus*
or GREEN COMMA
Pl. 21

Usually uncommon to rare, the Rustic Anglewing is found spottily in cool, moist forest near the coast. Like the other anglewings it hibernates as an adult and is territorial in sunflecks. Males have conspicuous dark green mottling on all wings below. Females are typically dull gray on the same surface. This is a poorly known species in our area and is seldom reported. Its numbers seem to vary on a fairly long "cycle," perhaps around 20 years.

DISTRIBUTION: Mainly coastal counties, in redwood, Douglas fir, and mixed forests; unknown in the Sacramento Valley and Delta.

HOST PLANTS: The literature usually cites California azalea (*Rhododendron occidentale*). I have never found it on this plant myself. All of my records are on willows (*Salix*).

SEASONS: Late winter, hibernators sometimes alive into May; fresh adults June to late July; occasional records from September and October. It is not clear whether we have one or two broods in our area.

SIMILAR SPECIES: The Satyr Anglewing (*P. satyrus*) is reddish brown below. The Zephyr Anglewing (*P. zephyrus*) is gray below with faint yellowish spots on all wings. The Oreas Anglewing (*P. oreas*) is very even, dark brown below. The shapes of the silvery "commas" on the hindwings differ (see pl. 21). The Rustic Anglewing has the most irregular wing margins. It and Oreas have the most extensive dark patterns above. The lichen-simulating underside of the male Rustic is both extraordinarily beautiful and unique in our fauna.

ZEPHYR ANGLEWING   *Polygonia zephyrus*
Pl. 21

This is yet another species of shaded woods, in this case mostly inland in canyons where the host plants (*Ribes*) grow. Although this is the commonest anglewing in the mountains of California, it is just as uncommon in our area as the rest of them. Habits of the genus.

DISTRIBUTION: Most Bay Area counties in riparian-canyon habitats. Absent from the Sacramento Valley and Delta.

HOST PLANTS: Currants and gooseberries, both in the genus *Ribes*.

SEASONS: Most records are for hibernators in late winter and early spring. Records of the summer brood(s) are surprisingly few, even in well-collected areas. Do adults live nearly a full year? In the Sierra Nevada they do, but there also appear to be two genetic species there lurking within this name. Only one of these occurs in our area.

SIMILAR SPECIES: See Rustic Anglewing. Note the combination of brown gray underside, faint yellow spots, narrow silver "comma" (more like the letter V), and rather sharply pointed wing projections of the Zephyr. The dorsal pattern is rarely as complete as in the Rustic and Oreas Anglewings.

OREAS ANGLEWING   *Polygonia oreas*
Pl. 21

This species either is or isn't conspecific with the Gray Comma (*P. progne*), depending on this week's reading of the goat entrails. It is another poorly understood species of cool, moist forest. In the redwoods and south of the Bay near the coast it can be the commonest anglewing, but that isn't saying much! Perhaps because it is less rare than the others, I have actually seen it nectaring (on tansy ragwort [*Senecio jacobaea*] and mints [*Mentha*]), but like most anglewings it prefers flowing sap and such. Habits of the genus.

DISTRIBUTION: All Bay Area counties, including San Francisco. Absent inland, including the Sierra Nevada!, but widespread in the northern mountains.

HOST PLANTS: Reportedly *Ribes divaricatum*, a gooseberry.

SEASONS: Records spotty, with records concentrated in late winter to early spring (hibernators) and midsummer; perhaps one brood, the adults living nearly a full year.

SIMILAR SPECIES: See Rustic Anglewing. The Oreas is our darkest anglewing both above and below; note the very large spot in the basal half of the hindwing above, the very well marked pale submarginal spotband enclosed by dark markings, and the dark, even brown ground color beneath.

CALIFORNIA TORTOISESHELL   *Nymphalis californica*
Pl. 22

Mass migrations of this butterfly, involving millions if not billions of individuals, enter not only the lore of California lepidopterology but the consciousness of the general public. And no wonder! Any butterfly that can stop traffic over Donner Summit is worth your attention.

The most recent great migration—by no means one of the

greatest on record—occurred in late July to early August 2004, when a swarm 65 to 80 km (40 to 50 mi) long (north to south) by 24 km (15 mi) wide (east to west) moved south along the Sierran crest. When the butterflies reached Lake Tahoe they blanketed Kings Beach, sipping water from the sand and entertaining myriads of children, dogs, and photographers. Where were they going, and why?

Adult California Tortoiseshells hibernate, mostly at low elevations in foothill canyons in both the Coast Range and the Sierra. They come out and sun themselves on warm midwinter days, returning to their previous shelter for the night. They have been known to hibernate in the interstices of talus slopes and rubble piles, under bark and wooden shingles, inside discarded beer cans, and in outbuildings of all kinds. When they emerge from hibernation in late winter the males are late-afternoon territorial perchers in classic nymphaline style. Eggs are laid on species of wild-lilac *(Ceanothus)* in March, right around the time of bud burst. The spiny, black-and-yellow larvae feed gregariously and, when abundant, can defoliate the plants over wide areas. When mature they usually pupate on the (now leafless) host, setting up a situation that once seen is seldom forgotten. The gray lilac pupae hang head-down on the branches, often almost as thick as leaves. On hot, sunny days—or if disturbed, as by a parasitoid or predator—they all begin twitching violently from side to side. The effect is…unique.

The adults of this brood emerge in late May or very early June. They mill around for a few days, often visiting mud puddles, and then migrate. Those in the Coast Ranges may go either north, toward the high Inner North Coast Range and Klamath Mountains, or east to northeast toward the Sierra Nevada. The latter movement takes them across the Sacramento Valley, where they can be spotted either in purposeful flight or visiting bird- or wasp-damaged fruit in orchards and gardens. By mid-June there are no California Tortoiseshells in the Coast Ranges south of Colusa or Mendocino County.

The areas where summer breeding takes place are scattered from the Pacific Northwest south through the High Sierra; in any given year the breeding is highly concentrated, and the venues change from year to year. This may be a response to climate, or it may be a way of confounding predators and parasites. In 2004 breeding was concentrated in far northeastern California. In the two prior years it was mostly in the Washington and Oregon Cascades. Eggs are laid on upland species of wild-lilac, especially tobaccobrush *(C. velutinus)*, which again may be completely defoliated. Wherever breeding may have occurred, the adults emerge in July. Rather than trying to breed again, they almost always pick up and migrate southward along the Cascade-Sierra axis to the High Sierra, where they estivate for the remainder of the summer. This is where they were going at the end of July 2004. Hikers above tree line in Yosemite, Sequoia, or Kings Canyon National Parks sometimes encounter masses of estivating Torties and emerge with tales of life-changing mystical experiences. They usually say the butterflies were Monarchs *(Danaus plexippus)*.

In late September the butterflies scatter, migrating down the west slope of the Sierra in search of hibernation sites. Some stop in the Sierran foothills but others press on across the Sacramento Valley again, showing up around Sacramento or Davis about October 5 every year en route to the Coast Range. Rarely, a few may overwinter on the valley floor. One male that did so on the Davis campus was out every sunny day that winter, sunbathing on the roof of his art studio-cum-hibernaculum.

The spring and fall migrations, unlike the summer one along the crest, are so inconspicuous that most people, including lepidopterists, don't realize they are happening. There are no Torties in the Coast Range all summer, and yet they reappear "as if by magic" in the same canyons next spring. Bizarrely, the mythology persists that this species has only one generation per year. It has one generation per year *in any one place.*

So why do they do it? The young larvae require young, tender foliage on which to begin feeding. Large larvae can eat mature foliage (which is more or less leathery and loaded with tannins). Wild-lilacs put on all their new growth at the beginning of the season, so by the time the first generation of adults hatches in May, no hosts in suitable condition are available in the foothills. Meanwhile, the high-elevation wild-lilacs have just melted out of the snow and are undergoing bud burst. Only by migrating can the California Tortoiseshell get in two broods a year. As in the case of the Painted Lady *(Vanessa cardui)*, population densities vary through several orders of magnitude, but the seasonal migrations go on whether tens or millions of animals are around. "Population pressure," pompously invoked in the literature, seemingly has nothing to do with it. However phylogeny does have something to do with it—as discussed under the other tortoiseshell species below.

Two species in the Old World *(N. polychloros* and *N. xanthomelas)* are very closely related to the California Tortoiseshell. They are both rare or endangered in most of their ranges.

Adult Torties are famous puddlers; during migrations hundreds may crowd onto a single puddle, and we have wonderful pictures of them doing so along wagon ruts back in the early twentieth century. They visit flowers, especially coyote mint *(Monardella),* and homopteran honeydew on various plants, and as earlier noted, they also frequent fruit. Torties often alight on people. The female has broader wings and is rather uniformly dark on the ventral surface; males are usually contrastingly paler on the outer half of these wings. There is remarkably little variation, though aberrations can be induced by cold-shocking the young pupae.

Migrating Torties, like Painted Ladies, are amply provisioned with yellow fat, making a distinctive splotch upon colliding with your vehicle.

DISTRIBUTION: All counties. In the Sacramento Valley breeding only on the uplands near Red Bluff, but crossing the valley floor twice a year in migration.

HOST PLANTS: Probably any species of wild-lilac (*Ceanothus,* Rhamnaceae).

SEASONS: In our area, late January to April (hibernators), late May to June (fresh), absent in summer with a few sightings most years in September to October.

SIMILAR SPECIES: Larger than all our anglewings, with no "comma" mark on the ventral hindwing.

COMPTON TORTOISE   *Nymphalis j-album (N. vau-album)*
Not illustrated

This boreal species was recorded once, bizarrely, in San Mateo County. It is otherwise unrecorded in California at all. The adult lives 11 or 12 months and is an excellent candidate to hitchhike in vehicles or commercial containers, but it is unlikely that it has bred in our area since the last Ice Age or will until the next one. The host plants, birches *(Betula),* do not occur here naturally.

MILBERT'S TORTOISESHELL   *Nymphalis (Aglais) milberti* or SMALL TORTOISESHELL

Pl. 22

This butterfly is yet another seasonal migrator with an annual cycle similar to that of the California Tortoiseshell *(N. californica)*, but not subject to population explosions in our area and usually considered rather rare. Adults hibernate at low elevation and breed in March to April on stinging nettle *(Urtica holosericea)*, mostly in riparian forest and tule marsh, but also in urban environments. The resulting adults migrate north- and eastward to breed at higher elevations in summer. They do so every year around Mendocino Pass in the high North Coast Range; in the Sierra Nevada they breed mostly on the east slopes and can be seen crossing the crest, even above tree line. Redispersal to the lowlands occurs in September and October. Since stinging nettles are available all summer, it is not obvious why this species (or the willow-feeding Mourning Cloak! *[N. antiopa]*) migrates as it does—unless it is to get away from the low-elevation heat, an objective with which many inland readers of this book may sympathize. This species visits flowers eagerly— coyote mint *(Monardella)* is a favorite—and puddles but does not seem to be territorial. There is no sexual dimorphism and little variation, mostly in the width and degree of contrast of the yellow band above.

DISTRIBUTION: Sporadic; apparently more frequent inland, including the Sacramento Valley, than along the coast. Larval colonies are sometimes found in places where adults have never been recorded.

HOST PLANTS: Both the native stinging nettle *(Urtica holosericea)* and the smaller introduced nettle *(U. urens)*. The larvae are gregarious and form a seething mass of spiny black-and-yellow caterpillars capable of defoliating entire clumps of the host, while nearby ones remain untouched.

SEASONS: Hibernators February to April; fresh adults late May to June and (rarely) September to October. Apparently absent in midsummer, when it is breeding upslope.

SIMILAR SPECIES: None in our area.

MOURNING CLOAK   *Nymphalis antiopa*
Pl. 22

Until very recently the Mourning Cloak was a very common butterfly throughout our area, even in cities. I remember the thrill experienced by a British postdoc in my lab upon seeing his first one. (As the Camberwell Beauty, this species is traditionally regarded as the rarest British butterfly. It has also declined dramatically on most of the continent of Europe since World War II.) But recently it has undergone a catastrophic regional population collapse, disappearing from many places and becoming very rare in others. This is tied in to its cycle of seasonal altitudinal migration, described below, and seems to have begun with a breeding failure in the Sierra Nevada in the summer of 2001. There have been stirrings of potential recovery in 2004/05, but only stirrings.

In northern California Mourning Cloaks are known to overwinter only as adults. Unlike many other hibernating nymphalids, they are rarely seen out and about before January 25, regardless of the weather. Breeding then occurs in spring; adults emerge in late May. Near the coast they appear to remain in place and there may be another brood in midsummer (and even a third, in September to October, some years). But inland this species disappears by early July and is not seen again until late September or October, if then. Earlier authors assumed the species estivated through the hot summer months, though there was no direct evidence of that. It turns out that there is plenty of circumstantial evidence to suggest they undergo altitudinal migration, just like the California Tortoiseshell *(N. californica)*. In fact, they can be seen migrating upslope individually along Interstate 80 in June! Their progeny can be seen at and above 1,500 m (5,000 ft) in August and September, hanging around willows *(Salix)* and sometimes visiting rabbitbrush *(Chrysothamnus)*. Some of these butterflies hibernate in the mountains, but others probably disperse downslope to repopulate the lowlands and hibernate there; we regularly see single Mourning Cloaks seemingly crossing the Sacramento Valley in October. Or we used to.

Hibernators are (or were) a conspicuous element in the late-winter landscape, being territorial along wood roads and nectaring on willow catkins. Old individuals—not necessarily hibernators—become very battered; the ground color fades to a dull dark red, and the yellow borders, or what is left of them, fade to cream. There is no sexual dimorphism or seasonal or local variation, but spectacular aberrations occur and can be reproduced by cold-shocking the young pupae. The most famous of these is called *N. a. "hygiaea"* and has the yellow border greatly expanded to cover the entire outer half of all wings, obliterating the row of blue spots. It has been taken in the wild several times in our area.

DISTRIBUTION: All counties, primarily in wooded and riparian habitats but often straying; breeding in older urban neighborhoods with mature landscaping. Now rare inland, but still fairly numerous near the coast.

HOST PLANTS: Willows *(Salix),* including the nonnative weeping willow *(S. babylonica);* hackberries *(Celtis);* and elms *(Ulmus).* Before the crash it was sometimes considered a pest for defoliating these trees in cultivation, though always spottily—a colony here, a colony there. Eggs are laid in large batches; the larvae feed colonially but do not spin a web. They are black and spiny, with conspicuous red spots down the back; it is easy to spot a colony from a passing car! They leave the host to pupate and, if near a building, often do so under the eaves. When each adult emerges, it produces a large drop of meconium that leaves a blood red spot on the ground below. In medieval Europe such "red rain" was taken as an omen and often stimulated civic disturbances and demonstrations of religious fanaticism. The pupa is grayish violet, finely dotted with black.

SEASONS: Hibernators late January to April. From Fairfield inland fresh butterflies in late May to June, then usually not seen again except a few individuals in late September to October. Near the coast, present all season, with two to three broods.

SIMILAR SPECIES: None.

AMERICAN PAINTED LADY or WEST VIRGINIA LADY

Vanessa virginiensis

Pl. 23

The least common of our three ladies, this one occurs in a great variety of habitats but is perhaps most frequent in riparian forest and tule marsh. It is reputed to migrate, but there is no solid evidence that it does so in our area; however, it is rarely seen in winter. Pattern aberrations are very rare in this species and are more difficult to produce by experimental "cold shock" treatment than in the other ladies. It is an eager visitor to a great variety of flowers, and particularly addicted to dogbanes *(Apocynum).*

DISTRIBUTION: All counties; occasional in cities.

HOST PLANTS: All species of everlastings *(Gnaphalium)* and pearly everlastings *(Anaphalis);* elsewhere also recorded on pussytoes *(Antennaria).* The larva webs up the host plant, incorporating dry chaffy material from the inflorescence into its rather messy nest. The caterpillar itself is beautifully banded in black, red, and white but can be seen only by opening up the nest—so the significance of its coloration is obscure. In Davis and Suisun City, larvae have been found feeding on the exotic bedding plant gazania *(Gazania),* and they are easily reared on it.

SEASONS: January to December, multiple broods (but unlike the other species of *Vanessa,* commonest in summer).

SIMILAR SPECIES: The very large ventral hindwing eyespots make this species unmistakable.

PAINTED LADY   *Vanessa cardui*

Pl. 23

The Painted Lady is the most cosmopolitan butterfly on the planet—and recorded throughout our area, including the Farallon Islands. It can be seen anywhere except under dense forest cover. The Painted Lady overwinters in the deserts along the U.S.-Mexico border, breeding on annual plants in the wake of the winter rains. The resulting adults then migrate northward in late winter. Although the occasional late larva survives in our area and manages to produce an adult in late winter, this appears to not be a factor in population dynamics; the species should be considered nonresident, as it is in virtually all of temperate North America. The northward migration occurs every year but varies in density by several orders of magnitude; when sparse it is nearly invisible, but in big years literally hundreds of millions of adults invade our air space on a broad front, astonishing spectators and tying up traffic. These butterflies have been programmed physiologically and are born with a large supply of yellow fat, which enables them to migrate single-mindedly from dawn to dusk without the need to stop and feed. This is what makes the familiar yellow splotch on your windshield when your car intersects their flight path. They fly from southeast to northwest "like bats out of Hell" and can make the trip from Bishop to Davis in about three days. They are not dependent on a tailwind, either.

As their fat reserves diminish, they begin to feed, stop migrating, and become sexually active. Males set up territories in the late afternoon. These may be on hilltops or, in flat country, in places open to the southwest and defined by a vertical backdrop (trees, walls, etc.). Mating occurs only in the very late afternoon, and pairs may remain in copula overnight. Eggs are laid singly, but in eruptive years larval densities can be enormous; host plants may be locally defoliated, and the larvae are forced to emigrate in search of food, again triggering public notice. The resulting adults usually emerge in late May to early June and themselves emigrate no later than their second or third day, continuing the movement northward. Individuals hatching in June show a decreasing migratory instinct, and after the middle of the month are likely to stay at home and attempt to breed locally. In some years the Painted Lady is absent in summer at low altitudes, while in others it maintains at least a low-level presence in our area. These facts suggest that migration is cued largely by photoperiod, and that as the Summer Solstice is approached the cues that program it cease to operate. Summer breeding is more likely in the mountains than near sea level.

Great Painted Lady years typically coincide with heavy winter rains in the desert, often but not invariably associated with El Niño. There is no objective yardstick for comparison, but by general agreement 2005 was one of the greatest Painted Lady years ever. The leading edge of the migration reached Sacramento on March 11, and wave after wave moved through for the next several weeks. One way to estimate numbers is to project an imaginary line to the horizon, perpendicular to the direction of movement, and make counts at intervals of the numbers crossing it per minute. At the height of the 2005 migration, counts reached 175 per minute, or nearly three per second. With some simplifying assumptions, one could calculate that the number passing through the Sacramento Valley in toto had to exceed a billion! The very first locally bred butterflies emerged and began emigrating in early May, and some were still emerging at the end of June. On certain days, concerted local movements in anomalous directions (toward the southeast or southwest) can be observed; these are not understood.

The same phenomena also occur in the Old World. The Painted Lady overwinters in North Africa and the Middle East, breeds in late winter, and migrates into Europe for summer breeding. Vast swarms may, for example, move through the Mediterranean island of Cyprus, typically at the same time they are moving across our part of California.

Different Painted Lady years differ in the geographic distribution of the dense streams. Sometimes the greatest numbers are in our part of California, but in other years numbers moving up the Great Basin east of the Sierra may be much greater. Few of the butterflies appear to cross the Sierran crest. They access the Central Valley via the Inyo and Kern Counties area and/or by crossing the Transverse Ranges.

A reverse (southward) migration occurs from late August through November. This is a much more casual affair. The butterflies, which were mostly born in the Pacific Northwest, are not provisioned with fat and must feed as they go. Most of them seem to follow Highway 395 south on the east side of the Sierra, where they can often be observed in numbers nectaring on the golden blossoms of rubber rabbitbrush *(Chrysothamnus nauseosus)*. (A good place to see them is Woodfords, Alpine County, in early October.) Large numbers occasionally come through the Delta and Suisun Marsh, where asters *(Aster)*, goldenrods *(Solidago)*, coyote brush *(Baccharis pilularis)*, and other fall flowers afford abundant nectar resources. Densities on October 10, 2001, at Suisun were comparable to a spring migration, but this is rare. I have marked dozens of flower-visiting Painted Ladies in gardens in Davis in October and have never had a recapture, suggesting that they move on quickly. There may be a little breeding in our area even after Halloween, however, resulting in the handful of fresh adults recorded in January and February. Most late larvae probably die during the winter.

Desert-bred Painted Ladies are generally small and dull colored. In the wettest desert years this is less noticeable (as in 2005). Those bred in our area and northward are usually large and richly pigmented with bright pinkish (male) and orange (female) hues. The largest locally bred specimens may have three times the wing area of the smallest desert-bred ones! Males have more pointed forewings than females.

Extreme and often very beautiful wing-pattern aberrations occur in this species. They are very rare in nature but are easily manufactured by subjecting young (8 hours) pupae to sustained chilling (3 weeks at 2 degrees C [35.6 degrees F]) — many will die or be crippled, but the survivors, which vary greatly among themselves, will be very striking. There are hints that certain individuals are genetically predisposed to produce these aberrations. The study of these phenomena began in Europe and America in the second half of the nineteenth century and was an important component in early evolutionary thinking and in the formulation of the science of developmental genetics.

The Painted Lady has the most inclusive larval diet of any butterfly in the world, though the vast majority of larvae will be found on thistles *(Cirsium, Carduus,* and *Silybum)*, the borage family (Boraginaceae), and the mallow family (Malvaceae). In outbreak years larvae can be found on truly bizarre hosts. When breeding occurs as late as June in our area, many of the preferred hosts have already senesced, and eggs are often laid on the noxious weed yellow star-thistle *(Centaurea solstitialis)*. Agricultural extension personnel are used to having larvae of the Painted Lady brought in by members of the public who hope they have found a biological control agent for the weed. (Alas, their impact on it is minimal.) The only record I have of Painted Lady larvae causing economic damage in our area was from an herb farm where they ate the crops of borage and comfrey *(Symphytum)* to the ground. It is not unusual in outbreak years for them to completely devour the popcorn flower *(Plagiobothrys)* in vernal pool sites in the Sacramento Valley; a dry vernal pool filled with hungry, writhing Painted Lady larvae presents a peculiar sight!

Genetically, our animals are indistinguishable from European ones. This perhaps is not surprising given the migratory capability of this butterfly, though it has never been documented as moving between Eurasia and North America. The great nineteenth century scientist Louis Agassiz was a creationist and never accepted Darwinism. He believed on religious grounds that God had created entirely separate biota on the various continents, and on that basis he explicitly argued that circumpolar butterflies such as the Painted Lady and Mourning Cloak *(Nymphalis antiopa)* must have been introduced to America by the early European colonists. It is possible that he may have been right on that one.

DISTRIBUTION: Everywhere.

HOST PLANTS: Very many. Borage family (Boraginaceae): fiddleneck *(Amsinckia)*, popcorn flower *(Plagiobothrys)*, borage *(Borago)*, comfrey *(Symphytum)*, hound's tongue *(Cynoglossum)*; sunflower family (Asteraceae): thistles *(Cirsium`* and the introduced plumeless thistles *[Carduus])*, Scotch thistle *(Onopor-*

dum), milk thistle *(Silybum marianum)*; artichoke and cardoon *(Cynara)*; cocklebur *(Xanthium)*, sunflower *(Helianthus)*, mule's ears *(Wyethia)*; vervain family (Verbenaceae): lippias *(Lippia, also called Phyla)*; plantain family (Plantaginaceae): plantain *(Plantago)*; Urticaceae: nettle *(Urtica)*, pellitory *(Parietaria judaica)*; pea family (Fabaceae): lupine *(Lupinus)*, sweet pea *(Lathyrus)*; mallow family (Malvaceae): mallow and cheeseweed *(Malva)*, alkali mallow *(Malvella leprosa)*, hollyhock *(Alcea)*, *Modiola*, checkerbloom *(Sidalcea)*, tree mallow *(Lavatera)* (herbaceous species only?), and so on and so forth. Although the Painted Lady seemingly never oviposits on woody plants, it can be reared easily on woody members of lineages it does eat, such as pride-of-Madeira *(Echium fastuosum)* and the various tree mallows. It is interesting to experiment to determine the limits of just what this species will and will not eat, but one should start with eggs, as it may be difficult to switch larvae to a new host after they have begun feeding. The larva lives in a rolled-leaf nest, often very sloppy and filled with frass. It makes a new nest after each molt. The vast majority of larvae are black, with an interrupted yellow line on each side and hints of orange at the base of the spines; the head is black. Rarely, larvae may be gray, greenish, or even orange. The pupa is fawn gray with iridescent golden markings and can often be found on the host. A generation takes about 28 days in warm weather.

SEASONS: January to December, multiple broods. Usually arriving from the desert between late February and late April, depending on the year; the first and sometimes only local brood emerges late April to late June, usually peaking in late May and emigrating. If there is a summer emergence, it is in mid-July to mid-August. Southward migrants begin appearing in late August and may straggle even into early December. Winter records are very rare.

SIMILAR SPECIES: The Painted Lady is decidedly pinkish in the male. Both sexes of the West Coast Lady *(V. annabella)* are orange. The male forewing of the Painted Lady is also more pointed. The band closing off the discal cell on the forewing upperside is white in the Painted and orange in female West Coast Lady. The black bar in the middle of the same cell is continuous in the West Coast Lady but broken into two spots in the Painted. The row of (usually) blue-centered spots on the hindwing upperside in the West Coast is set off by a row of curved black lines basad to it. In the Painted Lady, the spots are often "blind" (not blue centered), especially in males, and basad of them is broad brown shading which merges into the brown along the anal margin. The West Coast Ladies average smaller than locally bred Painted Ladies but larger than the small, pale desert ones. The American Painted Lady *(V. virginiensis)* is immediately recognizable by the very large eyespots beneath.

WEST COAST LADY   *Vanessa annabella*
Pl. 23

This very common species was formerly lumped with the nearly identical South American species under the name *V. carye.* The two differ in male genitalic morphology and have statistical differences in both wing pattern and molecular genetics but are extremely closely related and can be hybridized experimentally. They have probably been geographically isolated for about two million years. Because the original description, by Jakob Hübner in 1812, did not provide a precise locality, the genitalic anatomy was not mentioned, and no type-specimen has been found, the correct application of the name (North or South American?) had to be inferred from minor details of Hübner's published illustration!

The West Coast Lady moves up- and downslope seasonally in mountainous terrain (as does *V. carye* in Argentina), but there is no evidence of latitudinal migration as in the Painted Lady *(V. cardui).* Males are territorial in late afternoon exactly as described for the Painted Lady. They modulate their posture very carefully to optimize body temperature—sitting with wings open in the dorsal or body-basking modes when cold, or with the wings closed over the back when hot. They may sit on the ground, on low vegetation, or on tree trunks or walls, again as a thermoregulatory tactic. Male-male chases are easy to watch (and great fun). At high densities, many individuals may take part as a result of some disturbance. The territories are reallocated daily and the "best" sites fill up first, with latecomers relegated to the margins—the territories are minimally compressible. Males almost invariably return to their own perches after a chase, and individual males may live as long as 28 days in summer—several months in winter—being territorial whenever weather permits, though not necessarily in the same territory each day. It is very striking that these butterflies ignore one another completely at flowers; males do not react at all to the presence of a virgin female in their midst outside the context of late afternoon territoriality.

The West Coast Lady produces aberrations precisely parallel to those of the Painted Lady, and they are rather more frequently encountered in nature. They are often taken in summer, when they would be unlikely to be due to cold shock, and in this species there is strong evidence of a genetic predisposition to their production. In the 1970s they occurred remarkably frequently around Fairfield and Suisun City, Solano County, but have now disappeared from that population.

DISTRIBUTION: Everywhere except deep woods (but often seen being territorial in sunflecks in riparian forest). Commoner in cities and weedy agricultural land than in wildlands.

HOST PLANTS: Many genera in the Mallow family (Malvaceae) (mallow *[Malva]*, alkali mallow *[Malvella leprosa]*, hollyhock *[Alcea]*, checkerbloom *[Sidalcea]*, Modiola, tree mallow *[Lavatera]*) and several in the nettle family (Urticaceae) (nettles *[Urtica]*, pellitory *[Parietaria judaica]*, and the recently naturalized false nettle *[Boehmeria]* in the Delta). Larva in a rolled-leaf shelter; on cheeseweeds *(Malva neglecta, M. nicaeensis, and M. parviflora)* they live at the base of the leaf on the upperside, constructing a silken canopy overhead, but in wet weather they live underneath and pull the leaf down to form a tent. Usual form black with yellow markings, occasionally gray, greenish, pinkish, or orange; pupa much less elongate than that of the Painted Lady, reddish gray with two white kidney-shaped dorsal spots.

SEASONS: All year; in our area larvae, pupae, and adults can usually be found all winter. Adults can be seen on any warm, sunny midwinter afternoon, beginning their territorial behavior earlier in the day when it is cool. (Winter adults often visit blooming rosemary *[Rosmarinus officinalis]* or escallonias *[Escallonia]* in gardens.) In the Sacramento Valley this species is rather infrequently seen in hot weather.

SIMILAR SPECIES: See Painted Lady.

RED ADMIRAL   *Vanessa atalanta*

Pl. 23

Like the Painted Lady *(V. cardui)*, the Red Admiral (originally the "Red Admirable" in England in the early 1800s) occurs in both

Eurasia and North America, and we do not know if it is native or introduced on our shores. In the Old World and in much of North America it is strongly migratory. In California there is no clear evidence for migration, and we know that it overwinters every year in our area. Indeed, just as with the West Coast Lady *(V. annabella)*, territorial males can be seen perched in their sun-flecks on any mild midwinter afternoon. The territorial behavior of the two species is identical, and they often share lek areas—sometimes with Painted Ladies, California Tortoiseshells *(Nymphalis californica)*, and rarely other nymphalids as well. Under these circumstances it is not surprising that the Red Admiral and the West Coast Lady occasionally hybridize. I have seen one hybrid in northern California; they are more frequent (for some reason) in the south. Wing pattern aberrations parallel to those of the ladies occur but are very rare in nature and difficult to induce experimentally.

This species visits flowers avidly but also is a regular visitor to flowing sap, damaged fruit, and less appealing substrates (dung, garbage, carrion). Adults are very tolerant of shade and can often be flushed from deep riparian shade in the heat of day; they also show up sporadically at lights at night. They also have a strong propensity to land on people! This is the most frequently seen butterfly in the urban Bay Area. Its host plant in such places is pellitory *(Parietaria judaica)*, which grows under freeway overpasses, up chain-link fences, and in cracks in the sidewalk. This plant is naturalized from the eastern Mediterranean.

DISTRIBUTION: Everywhere; equally common in urban and wildland settings; found in both open and shaded environments.

HOST PLANTS: Nettle family (Urticaceae): in riparian habitat, native stinging nettle *(Urtica holosericea,* now sometimes considered a subspecies of *U. dioica);* in gardens, baby's tears (*Soleirolia,* also called *Helxine,* a ground cover used in cool, moist, shaded sites), and cultivated species of pilea *(Pilea);* in weedy, urban situations on the small stinging nettle *Urtica urens* but mostly on pellitory *(Parietaria judaica)* in the fog belt—this plant is not found inland. Larva solitary, black with white or yellow markings, in a rolled-leaf nest, but feeding in the open on the tiny-leafed baby's tears; pupa purplish gray, more elongate than other *Vanessa* species. Unlike the West Coast Lady, does not use mallows *(Malva)* as hosts.

SEASONS: All year, seemingly commonest October to April, especially inland.
SIMILAR SPECIES: None.

BUCKEYE   *Junonia (or Precis) coenia*
Pl. 22

The genera *Junonia* and *Precis* are very closely related and form a pantropical group. Many of these species have sharply differing seasonal forms in the seasonal (wet/dry) tropics. In keeping with its tropical roots, our species is intolerant of cold weather; it was extirpated from our area in the extreme cold wave of Christmas week 1990 and is only known to have survived for certain on San Bruno Mountain. The pattern of recolonization beginning in 1991 suggested that it survived sporadically in various places and spread out from them, rather than reinvading from the south along a broad front. However, apparent waves of immigrants can be observed in the Sacramento Valley in most years at some time between April and late June. Unusually large influxes of this sort occurred in May and June 2003 and 2005, moving from southwest to northeast. In 2005 the immigrants flooded the Sierra Nevada as well; one was even found flying over snow at about 3,000 m (9,900 ft) in Sequoia National Park on May 14! We do not know where these migrating masses originate.

The Buckeye rarely overwinters successfully on the floor of the Sacramento Valley, where most of its habitats are vulnerable to flooding. As the season progresses, it recolonizes the valley between March and May, depending on the year. It always survives in foothill canyons (except in 1990!) but the adults rarely appear before late January to mid-February. Supposedly these have hibernated, but this is far from certain.

Breeding occurs most of the year, with overlapping generations. Typically the species increases in abundance with each successive generation, peaking in fall, at which point females become very dispersive and scatter widely across the landscape—presumably a "risk-spreading" overwintering strategy. In years with a strong migratory influx in late spring, however, the seasonal peak may occur early, and natural enemies then assert control and keep the population in check the rest of the season.

The Buckeye displays seasonal variation, which is dramatic

enough, but does not begin to approach what its relatives do in the tropics. Beginning in late August and continuing into November, many Buckeyes are large and richly colored. The ventral hindwing is suffused with deep purplish to brick red. This very striking form, informally called *J. c. "rosa,"* is also produced in fall in the East. The last emergents, in November, are very small and a plain clay color beneath. It is these that hibernate; "*rosa*" does not. Since the "*rosa*" form is not produced in spring, absolute photoperiod cannot be responsible. Rather inconclusive experimental studies suggest a complex photoperiod-temperature interaction is at work. In any case, the phenomenon is predictable. Beautiful examples of the "*rosa*" form, especially females, are usually abundant in the Suisun Marsh in October, visiting asters *(Aster)* and other fall flowers.

The Buckeye visits a great variety of flowers, ranging from small ones at ground level (lippias *[Lippia]*, heliotrope *[Heliotropium]*, clovers *[Trifolium]*) to large shrubs (it often swarms over coyote brush *[Baccharis pilularis]*, in fall—much preferring the male plants!). It has no special relationship with its namesake, the California buckeye tree *(Aesculus californica)*. Males are territorial, usually in hot, dusty places in full sunshine—often in paths or roads. They perch on the ground. In hot weather they keep the wings tightly closed over the back, minimizing the surface exposed to incoming solar radiation, but they still sit out there in the middle of the road. If you look closely, you will see they are poised on tiptoe.

The eyespots are believed to dazzle and distract predators. Up to 10 percent of Buckeyes may show wing damage or beak marks indicative of bird strikes at the eyespot area. Such damage seems to be most frequent in early winter and early spring, when alternative prey may be scarce.

As with the genus *Vanessa*, Buckeyes are prone to color and pattern aberrations, ranging from minor to very extreme. One of the more frequent ones makes the white forewing band nearly disappear, filled in with brown. It thus resembles a related subtropical species (that strays regularly into southern Arizona). Such aberrations were fairly frequent in our area in the 1970s but have seemingly disappeared. As in *Vanessa*, they can be induced by temperature shock, but their seasonality in the wild, as well as breeding experiments, suggests the phenomenon is at least partly genetic. The Buckeye has been used for extensive studies of pattern determination and development by H. Frederik Nijhout of Duke University.

DISTRIBUTION: Everywhere in open country, including cities. Absent from closed-canopy forest. Generally less common in fog belt.

HOST PLANTS: The larvae feed in the open on plants containing iridoid glycosides, a group of very bitter compounds (you can get a harmless taste by chewing on a leaf or stem of English plantain [*Plantago lanceolata*]). In our area these include lippias (*Lippia*, Verbenaceae); plantains (*Plantago*, Plantaginaceae; all species, but English plantain—bitterest—seemingly preferred); snapdragons (*Antirrhinum*), monkeyflowers (*Mimulus*), bee plants (*Scrophularia*), and fluellin (*Kickxia*), all in the figwort family (Scrophulariaceae), and probably others. As with the Painted Lady (*V. cardui*), this butterfly never oviposits on trees or shrubs, but it can be reared on chemically suitable tree foliage, including catalpa (*Catalpa*, Bignoniaceae) and princess tree (*Paulownia*, Scrophulariaceae). Interestingly, a subtropical sibling species of the Buckeye feeds only on mangroves, which are of course trees. The larva is black and white and spiny, with a bright orange head. It is probably warningly colored, but the pupa is cryptic (resembling a bird dropping) and presumably edible. We know experimentally that the adults are.

SEASONS: Recorded in all months, but mostly February to November, earlier in the Bay Area and foothill canyons, later in the Sacramento Valley and Delta, where it may not overwinter. Usually commonest everywhere in fall.

SIMILAR SPECIES: None in our area.

White Admirals and Allies (Limenitinae)

LORQUIN'S ADMIRAL   *Limenitis lorquini*
Pl. 24

Lorquin's Admiral, named for California's first known butterfly collector, is a quintessential riparian species. Except for strays, it is found exclusively along streams lined with willows (*Salix*). It is still common in much of the Bay Area but has disappeared from many of its old haunts in the southern Sacramento Valley. Coast Range specimens tend to have more orange on the lower surface than those from the valley, which are like most Sierran ones in having a lot of violet blue or bluish white below. Mountain specimens average larger than those from the lowlands.

Males are territorial, perching on or near willows but also flying back and forth along streamsides. They are avid puddlers. Both sexes visit flowers, including California buckeye *(Aesculus californica)*, yerba santas *(Eriodictyon)*, dogbanes *(Apocynum)*, milkweeds *(Asclepias)*, coyote mint *(Monardella)*, buttonbush *(Cephalanthus occidentalis)*, thistles *(Cirsium, Carduus, and Silybum)*, asters *(Aster)*, and goldenrods *(Solidago)*. Except in hot weather, this species often rests with the wings partially opened, displaying the upperside pattern.

This is the only member of its genus with orange on the forewing apex above. The resemblance to the California Sister *(Adelpha bredowii californica)*, with which it is largely sympatric, has long been seen as more than coincidental. The hypothesis that it was a mimic of that species was only tested in 2001 (and sustained; the Sister proved distasteful, the Admiral edible to captive jays). The flight seasons of the two species tend to match almost perfectly. At one long-term monitoring site in Solano County, they usually emerge in spring on the same day! We do not know the chemical basis for the unpalatability of the Sister. Recent work on the Viceroy *(L. archippus)*, the famous mimic of the Monarch *(Danaus plexippus)* and a relative of Lorquin's Admiral, has shown that it is significantly distasteful as well—meaning that relationship is more Mullerian than Batesian. The one published study of the Admiral/Sister case so far argues for its being Batesian.

The larva of Lorquin's Admiral resembles a bird dropping. When young it sits on the midrib of a willow leaf and eats it back from the tip. Fall larvae construct a hibernaculum from a leaf, which is tied to its branch so it does not fall off.

DISTRIBUTION: All counties; formerly common, now scarce and local in the southern Sacramento Valley.

HOST PLANTS: Willows *(Salix)*, mainly the glabrous (nonpubescent) species. Other hosts are recorded in the literature, but not locally.

SEASONS: Generally three broods, late March to October (rarely to December in the Bay Area).

SIMILAR SPECIES: The California Sister is usually larger, with broader wings. The orange on Lorquin's Admiral runs along the margin of the forewing, from the apical area at least halfway down. On the California Sister it forms a discrete roundish spot, edged outwardly with black brown and never touching the outer margin. Note that the white band on the Sister is substantially narrower than on Lorquin's Admiral, especially on the hindwing, where it tapers nearly to a point near the anal angle.

CALIFORNIA SISTER *Adelpha bredowii californica*
Pl. 24

This common butterfly of oak woodland is a member of a tropical group, and its regal flight, gaudy conspicuousness, and adult feeding preferences all bear the stamp of the tropics. It soars effortlessly back and forth along roads, flapping its wings as infrequently as possible—or so it seems. Both sexes puddle. Although it visits California buckeye *(Aesculus californica)*, thistles *(Cirsium, Carduus,* and *Silybum)*, coyote brush *(Baccharis pilularis)*, and so forth, it is more likely to come to flowing sap, bruised fruit, dung, roadkill, and other unethereal attractants. Marked adults often live a month and seem unafraid of predators; beak-marked individuals are seldom seen. As described for the Lorquin's Admiral *(Limenitis lorquini)*, it has now been shown to be distasteful to at least one bird species, and we have at least a prima facie case for a mimicry relationship between the two butterflies. They are perhaps easier for us to tell apart than they are for birds.

Males are territorial, perching on twigs and often holding the wings at least partially open (at puddles they are usually fully open). This is one of the least variable butterflies in California; despite the complexity of the wing pattern, aberrations are virtually unknown. Warningly colored species are often invariant; in our fauna the Monarch *(Danaus plexippus)* and the Pipevine Swallowtail *(Battus philenor)* are further examples. Presumably variants would not be recognized and would be sampled by predators. So they are selected against.

Females of the California Sister become very dispersive in late spring and in fall and are often seen far from home—in cities and in high mountains above the upper limit of host plants. This seems to be a regular seasonal phenomenon, but its significance is not understood.

DISTRIBUTION: General in oak woodland; less common in the Sacramento Valley. Occasionally seen even in central cities.

HOST PLANTS: Oaks, especially the various live oaks *(interior live oak [Quercus wislizenii], coast live oak [Q. agrifolia], golden oak [Q. chrysolepis])*.

SEASONS: Two or three broods, March to November.

SIMILAR SPECIES: See Lorquin's Admiral.

Meadow Browns and Satyrs (Satyrinae)

More than 2,000 species of meadow browns and satyrs are distributed over most of the world. The group is most species rich in grassland and steppe environments and is the dominant butterfly lineage from eastern Europe through central Asia, and in Patagonia. There have been radiations on both the Old and New World bamboos, but most species feed on herbaceous grasses and sedges. The adults have swollen veins at the wing bases, which have been shown to contain hearing organs. They tend to have relatively simple wing patterns involving rows of eyespots, though plenty of tropical species have gone off in wildly different directions, and some even participate in mimicry rings. Several genera adapted to cold climates appear to be undergoing very rapid evolution since the last glaciation. This is one of the two butterfly groups in which speciation by polyploidy (chromosome doubling) seems to occur. Tropical species often have pronounced seasonal polyphenisms (compare our California Ringlet *[Coenonympha tullia california]*).

The larvae tend to be cryptic green or brown (often changing from green to brown when in diapause), with lengthwise markings contributing to their crypsis, and often with red-tipped horns at both ends. The pupae are smooth, often green, and hanging upside-down, but in steppe and tundra climates they are brown and anchored in a rudimentary cocoon. Their flight is distinctive, with a sharp downstroke that thrusts them up, followed by a long drift downward on the upstroke. Among our species, only the Great Arctic *(Oeneis nevadensis)* is a very energetic flier, as advertised by the wing shape of the male. Sexual dimorphism is usually weak in this group, but females are consistently larger and carry the male when mating.

CALIFORNIA RINGLET *Coenonympha tullia california*
Pl. 25

The California Ringlet is our representative of a species or species complex that reaches completely around the Northern Hemisphere, mostly in higher latitudes. Our subspecies is very different in appearance and life history from most others, including the one directly across the Sierra Nevada, but molecular genetics suggests all are a single biological species. In North America, the only other subspecies approaching ours in coloration is the Siskiyou Ringlet *(C. t. eryngii)* of western Oregon and northwestern California. There are populations of the *C. tullia* complex in Eurasia with somewhat similar coloring, but these are almost certainly convergent, not closely related.

The California Ringlet is still abundant in many areas of the Coast Ranges and Bay Area in oak woodland and grassland but has been in decline in the Sacramento Valley for years for reasons that we do not understand.

First-brood adults—produced from overwintered (third-instar) larvae—are pearly or even silvery gray above and heavily infuscated with blackish brown on the ventral hindwing (this is probably thermoregulatory melanism). They emerge in early spring and breed immediately. Second-brood adults are very pale reddish buff (exactly the color of dry grass) with little or no darkening of the ventral hindwing. The actual factors controlling the seasonal polyphenism are unknown. Summer adults are physiologically quite different, as well. They undergo estivation with reproductive diapause for as long as 2 to 3 months and do not begin laying eggs until August or even early September. While estivating they remain quiescent, preferably in shade, but can be kicked up; they may fly around on cool afternoons. Late-summer eggs hatch quickly, but the larvae usually have nothing to eat and remain inactive until new growth begins. They then feed, reach the third instar, and again become quiescent until early spring. Neither generation lays its eggs preferentially on grasses; the larvae are left to forage for themselves.

The adults don't seem to do very much. They don't puddle, are not territorial (though male-male chases can be seen, generally rather desultory ones!), and are not very avid flower visitors either—though they seem broad in their floral tastes (fiddlenecks *[Amsinckia]*, yellow star-thistle *[Centaurea solstitialis]*, milk thistle *[Silybum marianum]*, yerba santas *[Eriodictyon]*, asters *[Aster]*, goldenrods *[Solidago]*…). Mostly, they either sit still or fly with an odd undulating motion among tall grasses.

DISTRIBUTION: All counties, but declining in the Sacramento Valley and around urban and suburban areas generally.

HOST PLANTS: Grasses; field preferences are unknown. This species can be reared on many grasses, including some weedy Mediterranean annuals, which it almost certainly does not use in nature. It can breed in the wild as well as in the lab on Bermuda grass *(Cynodon dactylon)*, but that ability has not allowed it to enter the urban fauna. Seemingly associated with both perennial bunchgrasses and stoloniferous or rhizomatous species.

SEASONS: Recorded from January (unusual) to November; usually a break between the two generations, in late spring.

SIMILAR SPECIES: None in our fauna. The pale coloration of the spring brood may lead to it being mistaken for a pierid, but the flight is very different.

OX-EYED SATYR or COMMON WOOD NYMPH
Cercyonis pegala boopis

Pl. 25

The name *boopis* literally means "ox-eyed," and it is apt. The type-locality of this subspecies has been fixed retroactively as Point Richmond, Contra Costa County, thereby preventing a potentially ghastly taxonomic wrangle in a group where the limits of subspecies are not well defined. Point Richmond seems to be where the pioneer Bay Area lepidopterist H.H. Behr found it commonly. Its distribution, while encompassing all the Bay Area counties, is quite spotty, and it is only locally common. It occurs mainly in grassland, also in oak woodland, but almost always near water—often along watercourses that may dry up in summer. It has a very distinctive arcing flight, bobbing along and giving the impression that its wings are almost always together over its back. It twists in and out of shrubs, including bramble tangles, and if pursued often goes into the middle of a shrub and sits there, out of reach. I have watched Rheas (the bird) on the Patagonia steppe, and their movements remind me a great deal of this butterfly.

Females often have a vaguely yellowish cloud on the outer third of the forewing upperside, surrounding the two large eyespots.

They are rarely observed nectaring (dogbanes *[Apocynum]*, coyote mint *[Monardella]*, yellow star-thistle *[Centaurea solstitialis]*) or chasing anything; they are rather torpid beasts...until you try to catch them, that is.

DISTRIBUTION: All Bay Area counties, usually along streams or sloughs or in relatively moist, open grassland or brush. Present locally in parts of the Delta but absent in the southern Sacramento Valley; occasional in the north, mainly following blue oak woodland down from the foothills.

HOST PLANTS: Presumably perennial grasses; eggs are laid haphazardly on living and dry grass and elsewhere; the newly hatched larva immediately enters diapause until spring. It thus may not eat for some 6 months! Once it does begin feeding, it is nocturnal and grows rather slowly. It remains in the pupa (at "room temperature") for an agonizingly slow 3 weeks.

SEASONS: One brood, males as early as May, females often not seen until June with some still flying in September or even October. Despite the extended female flight period, they have not been shown to estivate.

SIMILAR SPECIES: The Great Basin Wood Nymph *(C. sthenele)* is consistently 20 percent smaller. On males, the lower eyespot on the underside of the forewing is almost always smaller than the upper in Great Basin; they are variable but usually nearly equal in the Ox-eyed Satyr. They are usually nearly equal in females of both. Also on the ventral forewing, the dark postmedian line (defining the basal edge of the pale zone around the eyespots) reaches the top of the upper eyespot in the Ox-eyed Satyr but usually fades out below that in the Great Basin Wood Nymph. These are weak and "statistical" characters, but in practice one hardly ever has any trouble telling the two species apart in our area.

GREAT BASIN WOOD NYMPH or WOODLAND SATYR *Cercyonis sthenele*

Pl. 25

The name "Great Basin Wood Nymph," currently in favor, is not exactly informative for our populations, which belong to subspecies *C. s. behrii* and should probably be called Behr's Satyr. It is smaller than the Ox-eyed Satyr *(C. pegala boopis)* and has a somewhat grayish cast. The characters usually cited for distinguishing the two are enumerated in the species account for the Ox-eyed Satyr, but usually they are no trouble. Behr's Satyr is found in grassland, coastal scrub, and oak woodland, arcing above and bobbing within tall grass. It is not particularly associated with water. It is a more avid flower visitor than the Ox-eyed, coming to various legumes (Fabaceae), milkweeds *(Asclepias)*, dogbanes *(Apocynum),* and sunflower family members (Asteraceae). One of its favorites, in my experience, is the common white buckwheat *(Eriogonum nudum).*

The first California butterfly known to go extinct—apparently due to habitat loss—was the endemic, nominate subspecies, the Sthenele Satyr *(C. s. sthenele),* which was confined to what is now the City and County of San Francisco. It was apparently done in by development before 1880. Many specimens were subsequently lost in the 1906 earthquake and fire, so it is rare in collections today. It differed from Behr's Satyr primarily in having a distinct white edge to the median band on the ventral hindwing. Some Marin County populations show a little of this today, and the endemic population on Santa Cruz Island, one of the California Channel Islands off Santa Barbara, is quite similar and was recently named subspecies *C. s. hypoleuca.*

DISTRIBUTION: Bay Area counties except San Francisco (today) and San Mateo; known in Solano County only as a stray. Absent from the Sacramento Valley, though the Sierran subspecies *C. s. silvestris,* which is much darker below, may stray into Blue Oak woodland northward and along the eastern edge.

HOST PLANTS: Grasses, but field preferences unknown. Eggs are laid on grass stems. The newly hatched larvae enter diapause until spring. The larva is largely nocturnal.

SEASONS: One brood, the males typically emerging a week before the females; May to early September.

SIMILAR SPECIES: See Ox-eyed Satyr. There is a third species in this genus, the Sagebrush Satyr *(C. oetus),* which does not enter the area covered by this book.

GREAT ARCTIC   *Oeneis nevadensis*
Pl. 25

The genus *Oeneis* is known as the "arctics" for good reason. Species of arctics fly on the shores of the Arctic Ocean, atop Mount Washington, New Hampshire, and widely in the most inhospitably cold places in the Northern Hemisphere. One species group of the genus, however, is not at all "arctic" in its preferences, and our species belongs to that group. It is also sometimes called the Nevada Arctic, which is doubly misleading since the species does not occur in Nevada! Its favorite habitat is moist Douglas fir forest, where it occurs along roads and in clearings, typically sitting on downed logs, occasionally on a standing tree. At rest it is absolutely invisible, and in sunlight adroit at posing so as to minimize or conceal any shadow. When it flies it provides a brief flash of butterscotch color and then disappears again. The males have a large, conspicuous olivaceous stigma (pheromone gland) on the forewing upperside. They are territorial and prone to prolonged chases. Females have much more conspicuous eyespots, generally of equal size, and some individuals are very large. There is no other butterfly in our area that can be confused with this. Our only Bay Area population is around Plantation, Sonoma County, at the very southern limit of the range near the coast. Inland, the species reaches Mendocino Pass (Glenn County) and the Feather River Canyon in the Sierra Nevada.

Adults rarely visit flowers. In the Feather River country I have seen them at self-heal *(Prunella vulgaris)*, ox-eye daisy *(Chrysanthemum leucanthemum)*, and black-eyed Susan *(Rudbeckia)*.

The life history of this beast is very strange. Most arctics take two full years to complete their metamorphosis, diapausing once as a second- or third-instar larva and again as a full-grown one. This makes sense in the very harsh climates favored by most of the species, where growth seasons may be only a few weeks long and freezes can occur at any time. It does not make sense in the climate where the Great Arctic lives, but that's what it does anyway.

But the strangeness doesn't end there. Quite a few boreal and alpine lepidopterans—both butterflies and moths—have 2-year cycles, and as a rule each species is at least regionally synchronized, such that adults fly *only every other year*. This is known as semivoltinism. It does not appear to be a response to competition, because where several such species co-occur, they are usually *all* synchronized. One would think that years of abnormally light or heavy snowfall would lead to desynchronization, so that given enough time the species would fly every year (though in reality it might be behaving as two semivoltine populations, temporally isolated and potentially free to diverge genetically). Yet this rarely happens. It has happened with our Great Arctic, though. There are records at Plantation in both odd and even years, but it is much commoner in the odd years. In other places it flies only in the even years. In Oregon and northwestern California most populations are even, a few are odd, and fewer still have flights in both. Why be semivoltine? It has been suggested that this life cycle makes it difficult for parasitoids to track their hosts. But why can't parasitoids become semivoltine, too?

The Sonoma County population is considered to belong to subspecies *O. n. iduna,* originally described from Mendocino County, which is slightly paler than the nominate subspecies. To observe the Great Arctic, drive north up the Coast Highway! Just be sure it's the right year.

DISTRIBUTION: Sonoma County and northward.

HOST PLANTS: Grasses; preferences in the wild unknown.

SEASONS: Mid-May to mid-July, one brood, mainly every other year.

SIMILAR SPECIES: None.

Milkweed Butterflies (Danainae)

MONARCH   *Danaus plexippus*
Pl. 24

Everyone knows the Monarch, or thinks he or she does. I have heard California Tortoiseshells *(Nymphalis californica)*, Red Admirals *(Vanessa atalanta)*, and even Western Tiger Swallowtails *(Papilio rutulus)* called Monarchs! Note that the Monarch is a large orange butterfly with black veins and white spots in the black marginal border. Nothing else in our area really looks like it. The male is brighter orange and usually larger than the female and has a conspicuous swelling on one of the veins on the hindwing upperside; this is a pheromone gland. The female is browner and more suffused with black, and the black vein lines are heavier.

The very conspicuous black, yellow, and white caterpillar of the Monarch, with its pair of black filaments at both ends, feeds on milkweeds *(Asclepias* and related exotic genera, sometimes in gardens). These plants produce toxic cardenolides, which are taken up and stored for the insect's defense. We now know an enormous amount about Monarch biology, largely through the efforts of Dr. Lincoln P. Brower and his associates. We know, for example, that different milkweeds differ greatly in their cardenolide content and that the toxicity and unpalatability of an individual Monarch depend on what it itself ate. We also know that the foulest-tasting cardenolides (they are, overall, very bitter) are deposited in the wings, so that a bird holding a Monarch in its beak by the wings will detect them at once and be likely to let it go. The most toxic cardenolides are stored in the abdomen and are only ingested when the predator has torn the butterfly to shreds. Many minutes later the predator is likely to feel ill and then vomit. In classical stimulus-response learning theory, it was postulated that such a delayed reaction would not be associated with its cause; in theory, then, Monarch avoidance could not be learned. But it turns out (thanks to the pioneering studies of John Garcia) that in fact when novel foods have delayed, nasty consequences, the association is made—perhaps by all vertebrates. This adaptive, hard-wired association is now called "food-aversion learning," and it makes the whole Monarch story work.

The Monarch is also famous for its long-distance seasonal migrations. Most North American Monarchs overwinter in the Transvolcanic Range near Mexico City. Ours do not. Monarchs from the western Great Basin and West Coast spend the winter along the California coast, from just north of the Bay (not all years) to Santa Barbara. In our area overwintering is attempted as far inland as Fairfield, Solano County, but is not usually successful. The best-known overwintering sites are on the central coast, and Pacific Grove has made Monarch overwintering the centerpiece of its tourism industry. It is very strange that there are no historic records of Monarch overwintering on the coast before the twentieth century. It has even been suggested that the species colonized California quite recently and that its migratory pattern here is also very recently evolved. There seem to be no significant molecular genetic differences between California Monarchs and eastern ones. Nonetheless, concern has been voiced that the release of Monarchs at outdoor events (weddings and such), now a popular practice, could pose a threat to the genetic integrity of our populations.

Coastward movement begins as early as late July some years and becomes pronounced by August, when Monarchs can be seen crossing the high Sierra from east to west. Breeding, however, commonly continues into October. The wintering clusters are in place along the coast (typically in groves of eucalyptus *[Eucalyptus]* or Monterey pine *[P. radiata]*) by late October and typically break up in late February. The overwintering adults feed opportunistically, drink water, and fly around on sunny, warm winter days. Just before the roost dissolves, there is a several-day orgy of mating. Both sexes disperse inland, reaching Lake Berryessa and the Davis-Sacramento area usually in early to mid-March. Usually the foothill purple milkweed *(Asclepias cordifolia)* is already in good condition when the butterflies arrive, but the common valley milkweeds (narrow-leaf milkweed *[A. fascicularis]* and broad-leaf milkweed *[A. speciosa]*) may not have broken ground yet. The females seem to "know" where they will appear, however, and may linger there until they do. In late years they may retreat westward to the Coast Range to oviposit. In some years, for reasons we do not understand, they may continue eastward and breed almost entirely on the east side of the Sierra Nevada.

In any case, numbers of Monarchs fluctuate dramatically from year to year for reasons that seem related to disease. They tend to not breed repeatedly in the same location, which may be an antiparasitoid tactic. They tend to avoid ovipositing on plants already occupied by the orange Oleander Aphid *(Aphis nerii)* or by the gregarious, brilliant blue green leaf beetle *Chrysochus cobaltinus.* There is no evidence that their probability of oviposition on a particular milkweed is related to its cardenolide content. Monarchs are, overall, much less common in our area than in the East and especially the Midwest.

The Viceroy (*Limenitis archippus,* Nymphalidae), the famous mimic of the Monarch, does not occur in our area. As with the Pipevine Swallowtail *(Battus philenor)*, the lack of mimics suggests the species may not have been here long enough for any to evolve.

Monarchs are avid flower visitors, particularly to their beloved milkweeds, but also to thistles *(Cirsium, Carduus,* and *Silybum)*, goldenrods *(Solidago)*, and a variety of garden flowers including zinnias *(Zinnia)*, marigolds *(Tagetes)*, and butterfly bush *(Buddleia)*. Males are not territorial and do not puddle. Females may drive off other females from clumps of the host plant—a most unusual behavior in butterflies.

To get involved in Monarch-related activities, check out www.monarchwatch.org!

DISTRIBUTION: All counties, general in open country and along roadsides in forest.

HOST PLANTS: Milkweeds *(Asclepias)* and exotic members of the milkweed family (Asclepiadaceae) in gardens.

SEASONS: Inland, March to December, several generations but rarely in the same place. Coastal wintering clusters October to late February. Massive concentrations of fall migrants, familiar in the eastern United States, are seldom observed here; most likely in October at Lagoon Valley and the Suisun Marsh.

SIMILAR SPECIES: The Queen *(D. gilippus)* is a very rare stray in our area.

QUEEN   *Danaus gilippus*
Not illustrated

Breeding in southern California, this subtropical species has been recorded once as a stray in Contra Costa County. It is likely to show up in our area again. It is slightly smaller than the Monarch *(D. plexippus)*, deep brown red above with black borders and white spots; the veins are black lined only on the underside of the hindwing. It is also a milkweed *(Asclepias)* feeder; the larva is marked similarly to the Monarch but has a third pair of black filaments on its back about one-third of the way aft.

PLATES

PLATE 1   **Swallowtails**

Large butterflies with *tails on hindwings.*

Pipevine Swallowtail *(Battus philenor)*   PAGE 88

Black above and below. Male with iridescent blue sheen on hindwing above; female vaguely purplish, with white spots. Underside of hindwing with blue iridescence and large orange, white-edged spots in both sexes. Common, mostly in riparian habitats, spring to fall.

Indra Swallowtail *(Papilio indra)*   PAGE 95

Small, *mostly black, all wings crossed by yellowish white spot-band* on both surfaces, sexes similar, *tails short.* Rocky, mostly serpentine-soil areas. Spring only.

Anise Swallowtail *(Papilio zelicaon)*   PAGE 91

Mostly yellow, with checkered, not striped pattern. Spring brood small, pale; summer broods often large, richly colored. Blue spots most prominent in spring (both sexes) and on summer females. Many habitats; common and widespread. Multiple broods, spring to fall, except only in spring in serpentine-soil areas.

Pipevine Swallowtail

underwing

male

female

Indra Swallowtail

underwing

Anise Swallowtail

underwing

PLATE 2   **Tiger-striped Swallowtails**

Large to very large butterflies with *striped patterns* and *tails on hindwings*.

Western Tiger Swallowtail *(Papilio rutulus)*   PAGE 96

Bright yellow above; black pattern more extensive than in Two-tailed Swallowtail. Females generally larger than males, with more blue above. Spring brood smaller and paler yellow than summer brood. Common and widespread, even in cities. Two to three broods.

Two-tailed Swallowtail *(Papilio multicaudatus)*   PAGE 97

Similar to Western Tiger Swallowtail, but *black pattern less extensive, stripes narrow,* creating an impression of a much *yellower* insect. Averages *larger* than Western Tiger, especially in summer. Note *prominent second and third tails,* but these may not be evident in flight. Usually uncommon, in riparian canyon habitats mainly inland; tends to stay high in trees. Two to three broods.

Pale Swallowtail *(Papilio eurymedon)*   PAGE 98

Wings rather *long and narrow,* the male forewing very triangular; ground color *white* in males, *pale yellow* in females; *black pattern more extensive* than in other species. *Conspicuous blue and orange near anal angle of hindwing.* Common in chaparral and canyons. Not found in Sacramento Valley or (usually) urban areas. One brood, spring to early summer.

Western Tiger Swallowtail

male

underwing

Two-tailed Swallowtail

underwing

Pale Swallowtail

male

underwing

female

PLATE 3 Atypical Mostly White Butterflies

Clodius Parnassian *(Parnassius clodius)* PAGE 86

Large white butterfly with rounded wings, the female semitranslucent with more extensive black pattern and larger red spots. Male body furry; female abdomen glossy black above, almost always with waxy pinkish white vaginal plug (sphragis) attached on lower surface as shown. Cool, moist forest north of the Bay; extinct in Santa Cruz Mountains since 1958. One brood, late spring.

Pine White *(Neophasia menapia)* PAGE 102

Male snowy white with crisp black pattern on forewing upper side. Female grayish to yellowish white with heavy dark pattern including vein lines and *pink along hindwing edges above and below.* Conifer forests north of the Bay; one brood in summer.

Clodius Parnassian

male

female underwing
(note sphragis)

female

Pine White

female

male

female
underwing

male
underwing

PLATE 4 **Checkered Whites**

Medium-sized mainly white butterflies with *checkered dark patterns,* more extensive in females. Often highly variable sexually and seasonally.

Checkered White *(Pontia protodice)* PAGE 104

Male much more lightly marked above than female. Spring brood small, lightly patterned above but with heavy black basal scaling on all wings and *heavy brownish or greenish scaling on veins of hindwing beneath.* Summer broods larger with more extensive pattern above in female; note *large squarish black spot* near hind margin. Summer male nearly immaculate white beneath; female with yellow on hindwing veins. Formerly common but rare in recent years; open grassland, disturbed habitats, all year, but now seen mostly in September and October.

Western White *(Pontia occidentalis)* PAGE 106

Very similar to Checkered White. Ground color chalkier; black pattern more extensive and contrasty, especially in summer. Spring brood very dark beneath with vein lines *strongly greenish;* autumn specimens often similar but not as dark. Summer broods usually with *some dark vein scaling in both sexes.* Rare in our area; strays in North Coast Range and to Sacramento Valley in autumn. Replaces Checkered White northward and at high elevations.

Spring White *(Pontia sisymbrii)* PAGE 103

Male white; female often yellowish. Black spot at end of forewing cell above *dumbbell shaped,* especially in male. Hindwing underside with vein lines always present but *broken by ground color in middle of wing.* Rocky areas and chaparral, especially on serpentine soils. One brood in spring.

Checkered White

spring male

summer male

underwing

underwing

spring female

summer female

underwing

Western White

spring male

summer male

underwing

underwing

summer female

underwing

Spring White

female

underwing

male

PLATE 5   C a b b a g e   a n d   G r a y - v e i n e d   W h i t e s

Medium-sized mainly white butterflies with *black patterns not appearing checkered. No black spot* at end of forewing cell above.

Cabbage White *(Pieris rapae)*   PAGE 108

Spring brood lightly marked above but with heavy black basal shading. *Underside of hindwing more or less suffused with gray, not heavier along veins.* Summer broods usually larger, brighter, upperside pattern crisper and more extensive, underside of hindwing *plain white or pale yellow* with few black scales. Females may be slightly buffy in all broods. Ubiquitous and abundant, often in cities. All year.

Gray-veined White (*Pieris "napi* complex")   PAGE 107

Spring brood with *dark scaling on veins of hindwing beneath* (and occasionally on upper surfaces of all wings in females). Later brood(s) *much paler,* often *immaculate white or pale yellow beneath.* Coastal fog-belt populations (*P. n. venosa* shown with an asterisk) larger and more boldly marked than inland ones *(P. n. microstriata).* Cool, moist forests near the coast and in canyons inland. One or two (rarely three) broods, late winter to late spring.

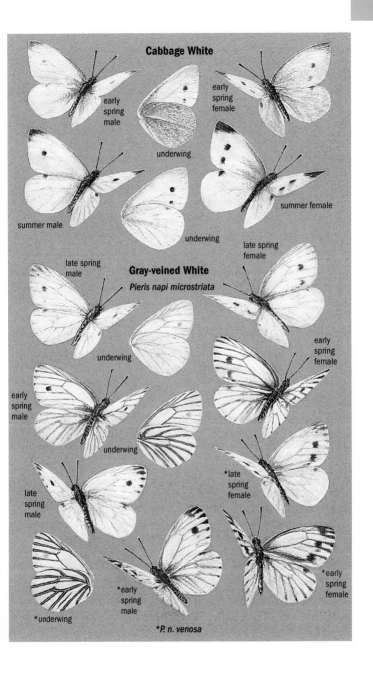

Cabbage White

early spring male

early spring female

underwing

summer male

underwing

summer female

Gray-veined White

Pieris napi microstriata

late spring male

late spring female

underwing

early spring female

early spring male

underwing

late spring male

*late spring female

*underwing

*early spring male

*early spring female

*P. n. venosa

PLATE 6 **M a r b l e s   a n d   O r a n g e t i p s**

Medium-sized white (rarely yellow or yellowish) butterflies with *green to purplish gray mottling on hindwing undersides*. Fly in spring to early summer only.

Sara Orange-tip *(Anthocharis sara)*   PAGE 112

Male with forewing border solid black; *orange spot large, UV reflective*. Female with border interrupted by spots of ground color; *orange spot smaller, UV absorbing, usually yellower*. Female ground color occasionally yellow. Second brood larger, less black above, mottling on hindwing underside reduced. Widespread and common except in Sacramento Valley. Spring to early summer.

Boisduval's Marble *(Anthocharis lanceolata)*   PAGE 113

Sexes similar, *plain white with dark spot at end of cell of forewing* and vague markings near apex, mainly in female. *Forewing pointed*. Hindwing underside finely mottled in grayish purple. Local and uncommon, North Bay inland. One brood in spring.

Small Marble *(Euchloe hyantis)*   PAGE 112

Always white above, sexes similar, female slightly broader winged. *Note large, more or less squarish black spot* at end of cell on forewing above; apical white spots *all of similar size;* heavy green mottling beneath with little or no yellow along veins; ground color *pearly above*. Local in bare, rocky sites, mostly on serpentine soils. Early spring.

Large Marble *(Euchloe ausonides)*   PAGE 110

Averages larger than Small Marble; male forewing more pointed. Note *thinner black spot at end of cell,* often in two parts as shown on male; *uppermost white apical spot much larger than others, rounded;* less intense green mottling, usually with *conspicuous yellow along veins;* ground color *matte white, not at all pearly,* female with broader and more diffuse apical markings than male; hindwing often tinged above with light orange (rarely all wings orange in second brood). Grassland, open country generally; often associated with weedy mustards; formerly common but rare inland in recent years. Normally does not occur together with Small Marble. One or two broods in spring to early summer.

Sara Orange-tip

spring male

spring female

underwing

summer underwing

summer male (second brood)

yellow female

Boisduval's Marble

male

male

female underwing

female

Small Marble

male

underwing

underwing

females

Large Marble

PLATE 7 Sulphurs

Medium-sized orange, yellow, or white butterflies with more or less black border above. An orange or yellow spot in center of hindwing above; beneath, this spot *silvery with purplish rim*. Males with border above solid black or slightly interrupted by yellow along veins. Females with border containing spots of ground color, sometimes fusing together.

Orange Sulphur *(Colias eurytheme)*   PAGE 114

Hypervariable. Summer forms large and richly colored, with bright hindwing undersides and pointed forewing in males. Cold-season forms small, the orange reduced to patches, the borders narrow, and the hindwing underside *heavily clouded with greenish gray,* silver spot on hindwing beneath usually *double rimmed.* Up to 50 percent of females white but otherwise identical to others. Abundant and generally distributed. Many broods, all year.

Western Sulphur   PAGE 117
(Colias occidentalis chrysomelas)

Bright *golden yellow* above (pale females rare); consistently larger than Orange Sulphur. Underside of hindwing warm orange ochre; silver spot with *one narrow purple rim.* Female border sometimes much reduced. Local in moist forests and forest edges north of the Bay. One brood, early summer.

Orange Sulphur

summer male

early spring male

summer female

underwing

underwing

white female

underwing

early spring
female

underwing

underwing

Western Sulphur

male

female

underwing

underwing

PLATE 8 Uncommon Sulphurs

California Dogface *(Zerene eurydice)*   PAGE 118

Male with unique forewing pattern and strong near-UV sheen. Male hindwing and female plain yellow. Note *sharply pointed forewing apex* and *large, round black spot* at end of forewing cell in female. Riparian habitats in canyons and Sacramento Valley. Most of year, irregular.

Cloudless Sulphur *(Phoebis sennae marcellina)* PAGE 119

Male *plain yellow, unmarked or nearly so.* Female ground color highly variable, often whitish or light orange, with dark spot at end of cell and variable spot-pattern as shown. Wing shape unlike other sulphurs. Sporadic stray from the south.

Dainty Sulphur *(Nathalis iole)*   PAGE 120

Tiny; unusual wing shape. Underside of hindwing variably clouded with greenish gray. Sporadic stray from the south.

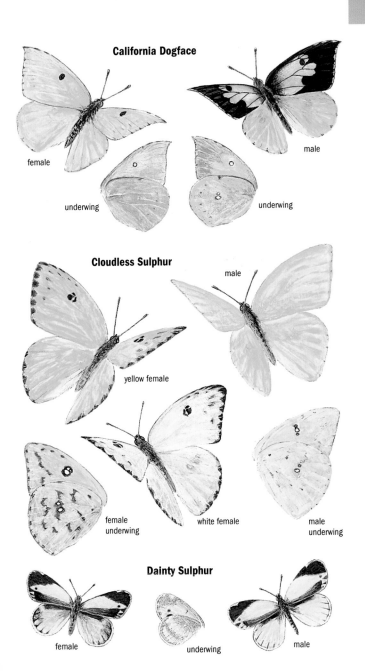

California Dogface

female

underwing

male

underwing

Cloudless Sulphur

male

yellow female

female
underwing

white female

male
underwing

Dainty Sulphur

female

underwing

male

PLATE 9 **Coppers**

Great Copper *(Lycaena xanthoides)*   PAGE 123

Our largest copper. Male *plain, even dark gray* above with a little orange near hindwing margin. Female variable, from mostly gray to mostly orange, with more complete spot-pattern. Underside with complex spot-pattern; size of spots quite variable. Marshes, riparian habitat, grassland, including Sacramento Valley. One brood, late spring to early summer.

Gorgon Copper *(Lycaena gorgon)*   PAGE 124

Male coppery red above. Female *straw yellow,* with *strongly checkered pattern above.* Underside of forewing tinged with orange, hindwing grayer, with sharply defined pattern of black spots, some of which may be *triangular.* Red spots strongly triangular. Mostly rocky habitats and chaparral; absent in Sacramento Valley. One brood, early summer.

Tailed Copper *(Lycaena arota)*   PAGE 123

Our only copper with *tails.* Male plain coppery red above; female orange with black strongly checkered pattern. Underside pattern complex, with curved black lines as well as spots and a *well-marked white band* basad of the marginal spots. Sunny spots in forest. Not in Sacramento Valley. One brood, summer.

Purplish Copper *(Lycaena helloides)*   PAGE 126

Small; male *strongly purplish* with line of submarginal red lunules above; female *mostly orange* with usual copper pattern. Underside gray brown (male) to orange tinted (female) with well-defined submarginal red line. Widespread; less common than formerly; especially frequent in marshes. Multiple broods, most of season; commonest in fall.

bright female

male

dull female

Great Copper

underwing

female

male

Gorgon Copper

underwing

female

male

Tailed Copper

underwing

female

male

underwing

Purplish Copper

PLATE 10   Hairstreaks

Mostly small butterflies with tails and (usually) a postmedian line crossing all wings.

Great Purple Hairstreak (Atlides halesus)   PAGE 128

Male *brilliant iridescent blue* above; female with outer half of wings mostly black. Underside *black* with red spots at base of hindwing and iridescent gold green spots near tails. Underside of abdomen *orange.* Our largest hairstreak. Riparian habitats, oak woodland, and older urban/suburban neighborhoods. Multiple broods, most of year.

Common Hairstreak (Strymon melinus)   PAGE 142

Slate gray above with large, bright red spot (sometimes two) at base of tail. Underside pearly gray with *postmedian line across all wings,* more or less edged with red. Common and general, mostly in disturbed habitats. All year.

Golden Hairstreak (Habrodais grunus)   PAGE 127

Plain golden brown above with little or no pattern. Underside *golden yellow* with postmedian line and small submarginal hindwing spots. Common around golden oak; absent elsewhere. One brood, summer.

"Inland" Bramble Hairstreak   PAGE 134
(*Callophrys "dumetorum complex"*)

Male grayish, female brown above. Hindwing underside *bright green* with faint white markings. Widespread inland, especially in chaparral. One or two broods, spring.

"Coastal" Bramble Hairstreak   PAGE 135
(*Callophrys "dumetorum complex"*)

Extremely similar to "Inland" Bramble Hairstreak, male darker gray above, hindwing underside *intense blue green,* often with more white pattern. Coastal fog belt only. Spring to early summer.

Great Purple Hairstreak

male

female

underwing

male

Golden Hairstreak

female

Common Hairstreak

female underwing

underwing

"Inland" Bramble Hairstreak

male

underwing

coastal underwing

"Coastal" Bramble Hairstreak

female

male

PLATE 11 Hairstreaks

Small butterflies, nearly unpatterned above; below with a *blue spot at base of tail*. Males with stigma along costal edge of forewing.

California Hairstreak *(Satyrium californicum)* PAGE 129

Male gray brown above with orange spots near tail. Female variable, mostly brown to mostly orange, with submarginal orange usually on all wings above. Underside gray with *sharply defined black spots* and a *usually complete or nearly complete row of red submarginal spots* on hindwing; blue spot may be capped with red. Often abundant where oaks are common. Now very rare in Sacramento Valley. One brood, late spring to early summer.

Sylvan Hairstreak *(Satyrium sylvinum)* PAGE 131

Very similar to California Hairstreak but underside paler, ash gray to nearly white; *submarginal pattern weaker, with little red;* postmedian spots often *much reduced or even absent;* variable. Formerly abundant, now declining, with willows in riparian habitats throughout. One brood, early summer.

Tailless Sylvan Hairstreak *(Satyrium dryope)* PAGE 131

Very similar to Sylvan Hairstreak but *tailless;* very pale below, usually with *much-reduced spot-pattern.* With willows in marshes and riparian woodland surrounded by arid grassland; Livermore Valley and Santa Clara County.

Gold-hunter's Hairstreak *(Satyrium auretorum)* PAGE 132

Rich brown above, lighter brown below, with postmedian line not strongly contrasting. No white frosting beneath. Sometimes common in oak woodland inland; nearly extinct in Sacramento Valley. One brood, late spring to early summer.

Mountain-mahogany Hairstreak PAGE 133
(Satyrium tetra)

Nearly tailless; charcoal gray above, lighter brownish gray below with *pronounced white frosting on hindwing;* postmedian line edged with white, mildly contrasting. Female substantially larger and more broad winged than male. Chaparral and rocky canyons. One brood, early summer.

Hedgerow Hairstreak *(Satyrium saepium)* PAGE 133

Coppery red above without pattern. Stigma black. Beneath, all wings *brown* with sharply defined postmedian line. Common in chaparral and coastal scrub. One brood, summer.

California Hairstreak

male

female

underwing

Sylvan Hairstreak

female

female underwing

Tailless Sylvan Hairstreak

underwing

Gold-hunter's Hairstreak

male

female underwing

male

Mountain-mahogany Hairstreak

female underwing

Hedgerow Hairstreak

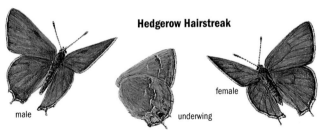

male

underwing

female

PLATE 12   Hairstreaks and Elfins

Small butterflies, nearly unpatterned above; tailed (hairstreaks) or not (elfins). Males with stigma along costal edge of forewing.

Thicket Hairstreak *(Mitoura spinetorum)*   PAGE 135

Steel blue above. Underside *rich chocolate brown* with *sharply contrasting white postmedian line.* Scarce, mostly in inland Coast Range, especially on serpentine soils. Several broods, all season.

Nelson's Hairstreak *(Mitoura gryneus nelsoni)* PAGE 138

Very similar to John Muir's Hairstreak but basal half of hindwing below *not significantly darker than distal half.* Coast Range north of the Bay, with incense cedar; one brood in spring.

Johnson's Hairstreak *(Mitoura johnsoni)*   PAGE 136

Rich brown above and below with *sharply contrasting white postmedian line.* Larger than *Satyrium* hairstreaks. Usually rare; Inner Coast Range. Several broods, most often seen in spring.

John Muir's Hairstreak *(Mitoura gryneus muiri)* PAGE 137

Dark brown (male) or more or less strongly orange-washed (female). Underside with basal half of hindwing very dark, *blackish;* area beyond the postmedian line *washed with lilac,* with several blue spots near tail. Mostly on serpentine soils with cypresses; rarely with California juniper in hot, dry grassland-steppe. One brood, late winter to early spring.

Moss's Elfin *(Incisalia mossii)*   PAGE 140

Several subspecies, differing in detail. *Fringes checkered;* outer (distal) half of hindwing below with *more or less gray frosting,* a usually distinct row of dark submarginal spots, and a fine white line along the wing margins. Around the Bay Area in local colonies associated with stonecrops growing on cool (usually north-facing) cliffs. The San Bruno Elfin is federally protected. One brood, late winter to spring depending on location.

Western Brown Elfin   PAGE 139
(Incisalia augustinus iroides)

Fringes *at most weakly checkered;* coppery red brown below, the basal half of the hindwing darker than the distal but with *no trace of gray frosting.* Common and widespread, except rare in Sacramento Valley. One or two broods in late winter to late spring. Many habitats.

Western Pine Elfin *(Incisalia eryphon)*   PAGE 141

Fringes *strongly checkered.* Hindwing underside with complex "Oriental rug" pattern. With native and planted two- and three-needle pines around Bay Area. One brood, spring.

Thicket Hairstreak

underwing

male

Nelson's Hairstreak

underwing

Johnson's Hairstreak

male

underwing

John Muir's Hairstreak

male

female

underwing

Moss's Elfin

female

underwing

Western Brown Elfin

underwing

Western Pine Elfin

female

underwing

male

PLATE 13   Blues

Small butterflies, predominantly blue above (at least in males), with complex spot-patterns beneath; occasionally with tails.

Marine Blue *(Leptotes marina)*   PAGE 150

Male violet blue above; female blue basally and purplish brown distally. Underside with *unique wavy-line pattern. No orange* in either sex. Sporadic stray from the south, occasionally breeding.

Western Pygmy Blue *(Brephidium exile)*   PAGE 152

Tiny; mostly brown above with more or less blue basally; checkered fringes; underside with complex pattern including a row of *iridescent marginal spots.* Often abundant, waste ground and saline and alkaline habitats. Nearly all year, but in most places seen only in second half of season.

Western Tailed Blue *(Everes amyntula)*   PAGE 146

Tailed; male rich blue above, female gray brown usually with much blue; underside ash gray, spots rather small. Cool, moist, mostly forested habitats in Coast Range. Absent from Sacramento Valley. Two broods, spring to summer.

Eastern Tailed Blue *(Everes comyntas)*   PAGE 144

Tailed; male rich blue above; spring females with much blue, summer females often all black. Usually *smaller* than Western Tailed Blue, the wings *less broad.* Underside pearly gray, spots usually larger and more contrasting than in western species. Sacramento Valley and adjacent foothills, infrequent in Bay Area. Disturbed sites, grassland, marshes. Multiple broods, all season.

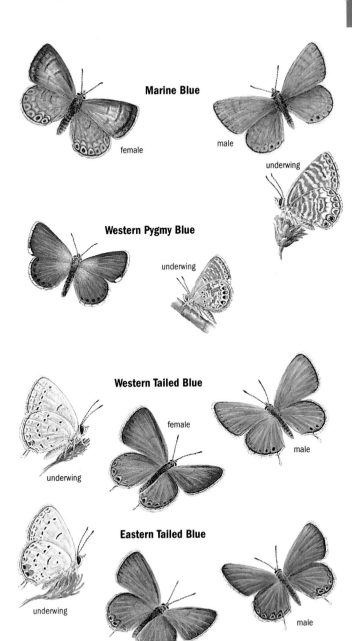

Marine Blue

female

male

underwing

Western Pygmy Blue

underwing

Western Tailed Blue

underwing

female

male

Eastern Tailed Blue

underwing

female

male

PLATE 14 **Blues**

Small butterflies, blue above (at least in males) with spotted patterns on undersides; no tails.

Greenish Blue *(Plebejus saepiolus)*   PAGE 160

Male *shiny, somewhat greenish blue above;* underside *chalky white with distinct black spots* with at most a tiny bit of orange. Forewing apex *pointed.* Female *rich orange brown* above with indistinct orange submarginal band on hindwing; underside *light brown,* heavily spotted. Moist meadows near the coast north of the Bay. One brood in spring.

Echo Blue *(Celastrina ladon echo)*   PAGE 156

Male violet blue, unmarked above except for checkering in fringes. Female paler, with broad black border. Undersides of all wings *chalky white with tiny spots,* more or less distinct; *no orange* in either sex. Abundant in Bay Area, many habitats; largely absent from Sacramento Valley. Two broods, late winter to early summer.

Xerces Blue *(Glaucopsyche xerces)*   PAGE 149

Rather large, the male lilac blue, female largely brown above. Underside brown to brownish gray, highly variable, with dark spots or with only the white rings surrounding them, as shown. Slightly larger than Silvery Blue. Formerly confined to San Francisco. *Extinct since 1941.*

male

Greenish Blue

female

male underwing

female underwing

male

Echo Blue

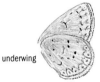

underwing

female

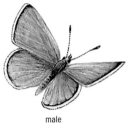

male

Xerces Blue

underwing variations

PLATE 15   Blues

Small butterflies without tails. At least one sex conspicuously marked with orange. Males blue above; females often brown.

Acmon Blue *(Plebejus acmon)* PAGE 157

Male blue, often violet tinged, above with conspicuous pink to orange marginal hindwing lunules (tending to disappear in fall). Female variable, mostly bright blue (early spring) to all brown black (summer) above, with strong orange hindwing lunules. Fringes *not strongly checkered.* Occasionally very small. Abundant and generally distributed. Multiple broods, all season.

Dotted Blue *(Euphilotes enoptes)* PAGE 155

Similar to Acmon Blue, slightly smaller with *less orange* (usually none in male); underside ashen, dark spots rounded, *usually with pale haloes;* forewing longer and narrower than in Acmon Blue. Very local, with wild buckwheat in scattered localities around the Bay. One brood in late spring to early summer.

San Bernardino Blue *(Euphilotes bernardino)* PAGE 154

Similar to Acmon and Dotted Blues. Underside *chalky white to pale gray,* black spots *heavy and more or less squarish, red spots distinct and contrasting.* Perhaps barely entering our area in Santa Clara County, associated with California buckwheat in chaparral. Late spring to early summer.

Sonoran Blue *(Philotes sonorensis)* PAGE 153

Bright shiny blue above with *conspicuous red markings in both sexes; strongly checkered fringes;* underside of hindwing *ash gray without red.* Very local, South Bay, in rocky areas. Late winter to early spring.

Acmon Blue

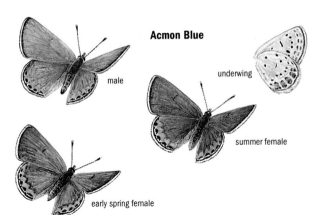

male

underwing

summer female

early spring female

Dotted Blue

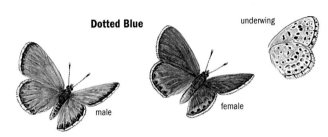

underwing

male

female

San Bernardino Blue

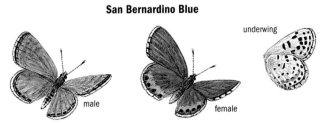

underwing

male

female

Sonoran Blue

underwing

male

female

PLATE 16 **Blues and Blue Copper**

Small bright blue (males) or bluish (females) butterflies with dotted undersides and little or no orange.

Blue Copper *(Lycaena heteronea)* PAGE 125

Male *brilliant sky blue above* with *pointed forewing apex.* Underside *chalky* white with distinct black pattern on forewing, but *little or none on hindwing.* Female brown with *more or less basal blue,* sometimes almost buff; submarginal pale orange lunules sometimes present above on hindwing; usual copper's black spot-pattern. Underside of female similar to male but yellowish. Restricted to a few cold coastal sites north of the Bay. One brood in summer.

Boisduval's Blue *(Plebejus icarioides)* PAGE 161

Several somewhat dissimilar subspecies. Male rich violet blue above with *dark submarginal spots on hindwing.* Female usually mostly brown, with some orange near margin on hindwing or all wings. Underside highly variable, light gray, some populations with greenish scales at base of hindwing, others not; *always with a row of dark spots along the hindwing margin.* Same size as Silvery and Arrowhead Blues; larger than Greenish Blue. Much of Bay Area; absent from Sacramento Valley. The Mission Blue *(P. i. missionensis)* is federally protected (see text); *P. i. pheres,* with white haloes but no black spots (like the usual form of the Xerces Blue), was confined to sand dunes in San Francisco and is *extinct* since 1950. Always with lupines; one brood, spring to early summer.

Silvery Blue *(Glaucopsyche lygdamus)* PAGE 147

Male bright blue above, unmarked except for narrow black border. Female usually mostly brown above, occasionally mostly blue. Color *less violet* than most populations of Boisduval's Blue. Underside pale gray (males) to brownish (females) always with iridescent blue green scales at base of hindwing and *never* with dark submarginal spots on any wings. Size of black spots highly variable. Generally common and widespread, including Sacramento Valley. One brood in spring.

Arrowhead Blue *(Glaucopsyche piasus)* PAGE 149

Male deep, rich blue; female with bases blue and outer part of wing dark above. Underside with *unique dark pattern;* fringes *very strongly checkered.* Rare and local in Bay Area; mesic forest. One brood in spring.

Blue Copper

female

male

underwing

Boisduval's Blue

male

female

underwing variations

Silvery Blue

male

underwing

female

Arrowhead Blue

female

male

underwing

PLATE 17 Unusual Fritillaries

See also pl. 18.

Unsilvered Fritillary *(Speyeria adiaste)*   PAGE 175

Compared to other fritillaries (pl. 18), *hindwing underside "washed out,"* the usual spot-pattern completely unsilvered. Increasingly rare; San Mateo and Santa Cruz Counties, mostly in redwood forest openings. One brood in summer.

Western Meadow Fritillary *(Boloria epithore)*   PAGE 169

Smaller than the true fritillaries *(Speyeria),* wing shape different. Underside of hindwing *mottled complexly in purple, white, yellow, and orange shades; no silver.* Local in moist forest in Santa Cruz Mountains. One brood, early summer.

Gulf Fritillary *(Agraulis vanillae)*   PAGE 168

Large, *bright red above,* forewing apex *elongate.* Underside of hindwing with many *long, narrow spots of brilliant silver.* Body also marked with silver. Mainly urban, in Bay Area. Many broods, all year. Especially common in Berkeley.

Unsilvered Fritillary

underwing

Western Meadow Fritillary

underwing

Gulf Fritillary

underwing

PLATE 18   True Fritillaries

Large orange-and-black butterflies with rounded wings and distinctive pattern of black spotting above and light, *often silvered,* spots on hindwing beneath. See also pl. 17.

Crown Fritillary *(Speyeria coronis)*   PAGE 171

Females quite large. Both sexes often *buffy. Heavily silvered beneath* with moderate see-through of spots on hindwing upperside. Bay Area, ranging widely. Long flight season, May to October.

Callippe Fritillary *(Speyeria callippe)*   PAGE 174

Smaller than others on this plate; populations near coast *strongly silvered,* inland less so; in all cases with *strong show-through.* Coastal populations appear *strongly checkered above* and may have *two-toned ground color.* Bay Area, now very local; late spring to early summer.

Zerene Fritillary *(Speyeria zerene)*   PAGE 172

Large, usually *bright reddish orange above* but golden to olivaceous in some coastal populations. Black spots on forewing underside usually large *and squarish.* Black pattern overall coarser than in Crown Fritillary; underside strongly silvered but with *little show-through.* Several named subspecies, some federally endangered. Spotty in Bay Area; one brood, summer to fall.

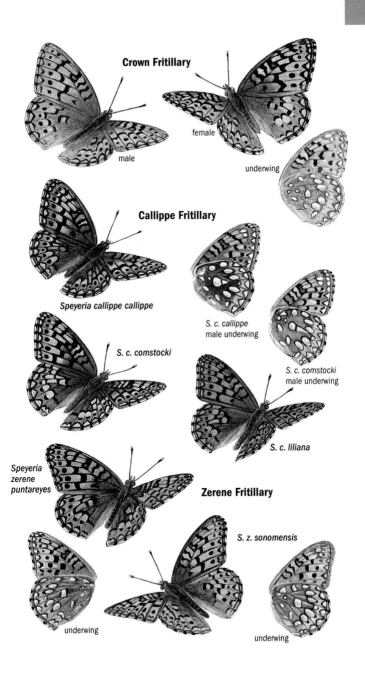

Crown Fritillary

male

female

underwing

Callippe Fritillary

Speyeria callippe callippe

S. c. callippe
male underwing

S. c. comstocki

S. c. comstocki
male underwing

S. c. liliana

Speyeria zerene puntareyes

Zerene Fritillary

S. z. sonomensis

underwing

underwing

PLATE 19 Crescents and Metalmarks

Small butterflies with "checkered" patterns in orange, black, brown, and white. Crescents have a crescent-shaped mark along hindwing margin below.

Mylitta Crescent *(Phyciodes mylitta)* PAGE 179

Variable. Male bright orange with more or less black pattern, *never two-toned*. Female with paler postmedian band and more extensive black pattern; larger. Hindwing underside with distinctly silvery crescent at margin. Antennal club orange. Abundant and general. Most of year.

Field Crescent *(Phyciodes campestris)* PAGE 176

Heavily marked above in both sexes; ground color above *usually two-toned*, paler in middle of wing. *Forewing apex rounded* (compare to other species on this plate). Hindwing underside often almost unmarked, light orange yellow in males; more heavily marked in females and in cool-season broods. Marginal crescent usually not very silvered. Locally common in wetlands and riparian habitat; very local in Sacramento Valley. Multiple broods, March to October.

California Crescent *(Phyciodes orseis)* PAGE 178

Larger and darker than Mylitta Crescent but otherwise quite similar. Male *slightly,* female *strongly two-toned above.* Antennal club orange. Looks like a large Mylitta Crescent with the coloring of a Field Crescent (!). Formerly, San Francisco and north. Now extinct in our area. Moist streamsides; one brood in late spring to early summer.

Mormon Metalmark *(Apodemia mormo)* PAGE 164

Note *white* markings and unique pearly brown gray hindwing pattern beneath. Lange's Metalmark (Antioch Dunes only) is federally protected; white pattern above reduced. Very local. One brood in second half of season.

Mylitta Crescent

light male

dark male

female

female underwing

male underwing

Field Crescent

California Crescent

underwing

male

male

female

female

underwing

Mormon Metalmark

A. m. mormo

underwing

Apodemia mormo langei

PLATE 20 Checkerspots

Medium-sized butterflies with "checkered" black-and-red or black-and-yellow patterns on both surfaces. Often highly variable geographically and individually.

Edith's Checkerspot *(Euphydryas editha)* PAGE 183

Abdomen without white spots; hindwing with a *complete row of postmedian red spots;* antennal club partly blackish; *apex of male forewing relatively rounded.* Grassland, chaparral, frequently on serpentine soils. Absent from Sacramento Valley. One brood, spring—usually earlier than other species. *Euphydryas e. bayensis* is federally protected.

Variable Checkerspot *(Euphydryas chalcedona)* PAGE 182

Larger than Edith's Checkerspot; *abdomen with white spots;* hindwing almost never with a complete row of red spots above; antennal club orange; *apex* of male forewing *pointed.* Widespread and common except in Sacramento Valley. One brood, midspring into early summer.

Northern Checkerspot *(Chlosyne palla)* PAGE 179

Size of Edith's Checkerspot. Male bright red orange with variable black pattern, the median spot-band sometimes lighter than the ground color. Females variable: entirely malelike, the median spot-band always light; or black and yellow; or intermediate. *Underside pattern constant.* Note on hindwing beneath that the first two (apical) postmedian spots are *not filled with red.* On black females there are no yellow spots on the hindwing above *basad of the one in the cell.* Widespread and common except in the Sacramento Valley. One brood, spring to early summer.

Leanira Checkerspot *(Thessalia leanira)* PAGE 181

Usually slightly smaller than Northern Checkerspot. Forewing *prolonged apically,* especially in male, with *red blotch at apex.* Ventral hindwing creamy white with black veins and *black postmedian line containing round white spots;* no red. Uncommon, local, usually in rocky areas; often on serpentine. Not in Sacramento Valley. One brood, late spring.

Edith's Checkerspot

underwing

Variable Checkerspot

male

female

underwing

Northern Checkerspot

male

females

underwing

Leanira Checkerspot

underwing

PLATE 21 Anglewings

Medium-sized butterflies with *very irregular wing margins.* Upper surfaces orange with black pattern; under surfaces brown to gray, cryptic, with a *silvery C- or L-shaped mark in the center of the hindwing.*

Satyr Anglewing *(Polygonia satyrus)* PAGE 186

Bright orange above; *wood brown beneath.* Cold-season form with more black and more irregular wing margins than summer form (shown). General in riparian and marsh habitats, but not common. Two to three broods.

Rustic Anglewing *(Polygonia faunus rusticus)* PAGE 186

Heavy dark pattern above. Male brown beneath, heavily marbled, with a series of *green blotches parallel to the wing margins.* Female underside rather even gray or gray brown, *no green or yellow.* Cool, moist coastal forests. Spotty records throughout year.

Zephyr Anglewing *(Polygonia zephyrus)* PAGE 187

Grayish brown beneath, usually with a series of faint *yellow spots* parallel to wing margins; wing "angles" sharply pointed; silver mark narrow, L shaped. Riparian canyon habitats, not common. Mostly late winter early spring.

Oreas Anglewing *(Polygonia oreas)* PAGE 188

Heavily marked above. Underside *very dark, even brown* with *narrow L-shaped* silver mark. Black spot near middle of leading edge of hindwing above *unusually large.* Cool, moist coastal forests, especially redwoods. Records spotty, much of year.

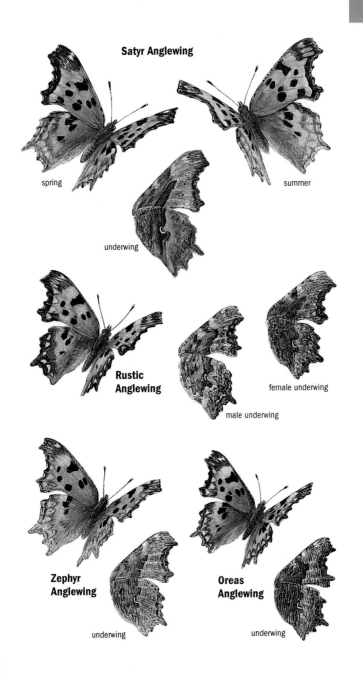

Satyr Anglewing

spring

summer

underwing

Rustic Anglewing

male underwing

female underwing

Zephyr Anglewing

underwing

Oreas Anglewing

underwing

PLATE 22 **Tortoiseshells and Allies**

Medium-sized to large butterflies with more or less scalloped wing edges and little sexual dimorphism; no silver "comma" marking on hindwing below. All the species on this plate are seasonally migratory.

Mourning Cloak *(Nymphalis antiopa)* PAGE 193

Deep *chestnut brown with yellow border* (fading to white) and *blue spots.* Widely distributed. Often seen in late winter.

Milbert's Tortoiseshell *(Nymphalis milberti)* PAGE 192

Smaller than California Tortoiseshell. *Broad orange-and-yellow band* on all wings above; dark borders containing *blue spots.* Basal half of hindwing upperside *solidly dark.* Underside *wood brown.* Uncommon, mainly late winter to early spring inland.

California Tortoiseshell *(Nymphalis californica)* PAGE 188

Larger than Milbert's Tortoiseshell. Truly "tortoiseshell" coloring above; basal half of hindwing *not solidly dark.* Small white spots near forewing apex above; marginal spots blue or bluish only near anal angle of hindwing. Underside variable, mainly brown; spots along margin greenish to bluish. Common to very abundant, except in Sacramento Valley and Delta, where occasional. A mass migrant. Often seen in winter.

Buckeye *(Junonia coenia)* PAGE 203

Not a tortoiseshell. *Large eyespots on all wings above; diagonal white band below forewing apex.* Female larger than male, with more rounded forewing. Hindwing underside seasonally variable, clay brown to (in fall) deep purplish red ("rosa"). Late winter and early spring individuals may be very small. Abundant in canyons and disturbed habitats; less frequent near coast. Many broods, all year.

Mourning Cloak

underwing

Milbert's Tortoiseshell

underwing

California Tortoiseshell

male underwing

female underwing

summer underwing

Buckeye

female

male

fall underwing

PLATE 23   **Ladies and Red Admiral**

Medium-sized orange- or red-and-black butterflies with complex, subtle hindwing patterns beneath. Some are strongly migratory.

Painted Lady *(Vanessa cardui)*   PAGE 195

Male pinkish, female more orange above. Hindwing below mottled in gray, brown, and white with triangular white spot at center. Eyespots on hindwing above and below *nearly equal in size;* may contain blue. *Black bar across cell of forewing above broken into two spots.* All spots near forewing apex *white.* Size and intensity of color highly variable. A mass migrant, occurring everywhere but not normally overwintering. January to November.

American Painted Lady *(Vanessa virginiensis)* PAGE 194

Male pinkish, female more orange above. Hindwing above with the two spots in the middle *smaller than the ones at the ends.* Hindwing below with *two very large eyespots.* Generally distributed but uncommon. Most of year.

West Coast Lady *(Vanessa annabella)*   PAGE 200

Both sexes orange. Bar at end of forewing cell above may be *orange or white.* Black bar across this same cell *continuous.* Hindwing pattern similar to Painted Lady beneath. Spots on hindwing *nearly equal in size,* often containing *blue;* on upper surface *bounded basad by a series of black crescents.* Averages smaller than Painted Lady. Extreme aberrations occur, of which an example is shown (upper and under surfaces). Exactly similar aberrations occur in the other ladies too. Common and general. All year.

Red Admiral *(Vanessa atalanta)*   PAGE 201

Above black *marked uniquely with red;* white spots near apex as in other vanessas. Underside of hindwing similar to other vanessas but darker. Common and general. All year.

Painted Lady

underwing

female

male

American Painted Lady

male

female

underwing

West Coast Lady

**Aberrant
West Coast Lady**

underwing

Red Admiral

underwing

underwing

PLATE 24 Various Large, Showy Brushfoots

California Sister *(Adelpha bredowii californica)* PAGE 207

Orange spot at apex of forewing *rounded and contained outwardly by a dark border, not reaching the margin. No white bar* at end of cell. White band on hindwing tapering sharply to a point. Larger than Lorquin's Admiral. Oak woodland, less frequent in Sacramento Valley. Two or three broods.

Lorquin's Admiral *(Limenitis lorquini)* PAGE 205

Orange spot at apex of forewing *elongate, reaching the margins. White bar at end of forewing cell above.* Hindwing white band narrowing gently toward anal angle. Riparian habitat throughout the area. Three broods, spring to fall.

Monarch *(Danaus plexippus)* PAGE 214

Orange with *black veins* and *black borders containing white spots.* Male with pheromone gland (black bulge on a hindwing vein above.) Everywhere; migrates to coast for winter. Inland, March to November.

California Sister

underwing

Lorquin's Admiral

underwing

male

female

Monarch

underwing

PLATE 25   Meadow Browns and Satyrs

Mostly medium-sized brown butterflies with eyespots on forewing and usually also on hindwing. Hindwing undersides cryptic.

California Ringlet
(Coenonympha tullia california)

PAGE 209

Small. Spring brood with *silvery luster above,* hindwing often *very dark below.* Summer brood *straw colored or buff* with little or no luster above, and basal half of hindwing *much lighter* than in spring brood. Eyespots variable, sometimes absent. Common and widespread but disappearing from the Sacramento Valley and intolerant of urbanization. January to November but usually commonest in spring.

Great Basin Wood Nymph *(Cercyonis sthenele)* PAGE 211

Noticeably smaller than Ox-eyed Satyr and usually *grayer* beneath. Eyespots relatively *small,* especially on male upperside. On forewing below, lower eyespot usually noticeably smaller than upper, especially in males. Grassland, oak woodland, and coastal scrub in much of Bay Area. Extinct in San Francisco (the endemic dune subspecies *C. s. sthenele,* exterminated by 1880). One brood, late spring to summer.

Ox-eyed Satyr *(Cercyonis pegala boopis)* PAGE 210

Larger than Woodland Satyr. Eyespots larger, especially in female; usually about equal in males, but the *lower spot larger* in females. Marshes, grassland, and oak woodland near water. Absent from most of Sacramento Valley. One brood, late spring to fall (!).

Great Arctic *(Oeneis nevadensis)* PAGE 212

Our largest Satyr. *Butterscotch* color above, the male with a large, greenish gray pheromone gland in middle of wing. Male with (usually) one and female with (usually) two eyespots. Male apex pointed, female rounded. Underside of hindwing gray, finely mottled. Cool, moist conifer forest north of the Bay. One brood in late spring to early summer; in many localities flies *only in alternate years.*

California Ringlet

summer

underwing

underwing

spring

Great Basin Wood Nymph

male

underwing

female

underwing

Ox-eyed Satyr

male

underwing

female

underwing

Great Arctic

underwing

male

female

PLATE 26 Dark-Colored Skippers

See also pls. 27 and 30.

Dreamy Duskywing *(Erynnis icelus)* PAGE 223

Forewing above *strongly grizzled, crossed by two dark-filled chain-like bands; no white spots* above. Hindwing dark, with two rows of light spots. Forewing *short and broad.* Rests with wings open. Cool, moist forest, near streams. North Bay only. One brood, late spring.

Sleepy Duskywing *(Erynnis brizo lacustra)* PAGE 224

Forewing grayish, *crossed by two open chainlike bands; longer and narrower* than in Dreamy Duskywing. Hindwing very dark below. Usually rests with wings open. Serpentine chaparral. One brood in spring.

Northern Cloudy-wing *(Thorybes pylades)* PAGE 222

Medium sized with *broad wings; two white spots near costa* of forewing; hindwing underside with *paler, slightly violet-tinged shade near margin;* this wing crossed by *vague, darker chainlike bands.* Various habitats in Bay Area, not in Sacramento Valley. One brood, late spring.

Common Sootywing *(Pholisora catullus)* PAGE 234

Glossy black with crisp, white dotted pattern, more extensive in female. Underside of hindwing *plain black. Face white.* Usually rests with wings open. Disturbed, weedy habitats throughout area. Multiple broods, spring to fall.

Roadside Skipper *(Amblyscirtes vialis)* PAGE 249

Dark brown with *conspicuous subapical white spot on forewing above* and *checkered fringes.* A *violet shade* along wing margins below. Usually keeps wing closed when at rest. Cool, moist woods, mainly north of the Bay. One brood, spring.

Dreamy Duskywing

Sleepy Duskywing

Northern Cloudy-wing

underwing

Common Sootywing

male

underwing

female

Roadside Skipper

underwing

PLATE 27   Duskywings

Medium-sized, very dark skippers that *usually rest with wings open.* Forewing upperside with *small white spots.* Two species *without* white spots are shown on pl. 26.

Propertius Duskywing *(Erynnis propertius)*   PAGE 225

Relatively large and strongly marked. Underside of hindwing with *two conspicuous white spots* near apex. Fringes *dark.* Widespread and usually common. Spring, with a partial second brood in summer.

Persius Duskywing *(Erynnis persius)*   PAGE 228

Small, more or less "*frosted,*" with *white dot at end of cell.* Many habitats, throughout area. Multiple brooded, spring to fall.

Pacuvius Duskywing   PAGE 227
(Erynnis pacuvius pernigra)

Small and very dark, usually *glossy; no white dot at end of cell.* Uncommon and local in Bay Area, mostly in chaparral. One brood in spring.

Mournful Duskywing *(Erynnis tristis)*   PAGE 225

Relatively large (size of Propertius Duskywing), somewhat *glossy, forewing not long and narrow,* fringes *white.* Common and widespread, often in cities. Most of year.

Funereal Duskywing *(Erynnis funeralis)*   PAGE 226

Similar to Mournful Duskywing, often with reddish tint beyond forewing cell; *forewing long and narrow.* Rare stray from the south, mainly in and near Central Valley.

Propertius Duskywing

male

female

underwing

Persius Duskywing

Pacuvius Duskywing

underwing

Mournful Duskywing

Funereal Duskywing

PLATE 28 Checkered Skippers

Small to medium-sized skippers with black-and-white checkered patterns.

Common Checkered Skipper
(Pyrgus communis)

PAGE 231

Male with *bluish sheen,* often conspicuous in flight. Female black and white without sheen. Underside bands gray (male) to olivaceous buff (female). Abundant and general. Most of year.

Two-banded Skipper *(Pyrgus ruralis)*

PAGE 229

Small; *grizzled,* with *strongly checkered fringes.* Hindwing pattern beneath *reddish.* Cool, moist habitats around the Bay. One brood, spring.

Least Checkered Skipper *(Pyrgus scriptura)*

PAGE 229

Smaller still than Two-banded Skipper. First brood very similar to Two-banded Skipper but hindwing pattern beneath *greenish gray.* Subsequent broods very small, mostly black, with white areas reduced to small spots. Alkaline areas and waste ground, mainly in Sacramento Valley. Most of year.

Northern White Skipper
(Heliopetes ericetorum)

PAGE 233

Largest of the group in our area. Male *mostly white;* female with checkered pattern *absent from middle of wings above.* Hindwing underside with *reduced, olivaceous pattern.* Migratory through our area inland in late spring and again in fall. Breeds sporadically in chaparral.

Common Checkered Skipper

male

female

underwing

Two-banded Skipper

underwing

Least Checkered Skipper

spring

summer

underwing

Northern White Skipper

female

male

underwing

PLATE 29   Mostly Branded Skippers

Small skippers that hold the wings partly closed at rest. Males (except Arctic Skipper) with a stigma on forewing.

Tilden's Skipper (Hesperia colorado tildeni)   PAGE 238

Hindwing beneath *golden to olivaceous,* males often with reduced pattern. Light markings near hindwing base beneath *form a C shape.* Forewing long, pointed in male (compare Woodland Skipper). Dry grassland and oak woodland in Inner Coast Range and Mount Hamilton Range. Local. One brood, summer to early fall.

Dodge's Skipper (Hesperia colorado dodgei)   PAGE 239

Similar to Tilden's Skipper but *much darker beneath* with pattern strongly contrasting. Coastal fog belt, summer to early fall.

Lindsey's Skipper (Hesperia lindseyi)   PAGE 240

Similar to *H. colorado,* hindwing below as light as Tilden's Skipper inland and as dark as Dodge's Skipper on the coast. Light spots beneath larger than in *H. colorado* and *squarish,* especially in female. *Basal spots not forming a C.* There may be some light scaling along veins below. Male stigma more slender than in *H. colorado.* Local in grassland, often on serpentine. One brood, April to June.

Yuba Skipper (Hesperia juba)   PAGE 237

Larger than related species. Body strongly *greenish.* Underside of hindwing dark olivaceous *green* with *large squarish silvery spots.* Uncommon and irregular; spring and fall. Not in Sacramento Valley.

Dogstar Skipper (Polites sonora siris)   PAGE 242

Rather small, dark above and below. Hindwing below *dark brown* with *distinctive curved band of yellow spots* and *dash near base.* Cool, moist habitats north of the Bay. Late spring to summer.

Columbia Skipper (Hesperia columbia)   PAGE 240

Underside of hindwing in both sexes *golden green* with *unique reduced pattern* of *silvered spots. No light spots near base.* Local, often on serpentine. Two flights, spring and fall.

Arctic Skipper (Carterocephalus palaemon)   PAGE 235

Strongly checkered pattern in dark brown and orange. Very local in cold, moist sites north of the Bay. Late spring.

Tilden's Skipper

male

female

male underwing

female underwing

Dodge's Skipper

female

female underwing

male underwing

male

Lindsey's Skipper

female

male underwing

female underwing

Yuba Skipper

female

underwing

Dogstar Skipper

male

underwing

female

Columbia Skipper

male

underwing

Arctic Skipper

underwing

PLATE 30 **Branded Skippers**

Mostly small skippers that hold the wings partly or completely closed at rest. Males usually with stigma.

Fiery Skipper *(Hylephila phyleus)* PAGE 236

Male bright golden yellow with *thin stigma.* Female mostly brown above with buff pattern. Underside of female hindwing olivaceous with a paler band enclosed by two rows of dark dots. *Antennae short.* Abundant in cities and suburbs, rare in wildlands. Most of year but commonest in fall.

Sachem *(Atalopedes campestris)* PAGE 243

Larger than Fiery Skipper. Male with large, broad, "smeary" stigma; hindwing underside not dotted. Female with a *large, squarish, hyaline spot near middle of forewing;* hindwing underside variable, in fall often chocolate brown with contrasting light spot-pattern. Antennae relatively long. Common, often in cities; scarce near coast. Most of year but commonest in fall.

Sandhill Skipper *(Polites sabuleti)* PAGE 241

Smaller than Fiery Skipper, male with relatively *thick, two-parted stigma;* female with light pattern above often contrastingly pale; conspicuous *yellow stripes on thoracic* "lappets." *Forewing short and broad.* Hindwing underside very variable, from yellow and nearly unspotted in summer males to "cobwebby" and contrasting in cool-season animals of both sexes. Common throughout. Multiple broods, spring to fall.

Umber Skipper *(Poanes melane)* PAGE 247

Mostly *rich brown* with light spots on forewing in both sexes. Underside of hindwing with unique, *subtle orange, lilac, and brown shading;* no distinct spots. Inland in riparian forest; an urban species in parts of the Bay Area. Multiple broods.

Eufala Skipper *(Lerodea eufala)* PAGE 249

Even brown with small white spots; male *without stigma.* Underside of hindwing *plain brown.* Grassland and marshes, Sacramento Valley and Delta, rare near coast. Multiple broods but commonest in fall.

Fiery Skipper

male

male underwing

female underwing

female

Sachem

male

female

Sachem underwings

male

light female

dark female

Sandhill Skipper

female

male

light

dark

Sandhill Skipper underwings

Umber Skipper

underwing

underwing

Eufala Skipper

PLATE 31 Mostly Branded Skippers

Mostly small skippers that hold the wings partly or completely closed at rest.

Woodland Skipper *(Ochlodes sylvanoides)* PAGE 244

Male with *triangular forewing and narrow stigma*. Hindwing underside golden brown or reddish orange (inland) to deep brown (coast) with *poorly defined pattern of squarish yellow spots*. Body hair of old specimens looks *greenish or bluish*. Abundant except in cities. Second half of season.

Farmer *(Ochlodes agricola)* PAGE 245

Small, with broader dark border than Woodland Skipper and shorter, slightly thicker male stigma. Underside of male hindwing usually *plain yellow, unmarked or nearly so;* female with *vague pattern and a purplish or lilac flush*. Abundant in wildlands; absent in Sacramento Valley. Late spring.

Yuma Skipper *(Ochlodes yuma)* PAGE 246

Larger than related skippers, *plain golden yellow with minimal pattern;* male with long, narrow stigma; *underside of hindwing plain yellow* with at most faint light spots. Suisun Marsh and Delta, local. Two broods, June and September.

Dun Skipper *(Euphyes vestris osceola)* PAGE 248

Dark brown with a golden luster; male with narrow stigma but *no markings*. Female with a few small and relatively indistinct light spots on forewing. Cool, moist forest north of the Bay and in Santa Cruz County. One brood, late spring.

Silver-spotted Skipper *(Epargyreus clarus)* PAGE 220

Large, with pointed forewing marked in amber; hindwing underside with middle third *iridescent silver*. Riparian canyon habitat north of the Bay. Formerly in Sacramento Valley, possibly extinct. Late spring.

Woodland Skipper

coastal underwing

interior underwing

male

female

Farmer

female underwing

male underwing

male

female

Yuma Skipper

underwing

male

female

Dun Skipper

male

female

underwing

Silver-spotted Skipper

underwing

Skippers (Hesperiidae)

The hesperiids form a large (more than 3,500 species) worldwide family of mostly small to medium-sized butterflies. This family has several subfamilies, but only three occur in our area: the spread-wing skippers (Pyrginae), skipperlings (Heteropterinae), and branded skippers or grass skippers (Hesperiinae). Most hesperiids have a distinctive recurved hook, the apiculus, at the end of the antennal club. Most have relatively large, muscular bodies relative to wing size, and rather broad heads. The common name "skipper" refers to their mode of flight, but many variations on the theme can be seen—some are very indolent, while others are among our swiftest fliers, engaging in dazzling aerobatics.

The spread-wing skippers are a diverse lot, but almost all of them sit with wings fully opened at their sides, hence the name. The Silver-spotted and Long-tailed Skippers (*Epargyreus clarus californicus* and *Urbanus proteus*) are exceptions, holding the wings closed over the back, and the cloudy-wings often hold them partly open. Most are dark in color, and members of the most species-rich genus, *Erynnis*—the duskywings—all look much alike. These skippers are primarily associated with hosts in the pea family (Fabaceae), the mallows (Malvaceae), and others including oaks (*Quercus*) and wild-lilacs (*Ceanothus*) (tannin-rich plants), and willows (*Salix*).

Among the skipperlings, we have only one species, the distinctively marked Arctic Skipper (*Carterocephalus palaemon*). These are small, thin-bodied, delicate hesperiids, usually with rather weak flight. They rest with the wings either open or closed over the back.

The branded skippers or grass skippers are mostly rather small and predominantly orange in color with brown markings. In some the female is mostly brown, and in a few both sexes are brown. They nearly all sit with the wings closed or with the hind wings widely spread but the forewings only slightly so—sometimes called the "jet plane" position. All of our species feed as larvae on grasses or sedges, but their field preferences are rarely known—in the lab they can usually be reared on almost any grass (Bermuda from your lawn, for example!). Grass skippers have a reputation as very difficult to identify. It helps to have an experienced colleague show you good series of mounted specimens. Although learning them entails a lot of work, these little butterflies are well worth it. Many of them are habitat specialists, and they may even serve as environmental quality indicators. (For example, the ongoing presence of the Columbia Skipper [*Hesperia columbia*] at Lagoon Valley Regional Park, Solano County, directly reflects the persistence of perennial bunchgrass there in a region where it has almost been extirpated.) Because we have much to learn about their life histories, hesperiids in general and grass skippers in particular offer excellent opportunities for the amateur to make a lasting contribution to science.

Most hesperiids have distinctive pheromone glands employed in courtship. Males of many spread-wing skippers have a costal fold. The costa is the vein along the front margin of the wing. On the forewing this appears rolled over to create a pocket or flap containing scent scales. Some species have hair pencils— tufts of erectile hairs like a powder puff, typically on the legs. Grass skipper males usually have a conspicuous scent gland, known as the stigma, in the middle of the forewing. This is the "brand." It consists of a black diagonal marking, sometimes interrupted, which partly or completely contains a pocket of silvery or velvety gray scent scales. The form of the stigma is usually diagnostic at the species level. Most male skippers are highly territorial, and some in our area hilltop.

Hesperiid larvae in our part of the world are unshowy, usually tan, brown, or green, with dark heads and often a dark, contrasting "cervical shield" behind the head. Most roll a leaf of the host plant, living in a tubular nest out of sight. Many pupate in the nest; some make a very rudimentary cocoon.

The Fiery Skipper (*Hylephila phyleus*) is the most distinctively urban butterfly we have—it is rarely found away from Bermuda grass (*Cynodon dactylon*) lawns, where it breeds and successfully dodges under the blades of the lawn mower. By way of contrast, one must go to the wildlands to see most of our "better" hesperiids, and they pose a challenge to collector and photographer alike.

Spread-wing Skippers (Pyrginae)

SILVER-SPOTTED SKIPPER *Epargyreus clarus californicus*
Pl. 31

This is the largest skipper in our fauna. Its flight is extremely fast, and the silver blotch on the underside of the hindwing flashes as the insect moves through dappled light and shade. It is difficult to follow in flight, but fortunately it is easily approached when feeding at flowers, especially California buckeye *(Aesculus californica)* and dogbanes *(Apocynum)*. The Silver-spotted Skipper is found in riparian canyon habitat, often along roadsides where males perch on foliage well off the ground. In the Gold Country, and to a lesser extent in the North Coast Ranges, it has benefited from the introduction of the ornamental tree black locust *(Robinia pseudacacia)* in the nineteenth century. This is the usual host plant of the eastern subspecies (typical *E. c. clarus*), and our subspecies quickly adopted it as a host as well. As a result, it is regularly seen in towns such as Downieville, Jackson, or Weaverville, where no native hosts occur. Unlike the eastern subspecies, ours is seldom very common.

DISTRIBUTION: North Bay counties, mainly inland. Formerly found with urban locust trees *(Robinia)* in Sacramento and Yolo Counties, but no recent records.

HOST PLANTS: All pea family (Fabaceae) plants including false indigo or lead plant *(Amorpha); Lotus crassifolius;* black locust *(Robinia pseudoacacia)* but apparently not honey or moraine locust *(Gleditsia triacanthos);* perennial vetches *(Lathyrus)* suspected. The rapidly spreading weedy riparian shrub Argentine flame pea *(Sesbania punicea)* is very likely to be adopted as a host plant in the next decade, potentially increasing the range and abundance of this species into the Sacramento Valley and Delta. Larvae tie several leaflets together to form solitary nests. The large larva has a brown head capsule with two bright orange eyelike spots; many people find its appearance clownish. Since the larva rarely leaves its nest in daylight, the function of these spots is unknown. If disturbed, the larva throws fecal pellets through the air, adding to the comical effect.

SEASONS: One brood, May to early July depending on locality. A very few late-summer records may indicate a rudimentary second brood. The eastern subspecies is multiple brooded.

SIMILAR SPECIES: None in our area.

DORANTES SKIPPER *Urbanus dorantes*

Not illustrated

Recorded once from Sebastopol, Sonoma County, this is a subtropical species with no business being in our area, the nearest resident populations being in northern Mexico near the Arizona border! The Long-tailed Skipper *(U. proteus)* is recorded intermittently in southern California and would seem a more likely stray here. But it *hasn't* been recorded. The Long-tailed Skipper has greenish lustrous hairs above and the Dorantes Skipper does not. Most distinctively, both have long tails. If you catch a long-tailed skipper in the area covered by this book, look in one of the field guides that covers the entire West to find out which one it is (there are several additional possibilities!) and communicate the record to Paul Opler, Ray Stanford, or me. It has been suggested that with global warming, such strays from the south may become more frequent and themselves serve as a barometer of ecological change.

NORTHERN CLOUDY-WING   *Thorybes pylades*
Pl. 26

This species is generally uncommon but widespread in our area in canyon riparian and mixed-mesic forest habitats, as well as foothill woodland and chaparral. Adults are most often seen at flowers (dogbanes *[Apocynum]*, yerba santas *[Eriodictyon]*, California buckeye *[Aesculus californica]*, brodiaeas *[Brodiaea]* and similar spring bulbs, vetches *[Vicia]*) along roadsides and near streams. The Northern Cloudy-wing is often confused with the larger species of duskywings, but with experience they are easy to distinguish; even in flight the shape of the forewing (more triangular in the cloudy-wing) is distinctive. Unlike the duskywings, the Northern Cloudy-wing never spreads its wings fully when at rest, making it a nonspreading spread-wing! Two other species of the genus occur in northern California, but not within the coverage area of this book. (Yes, there is a Southern Cloudy-wing *[Thorybes bathyllus]*, but it's an eastern species, while the Northern Cloudy-wing spans the continent.)

DISTRIBUTION: All Bay Area counties but San Francisco. Unrecorded in the Sacramento Valley.

HOST PLANTS: *Lotus crassifolius,* a robust herbaceous perennial, is most often used along roadsides. Also on perennial vetches and sweet peas *(Lathyrus)*, native species only, and scurf-pea *(Psoralea)*, a native perennial resembling a giant alfalfa plant. Perhaps other pea family (Fabaceae) plants. The larva constructs a silk nest, tying leaflets together.

SEASONS: One brood, mid-March through July depending on locality; typically a late-spring flier.

SIMILAR SPECIES: Compare to the larger duskywings. The Northern Cloudy-wing has a triangular, "high-angled" forewing and a distinctive barred pattern on the hindwing beneath, quite unlike any duskywing, and it rests with the wings half-open.

Duskywings (*Erynnis*)

These dark hesperiids have a justified reputation for difficulty in identification. Size, habitat, and seasonality are all good clues, but a few individuals will require examination of the male genitalia for confident determination. An oddity of duskywing genitalia, noted by Scudder and Burgess in the nineteenth century, is that the right and left sides are consistently asymmetrical. Rarely, one finds an individual with two "right" or "left" sides. This does not seem to serve as a reproductive isolating mechanism, and its function is not understood. All our species, whether uni- or multivoltine, apparently overwinter as large larvae. They are mimicked by day-flying noctuid moths of the genus *Euclidea*, which may be very difficult to tell from the duskywings in flight. Other species that may be confused with duskywings are the Common Sootywing *(Pholisora catullus)*, the summer brood of the Least Checkered Skipper *(Pyrgus scriptura)*, the Roadside Skipper *(Amblyscirtes vialis)*; and the Eufala Skipper *(Lerodea eufala)*.

DREAMY DUSKYWING　　　　　　　　　　　　*Erynnis icelus*
Pl. 26

We have only one confirmed record of this species in our coverage area—in Sonoma County—but it is included and illustrated because I am convinced it is underreported and probably occurs much more widely. Any small duskywing found in a cool, moist riparian habitat should be suspected of being the Dreamy Duskywing. Elsewhere in its range it is usually observed at mud puddles or at dogbanes *(Apocynum)* or other small woodland flowers.

DISTRIBUTION: Known only from Sonoma County. Possible throughout the Bay Area.

HOST PLANTS: Elsewhere on willows *(Salix)* and poplars *(Populus)*; in California generally on willows.

SEASONS: One brood, May to June.

SIMILAR SPECIES: Most resembles the Sleepy Duskywing *(E. brizo lacustra)* in lacking white spots above (except at best a hint of a subapical spot on the forewing). Differs from the Sleepy Duskywing in the markedly shorter forewing, and the outer half of the forewing is frosted with gray, crossed by two parallel bands of chainlike spots. Elsewhere, the Dreamy may co-occur with the Sleepy Duskywing on serpentine soils, but unlike that species it is not at all restricted to them. It generally rests with wings open.

SLEEPY DUSKYWING
Erynnis brizo lacustra
Pl. 26

Except for very rare strays, this species is a strict serpentine endemic in our area. It occurs in chaparral with its only known Bay Area host plant, leather oak *(Quercus durata),* also a serpentine endemic. It visits mud puddles and various spring flowers. Otherwise it can be seen flitting through the serpentine chaparral with a twisting, evasive flight that is almost impossible to follow—disappearing into one thicket and reappearing as if by magic from behind a different one. It is not usually common.

DISTRIBUTION: Serpentine barrens throughout the Bay Area. Unknown in the Central Valley.

HOST PLANTS: Leather oak *(Quercus durata).* This shrub is extremely variable and may be easily confused with other species. South of the range of leather oak in both the Coast Ranges and the Sierra Nevada, this spread-wing skipper has been recorded on a variety of shrubby oaks not confined to serpentine. This poses an eco-evolutionary mystery. Why the restriction in the north?

SEASONS: One brood, March to June depending on locality.

SIMILAR SPECIES: The habitat alone may be enough to identify this species, but if not, note the relatively long, narrow forewing, the lack of white spotting above, the chain-dotted bands crossing the forewing (generally more contrasting than in the Dreamy Duskywing *[E. icelus]*), and the very dark, nearly or quite solid black brown hindwing beneath. Outside our area, this species has been taken with all the other small duskywings at one time or another. Puddling males may hold the wings over the back.

PROPERTIUS DUSKYWING   *Erynnis propertius*
Pl. 27

At the time this species was named, there was a fashion for naming hesperiids after Greek and Roman poets, dramatists, orators, and such. Sextus Propertius was a major Roman poet of the first century B.C. The name of the lover to whom he addressed many poems, Cynthia, is another generic name for *Vanessa,* and the specific epithet of our Silvery Blue, *Glaucopsyche lygdamus,* commemorates the slave, Lygdamus, who according to Propertius treacherously murdered her. In any case, the Propertius Duskywing is our commonest and most widespread duskywing and the only large one with a dark hindwing fringe. Females are much more contrastingly marked above than males. It is usually common, sometimes abundant anywhere in the hills where oaks *(Querus)* grow. Males puddle avidly, and both sexes visit flowers, including brodiaeas *(Brodiaea)* and related spring bulbs, vetches *(Vicia),* dogbanes *(Apocynum),* yerba santas *(Eriodictyon),* wild onions *(Allium),* California buckeye *(Aesculus californica),* verbenas *(Verbena),* and so forth.

DISTRIBUTION: All counties, but uncommon to rare today on the floor of the Sacramento Valley.

HOST PLANTS: A variety of oaks, including coast live oak *(Quercus agrifolia),* interior live oak *(Q. wislizenii),* valley oak *(Q. lobata),* Oregon white oak *(Q. garryana),* black oak *(Q. kelloggii),* and blue oak *(Q. douglasii)*; perhaps others. Larva is a leaf roller, usually on young foliage near shoot tips.

SEASONS: One brood, March to June, with a partial second brood some years in inland localities, June to July (despite the assertion in most books that this species is always univoltine).

SIMILAR SPECIES: The other two large duskywings in our area have white hindwing fringes. The two white spots on the ventral hindwing distinguish this species from all our other duskywings. The Northern Cloudy-wing *(Thorybes pylades)* has a different wing shape and many pattern differences.

MOURNFUL DUSKYWING   *Erynnis tristis*
or SAD DUSKYWING
Pl. 27

This is the only duskywing routinely observed in cities and suburbs in both the Bay Area and the Sacramento Valley. In wildlands it occurs with oaks *(Quercus)*, often in riparian situations but also in chaparral. Males hilltop in midafternoon. This species is less of a puddler than most duskywings. It visits tall flowers such as tall blue verbena *(Verbena bonariense)*, milkweeds *(Asclepias)*, dogbanes *(Apocynum)*, yerba santas *(Eriodictyon)*, mints *(Mentha)*, and California buckeye *(Aesculus californica)*, and in gardens it is partial to butterfly bush *(Buddleia)*. There is little variation except between the sexes.

DISTRIBUTION: All counties.

HOST PLANTS: Oaks *(Quercus)*; in summer utilizes tender young growth, mainly at shoot tips; often oviposits on young trees and saplings. Recorded in our area from both evergreen species (coast live oak *[Q. agrifolia]*, interior live oak *[Q. wislizenii]*) and deciduous ones (valley oak *[Q. lobata]*) and on introduced species including cork oak *(Q. suber)* from the Mediterranean and eastern black oak *(Q. velutina)* from the eastern United States.

SEASONS: Three to four broods, January (rare) to October.

SIMILAR SPECIES: The Propertius Duskywing *(E. propertius)*, with which the Mournful Duskywing often occurs, has dark fringes and the two diagnostic white spots on the ventral hindwing below. The Funereal Duskywing *(E. funeralis)* has white fringes but the forewing is longer and narrower, with a somewhat frosted, sometimes vaguely bronzed or reddish appearance, and occurs mostly in a different sort of habitat and is rare in our area.

FUNEREAL DUSKYWING　　　　　　　*Erynnis funeralis*
Pl. 27

This species ranges from the San Joaquin Valley and central California coast south to Chile and Argentina. In most of this area its main host is alfalfa *(Medicago sativa)*, but it never seems to rise to the level of causing economic damage. For us it is a rather rare stray (recorded in scattered counties, mainly as singletons). Unlike other duskywings, it favors hot, dry, sunny habitats (such as agricultural land). Its flight is fast and the males are highly territorial.

This species is often listed as a subspecies of the Zarucco Duskywing *(E. zarucco)*. Its separation as a distinct species reflects apparent distributional overlap (without interbreeding) in central Texas.

DISTRIBUTION: Widely scattered records in both the Bay Area and Sacramento Valley; most likely along the eastern edge of the Bay Area and the west edge of the valley.

HOST PLANTS: Alfalfa *(Medicago sativa)*; occasionally on wild peas (Fabaceae) such as *Lotus* and probably others.

SEASONS: Multivoltine, February to November.

SIMILAR SPECIES: The only other white-fringed duskywing in our area is the Mournful Duskywing *(E. tristis)*, with a broader and less-pointed forewing, a less pronounced reddish subapical patch on forewing, and different habitat and habits. Almost all white-fringed duskywings seen in the area covered by this book will be the Mournful, at least until global warming really "kicks in."

PACUVIUS DUSKYWING   *Erynnis pacuvius pernigra*
Pl. 27

The Pacuvius Duskywing is local and usually uncommon, found mainly in chaparral, occasionally in moist riparian woodland. Despite the ubiquity of wild-lilac *(Ceanothus)* species—the host plant—this duskywing is a good find; its ecological and host preferences are poorly understood. An avid puddler; often seen on low flowers (vetches *[Vicia]*, pink dogbane *[Apocynum androsaemiifolium]*, mints *[Mentha]*) near the host plants.

DISTRIBUTION: Spotty in the Bay Area; absent from the Sacramento Valley. I have found it most often in areas burned or cut over several years before, where low-growing species of wild-lilacs are prominent in the successional vegetation.

HOST PLANTS: Wild-lilac *(Ceanothus)* species; in my experience associated with *C. leucodermis* and *C. oliganthus* in our part of the world.

SEASONS: One brood, April to July, with a flight season of about 3 weeks.

SIMILAR SPECIES: Distinguishing this species from the Persius Duskywing *(E. persius)* without recourse to the male genitalia is at best difficult and sometimes impossible. In our area the Pacuvius Duskywing has a dark, sometimes glossy look, while the Persius is more "matte." The white dots on the Pacuvius are usually confined to the subapical region; the Persius is more likely to have a conspicuous light dot at the end of the forewing cell. The underside of the Pacuvius is more uniformly dark than the Persius. Because of different host associations and seasonality, identification may be easier using collection data rather than wing characters! Distinguished from other small duskywings (Dreamy Duskywing *[E. icelus]* and Sleepy Duskywing *[E. brizo lacustra]*) by not having the outer half of the forewing frosted above or crossed by chainlike bands.

PERSIUS DUSKYWING   *Erynnis persius*
Pl. 27

This is our commonest small duskywing and the only one that is multiple brooded, with a long flight season. It occurs in a variety of habitats, most often streamsides and dry stream beds but also forest clearings, clear-cuts, and the like. It is quite mobile and seems to have high rates of local colonization and extinction. In the Sacramento Valley it is found irregularly in annual grassland, as often as not along roadsides where Spanish lotus *(Lotus purshianus)* grows. The first and later broods differ in details of wing pattern. This may complicate efforts at identification! Males "fly a beat," are highly territorial, and also visit puddles. Both sexes are especially addicted to the flowers of pink dogbane *(Apocynum androsaemiifolium)*.

DISTRIBUTION: Most area counties, including the Sacramento Valley and Delta. Populations tend to be ephemeral, so old records may not indicate the current distribution.

HOST PLANTS: In the pea family (Fabaceae), *Lotus,* including annual species, especially Spanish lotus *(L. purshianus),* as well as large perennials such as *L. crassifolius;* so far unrecorded from the widely naturalized alien birdfoot trefoil *(L. corniculatus).* Native perennial vetches *(Lathyrus)* suspected also. Reports on willow *(Salix)* are based on eastern U.S. populations that are probably not conspecific with ours.

SEASONS: Three to four broods, March to October.

SIMILAR SPECIES: As noted in the Pacuvius Duskywing *(E. pacuvius)* account, these two may be impossible to differentiate without recourse to the male genitalia. Specimens taken in association with *Lotus* or in the second half of the season are likely to be Persius Duskywings—and those may be the best "taxonomic characters" you will get!

TWO-BANDED SKIPPER   *Pyrgus ruralis*

Pl. 28

This is a species of cool, moist habitats, disjunctly distributed between the Coast Ranges and the west slope of the Sierra Nevada. Its typical habitats are forest-meadow ecotones, coastal scrub-meadow ecotones, and the margins of ponds and bogs, where it can be found flying just above the ground near its host plant. Local and spottily distributed, it is usually uncommon in our area. Globally, this may actually be more than one genetic species, but there is no suggestion of more than one within the area covered by this book. Males perch on the ground and visit puddles.

DISTRIBUTION: Most Bay Area counties, but very localized. Absent from the Delta and the Sacramento Valley.

HOST PLANTS: Horkelia *(Horkelia)* species (in the rose family [Rosaceae]). Species of the related cinquefoils *(Potentilla)* and strawberries *(Fragaria)* have been recorded outside our area.

SEASONS: One brood, March to mid-July, typically flying for about 3 weeks in a place.

SIMILAR SPECIES: It is smaller than the Common Checkered Skipper *(P. communis)*, with more grizzled pattern and strongly checkered fringes, and the underside hindwing pattern is reddish, not gray. It is similar to the spring brood of the Least Checkered Skipper *(P. scriptura)*; again, the underside pattern is reddish, not mustard brown to gray. The summer brood of the Least Checkered Skipper is very different. The habitats of the Two-banded and Least Checkered Skippers are mutually exclusive; they never occur together, though both co-occur with the Common Checkered Skipper.

LEAST CHECKERED SKIPPER   *Pyrgus scriptura*
or SMALL CHECKERED SKIPPER

Pl. 28

Formerly common in its special habitat, this little skipper is the only Sacramento Valley butterfly that collectors elsewhere are eager to obtain. It is an extreme host specialist, rarely straying from patches of alkali mallow *(Malvella leprosa)*. The summer form is tiny, with much-reduced white markings above, and really looks like a different species. The flight is "busy" like that of a small bee, usually just above the ground. Males perch but can often be seen patrolling in linear habitats; they also dive into dense stands of the host, apparently looking for newly emerged females. This species visits flowers with shallow corolla tubes, including heliotropes *(Heliotropium)*, lippias *(Lippia)*, and, oddly, narrow-leaf milkweed *(Asclepias fascicularis)*, but not its host plant or its other most consistent associate, *Cressa* (Convolvulaceae).

Jerry Powell is quoted by Steiner (1990) as stating that the Least Checkered Skipper "invaded the Central Valley via weedy habitats and was discovered in the Delta area in 1956–57." If this is correct, it presumably came from the south. However, when Emmel, Emmel, and Mattoon fixed the type-locality of the species as Sacramento (see "California's First Lepidopterist"), they ignored this claim and assumed the species was present in Lorquin's time. The valley has been so poorly collected that it may well have been there all along, and the habitat is not one that generally attracts collectors.

DISTRIBUTION: Sacramento Valley, north at least to Butte County, and Delta portions of Bay Area counties; a few records from Napa, Sonoma, and San Mateo Counties. Found on heavy, often alkaline, clay soils with the host plant; also on railroad embankments and on waste ground and agricultural or ranch land. Commonly associated with the Sandhill Skipper *(Polites sabuleti)* and Western Pygmy Blue *(Brephidium exile)*. Colonies of this butterfly (like the host plant) are very persistent; its distribution in greater Sacramento has not changed significantly in over 30 years, but it disappeared in 2006 from many sites and it remains to be seen if it will recover. Many large patches of the host remain undiscovered by it, thus unoccupied.

HOST PLANTS: Alkali mallow *(Malvella leprosa,* formerly known as *Sida hederacea)*. This plant has creamy white flowers in our area. It occurs disjunctly near Mexico City and in eastern Patagonia, Argentina, and near Buenos Aires; in these areas it has lemon yellow flowers. In Patagonia a different species of *Pyrgus* (*P. seminigra*), which is not closely related to our Least Checkered Skipper, also specializes on this plant and is almost indistinguishable, superficially, from our species—a striking case of convergent evolution! Although the Least Checkered Skipper has never been known to breed on any of the common weedy mallows *(Malva)* in nature, it is easily reared on them and in fact may actually grow faster and larger on them than on its own host.

SEASONS: February to October; several broods. The first brood resembles the Two-banded Skipper *(P. ruralis);* the second is variable and the third and subsequent broods are very small and dark. Control of the seasonal polyphenism has not been studied. In drought years this species may enter diapause as early as July and "disappear" until the following spring, apparently skipping summer breeding.

SIMILAR SPECIES: The Two-banded Skipper has the underside pattern distinctly reddish, and it occurs in moist habitats near the coast.

COMMON CHECKERED SKIPPER   *Pyrgus communis*
Pl. 28

This species lives up to its name; it is common and widely distributed throughout our area in open, sunny habitats. It is often seen in city parks, gardens, and urban vacant lots, as well as in rural areas. Males are territorial and typically perch well off the ground, up to waist height. They perform spectacular aerial male-male chases, typically returning to their previous perches by executing a wide circle near human eye level. Males flash bluish in flight. Females usually fly close to the ground and lack the bluish sheen. Both sexes visit a great variety of flowers, with mints *(Mentha),* including garden lavender *(Lavandula),* and birdfoot trefoil *(Lotus corniculatus)* among their favorites.

The Common Checkered Skipper seems to have benefited from the human transformation of lowland Californian landscapes and vegetation. Our populations are multivoltine. Assuming they were multivoltine before Europeans came, they had to have been restricted to a handful of native host plants whose seasonality would support multiple broods: alkali mallow *(Malvella leprosa)* in alkaline clay soils in the Central Valley, and various species of checkerbloom *(Sidalcea)* in tule marshes and moist meadows. These are still used, but the vast majority of eggs can be found today on the introduced weedy mallows *(Malva)* in disturbed habitats. It seems certain that the Common Checkered Skipper must be much more abundant now than it could have been 200 years ago.

From the Transverse Ranges southward throughout southern California, the Common Checkered Skipper is replaced by an animal called the White Checkered Skipper *(P. albescens).* The common and scientific names are both deceptive, because the White on average is no whiter than the Common; in fact, the two are reliably distinguishable only by the male genitalia. (See John M. Burns, "*Pyrgus communis* and *Pyrgus albescens* are separate transcontinental species with variable but diagnostic valves," *Journal of the Lepidopterists' Society* 54:52–71, 2000.) Their biology is also identical. Only a few individuals with *albescens* genitalia have been detected north of the Transverse Ranges—all south of our area.

Are these really good species? Molecular studies, still in progress at this writing, present a confusing picture. At the molecular level, southern California White Checkered Skippers are consistently different from northern California Common Checkered Skippers—except in the Sierra Nevada. Sierran genitalic Commons (from above 1,200 m [4,000 ft]) are univoltine and feed only on native checkerbloom (*Sidalcea*), often on the floor of open coniferous forest, sometimes on meadows. Molecularly, these are White Checkered Skippers! Along the Sierran east slope and in the western Great Basin are multivoltine populations feeding on weedy mallows, and these are indistinguishable from our Commons. There thus appear to be three entities: multivoltine low-elevation and Great Basin morphological and molecular *P. communis;* multivoltine southern Californian morphological and molecular *P. albescens;* and Sierran univoltine, morphological *P. communis*–molecular *P. albescens.* Clearly, there is a great deal we do not understand about these critters.

We know, thanks to Burns, that the White Checkered Skipper has expanded its range eastward all the way to the Atlantic coast in recent decades; it shows up abruptly in collections at specific times. It is possible that it is also moving into lowland northern California! Because the species is considered so ubiquitous, this could easily slip under our radar. No one is seriously suggesting that the genitalia of every male in our area be checked, but users of this book should keep in mind the possibility of a White Checkered Skipper invasion and—if so inclined—take a peek at the form of the valves from time to time (see Burns' article for drawings).

DISTRIBUTION: All counties, everywhere except deep shade.

HOST PLANTS: Herbaceous mallows (Malvaceae), including alkali mallow (*Malvella leprosa*), checkerblooms (*Sidalcea*), garden hollyhocks (*Alcea*), and all species of the weedy mallows (*Malva*), for example the cheeseweeds (*M. neglecta, M. nicaeensis,* and

M. parviflora). The use of velvet leaf *(Abutilon theophrasti)* also called Indian mallow, in our area needs to be confirmed. In the Sacramento area, now using the introduced lawn and golf course weed *Modiola caroliniana.* Eggs often laid on prostrate mallows in mowed lawns, seemingly in preference to tall, robust specimens of the same species. Larva in a rolled-leaf nest, resting in a fishhook posture, apple green to purplish brown, with a dark line down the back and a very dark, blackish brown head. Hibernates as part-grown larva, usually pupating in the last larval nest. Pupa buff with brown markings, attached by the cremaster.

SEASONS: Continuous broods February to November (rarely December).

SIMILAR SPECIES: See discussion for White Checkered Skipper, not known in our area. The Two-banded Skipper *(P. ruralis),* is about 75 percent the size of an average Common Checkered Skipper, with a distinctly grizzled look, no blue sheen, and a complex reddish ventral hindwing pattern; it is very local with the host *(Horkelia),* in moist habitats near the coast. The Least Checkered Skipper *(P. scriptura)* is 65 percent (early spring) to only 50 percent the size of the Common; it has more black in the wing pattern, especially in summer broods where white above may be reduced to small dots; the underside hindwing pattern is grayish and different in detail from that of the Common (see pl. 28). It prefers hot inland areas, often alkaline, with alkali mallow. Both species may co-occur with the Common Checkered Skipper. The Northern White Skipper *(Heliopetes ericetorum)* averages 115 percent the size of the Common Checkered Skipper; the male has the upper surface nearly all white except near the margins; the female has a checkered pattern that is incomplete in the middle of the wings; the underside of the hindwing has a wavy brown line in the middle and is not checkered. It is found in sunny, open country, especially chaparral, and is scarce.

NORTHERN WHITE SKIPPER   *Heliopetes ericetorum*
or LARGE WHITE SKIPPER
or LARGE CHECKERED SKIPPER
Pl. 28

This species is poorly understood in our area. It seems to be a seasonal migrant, passing through at low density in late spring and again in fall, but we do not know where it is coming from or where it is going. Aside from a record on the Peninsula, it occurs mainly in the Central Valley and along the east edge of the Inner Coast Range. I have collected it from sea level to 2,100 m (7,000 ft) in northern California but very rarely see a breeding population. A swift, strong flyer, it is easy to spot but difficult to catch. Fortunately it is an avid flower visitor; I once caught a male by hand on a flowering privet in a neighbor's garden in Davis. What was it doing in Davis, and where was it going when I interrupted its journey?

DISTRIBUTION: Central Valley, Delta, and interior Coast Range counties. I have found it breeding in Mount Diablo State Park, on the south slope of the Sutter Buttes, and in interior Napa County (once each!).

HOST PLANTS: Recorded by Coolidge on a variety of mallow family (Malvaceae) plants in southern California. I have found it only on bush mallows *(Malacothamnus)* in the northern part of the state. Incredibly, I once found a female nectaring on a specimen of this plant in the University of California at Davis Arboretum!

SEASONS: Recorded from April through October, but most records are from May to June and September to early October, strongly suggesting seasonal migration.

SIMILAR SPECIES: Females are vaguely similar to the Common Checkered Skipper *(Pyrgus communis)*, but substantially larger; the ventral hindwing is distinctive.

COMMON SOOTYWING   *Pholisora catullus*
Pl. 26

Catullus was a Roman love poet. This gloomy-looking creature seems an odd choice to name for him. The "Common" Sootywing is not so common any more. In fact, at this writing it has become rare and has disappeared from many places in our region—which is interesting since its hosts are abundant urban and rural weeds, and it formerly bred in numbers in vacant lots and untidy farm yards all over northern California. As a demonstrably successful human associate, it would seem an unlikely candidate for catastrophic decline. Yet…

The Common Sootywing flies just above the ground; males have a characteristic zigzag motion as they "fly a beat" along roadsides. They nectar at lippias *(Lippia)*, heliotropes *(Heliotropium)*, and similar low, shallow-tubed flowers, and are the only butterflies I have seen to nectar at yellow oxalis *(Oxalis)* in lawns (the small summer-flowering species that go under mowers, not Bermuda buttercup!). The female has a more complete white pattern above; both sexes are completely dark on the hindwing below. This is strictly a species of full sun, but it usually sits with the wings open.

DISTRIBUTION: General in disturbed, weedy habitats, including the Mission District and industrial areas in the East Bay, but in rapid decline in most areas. Between 1995 and 2005 it largely disappeared from Yolo, Solano, and Sacramento Counties.

HOST PLANTS: In our area reported reliably only on pigweeds (numerous species of *Amaranthus*) and the closely related garden cockscomb *(Celosia),* both in the amaranth family (Amaranthaceae). Reports on the related goosefoot family (Chenopodiaceae) need confirmation in our area. Larva is apple green with a black head; lives in a rolled-leaf shelter. Solitary, but larvae tend to be clumped on one or adjacent plants. Hibernates as mature larva.

SEASONS: Multiple broods, February to November.

SIMILAR SPECIES: The small duskywings are all larger and (except the Pacuvius Duskywing *[E. pacuvius]*) less glossy. The white markings on the Common Sootywing are extremely sharp; the hindwing is unmarked below. The white face is unique. The Roadside Skipper *(Amblyscirtes vialis)* holds the wings over the back at rest, has a violet shade along the outer edge of the ventral hindwing, has checkered fringes, and occurs in cool, moist woods, usually far from human activity.

Skipperlings (Heteropterinae)

ARCTIC SKIPPER   *Carterocephalus palaemon*
Pl. 29

Not truly arctic, but boreal, this delicate little skipper occurs in a few cool, damp localities north of the Bay. Like the Dreamy Duskywing *(E. icelus),* it may be overlooked easily, as it is inconspicuous and retiring in its habits, but unlike that species, it can be confused with nothing else in our area. It flies in dappled light and shade along streamsides through forest. Males perch in sunflecks with the wings opened slightly. They also fly short loops around the perimeter of their sunfleck and have been known to puddle. Their flower-visiting behavior is very discreet: they usually visit flowers of perennial vetches *(Vicia)* and sweet peas *(Lathyrus)* by pitching up onto them from below. The Arctic Skipper is usually a low-density species but has infrequent years of abundance, when it can be found in places where its existence was unsuspected; it ranges completely around the Northern Hemisphere.

DISTRIBUTION: Very local in North Bay counties, reaching its southern limit in Marin. Inland it reaches at least the Mendocino Pass area in Glenn County, with a few very isolated populations in the northern Sierra to near Interstate 80.

HOST PLANTS: Grasses; no specific records in our area.

SEASONS: One brood, May to June.

SIMILAR SPECIES: None.

Branded Skippers or Grass Skippers (Hesperiinae)

FIERY SKIPPER   *Hylephila phyleus*
Pl. 30

This is our most urban butterfly, rarely found far from mowed lawns. After suffering apparently heavy mortality over winter, it is usually rare and spotty early in the season but becomes abundant by late July. By September and October it is abundant in gardens, visiting zinnias *(Zinnia)*, marigolds *(Tagetes)*, lantana *(Lantana)*, and other flowers. It tends to overwinter in the same places every year; these locations presumably serve as foci from which successive broods spread out to blanket the region. Adults are very active and agile; male-male chases can be marvelously aerobatic. This species belongs to a mostly Andean genus; the Fiery Skipper itself has a huge range, from northern California to central Argentina and Chile. It is not known whether it is native to California or was introduced from the tropics or subtropics. It may not have been in our area before the 1930s.

Specimens collected in cool weather tend to have heavier spot-patterns beneath, and the female ground color beneath is sootier. Otherwise, there is little variation.

DISTRIBUTION: All counties, almost never in wildlands.

HOST PLANTS: Mostly Bermuda grass *(Cynodon dactylon)*, occasionally on other lawn grasses such as Dallis grass *(Paspalum)*; its native host may be alkali grass *(Distichlis spicata)*, which occurs naturally nearly throughout its hemispheric range.

SEASONS: Records exist for all months, but mostly March to early December; rarely common before July.

SIMILAR SPECIES: The Fiery Skipper has a sharply pointed fore-wing and short antennae; the ventral hindwing pattern has discrete spots not forming connected bands; the stigma is slender, the gray in center scarcely noticeable; the male has triangular dark spots along the margins. This set of traits should distinguish the Fiery Skipper from all the other similarly colored skippers.

YUBA SKIPPER   *Hesperia juba*
Pl. 29

The Yuba Skipper is seldom encountered in our area and is probably not resident below about 1,000 m (3,000 ft). It is a very distinctive, large skipper with strongly greenish pelage and a dark green hindwing below with bold, squarish silvery spots in both sexes. This animal is seen in a great variety of habitats but above all is a strong flower visitor.

DISTRIBUTION: Sporadic in most Bay Area counties. Unrecorded in the Sacramento Valley except at the far north end and (very rarely) in the Delta.

HOST PLANTS: Perennial bunchgrasses. Preferences in our area unknown.

SEASONS: Two flights, early spring and late summer to fall, the latter typically much more abundant in its breeding localities in the mountains. Circumstantial evidence suggests this species overwinters as an adult, which would imply only one brood per year.

SIMILAR SPECIES: The Yuba Skipper is at least 15 percent larger than other species of *Hesperia* in our area. The only other similarly colored skippers approaching it in size are the very different-looking Yuma Skipper *(Ochlodes yuma)* and Sachem *(Atalopedes campestris)*.

Comma Skipper Complex
(Hesperia comma, H. colorado)

The comma skipper complex occurs all around the Northern Hemisphere, breaking into many named entities in western

North America, but far fewer in Eurasia. The greatest taxonomic diversity is right here in California. Studies of DNA phylogeography mirror this pattern: molecular diversity is greatest in California and lowest in the Old World. The DNA suggests that the group originated in North America, crossed the Bering Strait into Eurasia, evolved for a while there, then recrossed from Siberia into far northwestern North America very recently. Older works usually call the whole group *H. comma,* the oldest name (based on European material), but recent authors separate most North American populations as a separate species *H. colorado* (some from British Columbia and north appear to be true *H. comma*). All of ours fall under *H. colorado.* Our two named entities are different in so many ways as to merit separate treatment here. They tend to be quite common where found, as well as highly variable within and among local populations. All are single brooded, but the flight season can be locally variable, with populations on serpentine often flying later than others. (Why?) There are surely many more populations in our area than have been recorded to date. Watch for them.

TILDEN'S SKIPPER *Hesperia colorado tildeni*
Pl. 29

This is a dryland entity, found in grassland, chaparral, and oak woodland. It is quite lightly marked in both sexes, the males sometimes nearly unmarked below, the females ranging from nearly unmarked to fully, but lightly patterned. This beast tends to be missed because it flies together with the ubiquitous Woodland Skipper *(Ochlodes sylvanoides),* which it strongly resembles. It nectars at yellow star-thistle *(Centaurea solstitialis)* and coyote mint *(Monardella)* among other summer flowers.

DISTRIBUTION: The classic Bay Area locality is the Mount Hamilton Range in Alameda and Santa Clara Counties. Should be looked for in dry inland Coast Range areas and along the west edge of the Sacramento Valley, even in areas heavily invaded by Mediterranean annual grasses. Similar populations occur in the inner North Coast Range from Colusa County northward.

HOST PLANT: Grasses; one definite record is one-sided bluegrass *(Poa secunda).*

SEASONS: June to October, depending on locality, but always only one brood. Overwinters as first-instar larva inside the egg. According to C. D. MacNeill, the larva of Tilden's Skipper has only five instars (Dodge's Skipper *[H. c. dodgei]* has six; a very unusual life-history difference).

SIMILAR SPECIES: It is extremely similar to Lindsey's Skipper *(H. lindseyi)* and sometimes sympatric with it. Lindsey's Skipper is slightly larger and has better defined, distinctly squarish spots on the ventral hindwing, especially in the female; the veins may be vaguely lighter than the ground color. Depending on locality, Lindsey's Skipper flies two to several weeks earlier than Tilden's Skipper. If a skipper that looks like this appears to be two brooded, it is probably a seasonal succession of both. The Woodland Skipper is slightly smaller and has a much shorter, less pointed forewing, especially in males. The hindwing underside has a more or less contrasting yellow spot-band; the edges of the spots are relatively diffuse.

DODGE'S SKIPPER   *Hesperia colorado dodgei*
Pl. 29

Dodge's Skipper is much darker than Tilden's *(H. c. tildeni)*, especially beneath, where the ground color may approach chocolate. The upperside borders are also wider and darker. This is a fog belt animal, recorded in a wide variety of habitats, from dunes and coastal prairie to roadsides through forest. It often visits mints *(Mentha)*, milkweeds *(Asclepias)*, dogbanes *(Apocynum)*, and the sunflower family (Asteraceae) and puddles when it gets the chance. It is not known whether the dark color is determined genetically, environmentally, or by an interaction of the two factors.

DISTRIBUTION: Coastal areas of all counties (except not recorded in San Francisco and Sonoma), plus part of Contra Costa County.

HOST PLANT: Grasses; recorded specifically on red fescue *(Festuca rubra)*.

SEASONS: June to October depending on locality. Hibernates as the first-instar larva inside the egg.

SIMILAR SPECIES: Populations of Lindsey's Skipper *(H. lindseyi)* north of the Bay and near the coast may be almost as dark as Dodge's Skipper. Like other populations of the Lindsey's Skipper, they fly early and have rather large, squarish spots on the ventral hindwing. Coastal Woodland Skippers *(Ochlodes sylvanoides santacruza)* are also rich chocolate to deep reddish brown below but differ from Dodge's Skipper in all the same ways as the nominate Woodland Skipper *(O. sylvanoides sylvanoides)* differs from Tilden's Skipper.

COLUMBIA SKIPPER
Pl. 29

Hesperia columbia

A very strong hilltopper with a preference for—but not a restriction to—serpentine soils, this species is seldom common. The Columbia Skipper is usually found in rocky, barren areas or chaparral, or else along roads or streams at the bottoms of canyons. A puddler, it also visits goldenrods *(Solidago)*, asters *(Aster)*, dogbanes *(Apocynum)*, milkweeds *(Asclepias)*, and rabbitbrush *(Chrysothamnus)*. This is our only unambiguously double-brooded *Hesperia*.

DISTRIBUTION: All Bay Area counties. Unknown in the Sacramento Valley. An isolated population in the Cement Hill range (Lagoon Valley Regional Park). Very rare and local in the Sierra Nevada foothills.

HOST PLANTS: Perennial bunchgrasses, specifically recorded on junegrass *(Koeleria)* but almost certainly not restricted to it.

SEASONS: Two broods, late March to June and late August to November.

SIMILAR SPECIES: Normally easily told from all other similarly colored skippers by the very unusual underside hindwing pattern. A very few female Tilden's Skippers's *(H. colorado tildeni)* hindwings may resemble the Columbia Skipper's hindwing pattern, but they are never greenish. In some localities this species flies in early fall with both Tilden's Skipper and the Woodland Skipper *(Ochlodes sylvanoides)!*

LINDSEY'S SKIPPER
Pl. 29

Hesperia lindseyi

Like the Columbia Skipper *(H. columbia)*, this species tends to be commoner on serpentine than elsewhere. Unlike it, it prefers serpentine grassland, often in places that turn pink with clarkia *(Clarkia)* flowers in late spring. Not a hilltopper, it flies low and often perches on bare ground. It frequently visits thistles *(Cirsium, Carduus,* and *Silybum)*, mule's ears *(Wyethia)*, and balsamroot *(Balsamorhiza)* and others of the sunflower family (Asteraceae). As noted under Tilden's Skipper *(H. colorado tildeni)*, it is often sympatric with that species and the two are difficult to tell apart. The early flight period is often the best clue.

DISTRIBUTION: Widespread in the Bay Area, mainly in native grassland, including coastal prairie. Unrecorded from the Sacramento Valley, but possible, especially north. Populations from near the coast from southern Marin County northward are darker both above and below and were recently designated subspecies *H. lindseyi macneilli;* the female may be very dark, and the light spots on the ventral hindwing, while all present, are smaller than usual.

DISTRIBUTION: All Bay Area counties except Solano, San Mateo, and Santa Cruz.

HOST PLANTS: Perennial bunchgrasses including fescues *(Festuca)* and oatgrass *(Danthonia)*. C.D. MacNeill famously discovered that eggs are laid on lichens on fence posts and trees; the larvae drop to the ground and forage actively. This odd trait, recorded in Marin County, may be a way to avoid detection by egg parasitoids that key in on host plant chemistry. It is not known how widespread the behavior is geographically. See if you can find out.

SEASONS: One brood, mid-April (warmest inland sites) to mid-June (fog belt), always flying before the sympatric Tilden's Skipper and Dodge's Skipper *(H. colorado dodgei)*.

SIMILAR SPECIES: See Tilden's Skipper. The large, squarish spots on the ventral hindwing, sometimes with pale scaling along the veins, are helpful especially in females. Coastal females (subspecies *H. lindseyi macneilli)* may be about as dark as Dodge's Skipper; note that the spots, while not as large as in nominate *H. l. lindseyi,* are still squarish.

SANDHILL SKIPPER   *Polites sabuleti*
Pl. 30

Our multivoltine Sandhill Skipper is closely associated with the native turfgrass alkali grass *(Distichlis spicata)* but has been known to breed in Bermuda grass *(Cynodon dactylon)* lawns in the East Bay and Sacramento Valley. Despite its name, this species is not particularly associated with either sand or hills. It typically occurs in saline or alkaline areas, including the drier parts of salt marshes, and even in hillside seeps, as at Cement Hill. It often co-occurs with the Least Checkered Skipper *(Pyrgus scriptura)* on compacted alkaline clay soils. It nectars at low flowers such as lippias *(Lippia)* and heliotropes *(Heliotropium)* but in fall frequents taller plants such as asters *(Aster)* and rabbitbrush *(Chrysothamnus)*. Males perch on the ground and dart out at other butterflies.

A bewildering array of subspecies occurs in California, from sea level to above tree line. Some of these are single brooded. More than one genetic species may be involved. Some North Bay populations would repay genetic study, as noted below. In multivoltine populations, cool-season animals are small and may be much darker than summer ones, but can always be distinguished from the high-altitude forms. The lightest summer males are virtually indistinguishable from the usual forms seen in desert oases in southern California and the Great Basin.

DISTRIBUTION: All counties; often abundant.

HOST PLANTS: Alkali grass *(Distichlis spicata)* and Bermuda grass *(Cynodon dactyon),* perhaps other turfgrasses. Other subspecies often use perennial bunchgrasses.

SEASONS: March to November, with multiple broods. Populations near the coast north of the Bay, for example, at Salmon Creek near Bodega Head, seem to be univoltine (September to October) and may represent the northern subspecies *P. s. aestivalis,* known from the high Interior North Coast Range (Colusa County) north into the Pacific Northwest.

SIMILAR SPECIES: The underside pattern, small size, and unusual male stigma make this species unmistakable in our fauna.

DOGSTAR SKIPPER   *Polites sonora siris*
Pl. 29

The Dogstar Skipper is a coastal fog belt representative of a species widely distributed in the Sierra Nevada and northern mountains. As usual, the ground color of the ventral hindwing is very dark. Also as usual, we do not know if this trait is genetically or environmentally determined. Apparently it is restricted to coastal prairie and openings in Douglas fir or closed-cone pine forest on the west side of Sonoma County. It would not be surprising if the range of this inconspicuous hesperiid were considerably wider than we now know. The Dogstar Skipper visits clovers *(Trifolium)* and other low flowers; males perch low and are territorial.

DISTRIBUTION: Northwestern Sonoma County and northward.

HOST PLANTS: Presumably perennial grasses.

SEASONS: One brood, May to August, reflecting the usual attenuation of the season observed in the fog belt.

SIMILAR SPECIES: The hindwing underside is very different from that of the Sandhill Skipper *(Polites sabuleti),* and its habitat is different. Coastal Woodland Skippers *(Ochlodes sylvanoides santacruza)* have a similarly dark hindwing, but the spot-band is much broader and has diffuse edges.

SACHEM or FIELD SKIPPER   *Atalopedes campestris*
Pl. 30

Common and widely distributed in urban and suburban contexts, the Sachem is less common to rare in wildlands. This species has been extraordinarily responsive to climate change. It is now emerging nearly a month earlier in the Sacramento area than it did 30 years ago, and its range has exploded both northward (to central Washington) and eastward (to Sierra Valley, Carson Valley, and the vicinity of Verdi, Nevada) in the same period. Fall individuals, especially females, may be very dark, chocolate brown beneath even in the hot interior. There is substantial variation on the under surface, but the upper pattern is quite constant.

This species is usually scarce early in the season but explodes in fall, often even outnumbering the Fiery Skipper *(Hylephila phyleus)* for a while in gardens. For many years the Sachem has been very abundant in Capitol Park in downtown Sacramento. It is less numerous coastside. Adults visit a great variety of flowers, particularly tall ones such as thistles *(Cirsium, Carduus,* and *Silybum)* and tall blue verbena *(Verbena bonariense),* as well as zinnias *(Zinnia),* marigolds *(Tagetes),* lantana *(Lantana),* and the like. Males perch and are highly territorial.

DISTRIBUTION: All counties in urban and suburban environments, most common inland; frequent in riparian and tule marsh habitats; sporadic and unpredictable in woodland, forest, and chaparral, mostly fly-ups from open country.

HOST PLANTS: Lawn grasses including Bermuda grass *(Cynodon dactylon)* and Dallis grass *(Paspalum)*; presumably others away from civilization. As usual for grass-feeding skippers, the larva lives in a rolled-leaf nest. In mowed lawns it passes unharmed under mower blades.

SEASONS: March through November, three to four generations, the last always commonest.

SIMILAR SPECIES: The only other similarly colored skippers as big as this are the Yuma *(Ochlodes yuma)* (nearly unmarked orange) and Yuba *(Hesperia juba)* (green ventral hindwing with silver spots). It is easily told from its usual companion, the Fiery Skipper, by the very unusual male stigma and the square hyaline spot near the middle of the female forewing, with a very dark area basal to it. It is not uncommon to find the Sachem, the Fiery Skipper, the Woodland Skipper *(O. sylvanoides)* and the Sandhill Skipper *(Polites sabuleti)* all nectaring at the same plant on the same September day. In the East Bay they are likely to be joined by the Umber Skipper *(Poanes melane)*!

WOODLAND SKIPPER   *Ochlodes sylvanoides*
Pl. 31

It may be common in woodlands, but the Woodland Skipper also occurs everywhere else except manicured lawns: forest openings, chaparral, grassland, and tule marsh. It is, however, retreating steadily from the advance of suburban subdivisions. Males bask in sunspots and chase one another vigorously, but such interactions never occur when nectaring. Males often remain at their perches until very late in the afternoon. They also puddle freely.

The Woodland Skipper visits many flowers, ranging from yellow star-thistle *(Centaurea solstitialis)* to tall blue verbena *(Verbena bonariense)*, but is unique in our fauna in its addiction to vinegar weed *(Trichostema lanceolatum)* flowers. Other similar-sized hesperiids almost never visit this plant, but the Woodland Skipper swarms over it and clearly is able to reach the nectar in its very oddly shaped and oddly balanced flowers. (The only other butterfly that routinely accomplishes this is the Pipevine Swallowtail *[Battus philenor]*.) It also is the only hesperiid to visit the pink, delicate flowers of *Epilobium brachycarpum* (usually called *E. paniculatum*). In both cases it triumphs by hanging upside-down. Although this species often occurs in cool, shady places under trees, it is also one of the three butterflies that can be found in hot, dry grassland in fall, visiting tarweeds (subtribe Madiinae, Asteraceae) (the other two are the Buckeye *[Junonia coenia]* and, when present, the Checkered White *[Pontia protodice]*).

The hindwing underside is extremely variable both within

and among populations. Coastal fog belt populations are dark brown below with sharply contrasting yellow markings (subspecies *O. s. santacruza*), but rather dark reddish females (rarely males) can turn up in inland populations as well. I have seen two males of this species apparently born without claspers. One wonders if such individuals can mate! I have never seen this trait in any other butterfly.

DISTRIBUTION: Throughout the area, but rare or absent in urbanized settings.

HOST PLANTS: Perennial grasses, including *Phalaris* and *Leymus*.

SEASON: June through November, apparently two broods, both in the second half of the season.

SIMILAR SPECIES: The Farmer *(O. agricola)* is substantially smaller, flies earlier, and has a differently shaped stigma; the underside pattern of light spots is usually very diffuse. From Tilden's Skipper *(Hesperia colorado tildeni)* it may be told by the short forewing and less distinct underside pattern. Substantially smaller than the Yuma Skipper *(O. yuma)*, which is plain unmarked yellow beneath.

FARMER or RURAL SKIPPER   *Ochlodes agricola*
Pl. 31

The common name "Rural Skipper," which is appearing increasingly in field guides, is merely confusing given the existence of a *Pyrgus ruralis* in our fauna, but at least it is ecologically valid: this is a species that eschews civilization. It is generally abundant in foothill canyons, flying exactly when California buckeye *(Aesculus californica)*, its principal nectar source, blooms. It also visits coyote mint *(Monardella)*, dogbanes *(Apocynum)*, and other flowers when available. Adults fly in and out of sunflecks and are undeterred by shade. Males perch but seldom remain in one place for long. This species usually co-occurs with the Woodland Skipper *(O. sylvanoides)* but disappears just as that species begins emerging, so seasonal overlap between them is very limited if it exists at all. One might speculate that their seasonality evolved to prevent interspecific competition, were there a resource they might compete for! Alternatively, it could be an anti-hybridization mechanism—but they are not known to hybridize anywhere.

DISTRIBUTION: All Bay Area counties in wildlands. Unrecorded in the Sacramento Valley except on the fringes mostly at the extreme north end; near Sacramento to Iron Point, Folsom.

HOST PLANTS: Perennial grasses; specific records are lacking.

SEASONS: Recorded mid-March to mid-August, but the vast majority of records are from mid-April to late June. One brood.

SIMILAR SPECIES: Smaller than the Woodland Skipper; it has a somewhat squarish forewing shape, and the form of the male stigma is distinctive; the ventral hindwing has no dark pattern, with at most a vague, diffuse medial pale pattern; the ventral hindwing often is distinctly purplish, especially in the female. The early flight period, shared only with the significantly larger species of *Hesperia,* is itself diagnostic.

YUMA SKIPPER   *Ochlodes yuma*
Pl. 31

This branded skipper occurs disjunctly in the desert Southwest, including the southern Great Basin, and in the Sacramento–San Joaquin Delta and Suisun Marsh. We do not know if this is an old relict pattern or if our populations represent an introduction. The oldest records are from near Modesto in 1938 and Antioch in 1954. Scott, Shields, and Ellis in 1977 suggested that this was a recent colonization (*Journal of the Lepidopterists' Society* 31: 17–22), but in 1999 George Austin named our populations as a new subspecies *O. y. sacramentorum,* noting that "the considerable distribution and phenotypic differences from Great Basin populations" argue otherwise. Emmel, Emmel, and Mattoon (1998) have further suggested that the name *Hesperia ruricola,* proposed by Boisduval in 1852 and presumably based on Lorquin's material, might be based on this species! The name has been variously associated (e.g., with the Dun Skipper *[Euphyes vestris],* a very different beast), and the type is lost. The *H. ruricola* angle can only be resolved by discovery of a genuine Lorquin specimen, but the provenance of the Delta *O. yuma* might conceivably be determined by molecular means.

This is our only other branded skipper as large or larger than the Yuba Skipper *(H. juba).* It is distinctively bright orange, with little or no pattern. Males perch, typically at about waist height, but sometimes on the ground, and engage in aerial chases too fast to follow long with the eye. Females are much less conspicuous. Despite the abundance of the host plant, the

Yuma Skipper is never very common. It has two well-marked broods a year.

DISTRIBUTION: Delta and Suisun Marsh, with the host plant. Very rare elsewhere, but recorded several times around Sacramento as an apparent stray.

HOST PLANTS: Common reed *(Phragmites australis,* also called *P. communis).* The larva lives in a rolled-leaf nest. Common reed in North America is actually two genetic entities. One is native to the West and is nonweedy. The other is of Old World origin and is an aggressive, invasive weed in the East and in some parts of the West, such as Utah. The weedy common reed has been adopted as a host by the large Eastern species called the Broad-winged Skipper *(Poanes viator),* which subsequently expanded its range and became extraordinarily abundant. In contrast, all of our populations of *O. yuma* are associated with the native strain. It has not been determined whether this butterfly has encountered the weed. It should be interesting to observe what happens if it does. The plants must be differentiated molecularly, although their invasiveness is usually a good indicator of their identity.

SEASONS: Two broods, late May to July, late August to October.

SIMILAR SPECIES: Much larger than the Woodland Skipper *(O. sylvanoides);* ventral hindwing usually unpatterned.

UMBER SKIPPER   *Poanes (Paratrytone) melane*
Pl. 30

Your image of the Umber Skipper depends on where you live. In the East and South Bay this is an abundant urban species, the commonest "lawn skipper" in Berkeley. Inland it is scarce, local, and confined to riparian habitat, where it flits in and out of bramble thickets in dappled light and shade. There is little sexual dimorphism. The adults are eager flower visitors, and the males perch in sunflecks and engage in very rapid aerobatic chases.

DISTRIBUTION: All counties, but in quite different ecological contexts inland versus near the coast.

HOST PLANTS: In the East Bay largely on Bermuda grass *(Cynodon dactylon)* in lawns, but inland presumably on native riparian grasses, not identified. Recently (2006) observed ovipositing on the naturalized grass *Rytidosperma racemosum* in Berkeley.

SEASONS: March to mid-December in the Bay Area, where multivoltine. Inland April to October, with two or three reasonably discrete generations. Hibernation as part-grown larva. This species usually pupates in the larval shelter, which is reinforced with a flocculent, waxy material.

SIMILAR SPECIES: None in our fauna.

DUN SKIPPER   *Euphyes vestris osceola*
Pl. 31

It seems odd that our subspecies of the Dun Skipper, endemic to the California coast, is named for a Seminole Indian chief from Georgia or Florida. It is an old name, dating from the late-nineteenth-century tradition of naming hesperiine skippers for Native Americans (and pyrgine skippers for classical authors!).

Restricted to cool, damp, even boggy spots along streams and roads in Douglas fir and redwood forests, this insect may well be more widespread than we think—it is very easy to overlook. Three subspecies are recognized in California, differing primarily in their degree of inconspicuousness. Populations placed (perhaps incorrectly) in the nominate subspecies *E. v. vestris,* sometimes called the Eastern Dun Skipper, occur primarily in pitcher plant *(Darlingtonia)* bogs in far northern California and in the Sierra Nevada south to the Feather River Canyon. Subspecies *E. v. harbisoni,* sometimes called Harbison's Dun Skipper, occurs, bizarrely, in desert oases in San Diego County and may represent a completely separate salient, derived from Texas via the desert Southwest.

Males perch and puddle. They are especially fond of visiting flowers of self-heal *(Prunella vulgaris),* also called heal-all, a small, inconspicuous mint of streamsides and forest edges. It is not uncommon in its specific localities and should be looked for elsewhere.

DISTRIBUTION: Santa Cruz and Sonoma Counties, and north of our area. Probably a fog belt endemic.

HOST PLANTS: Sedges *(Carex* or *Cyperus).*

SEASONS: One brood, May to July, attenuated as usual for the fog belt.

SIMILAR SPECIES: The Roadside Skipper *(Amblyscirtes vialis)* has a more truncate forewing and a violet shade on the ventral hindwing. The Eufala Skipper *(Lerodea eufala),* found mainly in the hot interior, is weedy and has sharply defined white spots. The female Dogstar Skipper *(Polites sonora siris)* has more gold above and a distinct pattern on the ventral hindwing below.

ROADSIDE SKIPPER *Amblyscirtes vialis*
Pl. 26

As its name suggests, the Roadside Skipper occurs along roadsides, but only in cool, moist, generally mixed forest, where it is often associated with the Arctic Skipper *(Carterocephalus palaemon)*, the Dreamy Duskywing *(Erynnis icelus)*, the Western Tailed Blue *(Everes amyntula)*, and the Western Sulphur *(Colias occidentalis chrysomelas)*. It is probably undercollected and its range in our area underestimated, as it is so easy to overlook. Males perch on leaves in sunflecks and visit puddles and various flowers, especially the native vetches (*Vicia* and *Lathyrus*). There is no phenotypic variation.

This is the only Californian species of a large genus whose metropolis is in the Southwest. It is the most widely distributed member of the genus, and much of its distribution appears relictual.

DISTRIBUTION: North Bay, plus Alameda County; local.

HOST PLANTS: Presumably grasses; undetermined in our area, but almost certainly native perennials of shaded habitats.

SEASONS: One brood, mid-April to June.

SIMILAR SPECIES: Its habitat and the violet shade below readily distinguish this species from the Eufala Skipper *(Lerodea eufala)*. It usually holds its wings up over the back at rest, unlike the small duskywings, with which it may occur, and the Common Sootywing *(Pholisora catullus)*. Easily told from the Dun Skipper *(Euphyes vestris osceola)* by wing shape and checkered fringes.

EUFALA SKIPPER *Lerodea eufala*
Pl. 30

This skipper is poorly understood in our area—is it resident or a migrant or what? The species ranges from the Sacramento Valley to central Argentina and Chile and looks the same everywhere. In our area it occurs primarily in the Sacramento Valley and Delta, though sporadic records exist for most Bay Area counties, and it seems to turn up regularly around Bodega Bay. There are a few

April records, but the Eufala Skipper usually first appears in July and becomes abundant from September through November in annual grassland, agricultural areas, urban waste ground, and roadsides. Adults visit yellow star-thistle *(Centaurea solstitialis)*, lippias *(Lippia),* heliotropes *(Heliotropium),* and various garden flowers. They will also pitch upward to visit off-season vetch *(Vicia)* flowers upside-down, and they frequently forage in shade. This species shows no sign of territoriality and always flies close to the ground. Several buff-colored individuals, presumably mutant, have been taken in our area. The Eufala Skipper has a unique behavior: upon alighting it "shrugs" its wings several times before folding them.

DISTRIBUTION: Sacramento Valley and Delta; sporadic elsewhere; not found in forest, though not averse to shade.

HOST PLANTS: Bermuda grass *(Cynodon dactylon),* Johnson grass *(Sorghum halepense),* Dallis grass *(Paspalum),* barnyard grass *(Echinochloa),* cultivated rice *(Oryza);* probably other grasses, annual and perennial.

SEASONS: March to December, but rarely seen before June; over-wintering status unknown.

THINGS TO DO
WITH BUTTERFLIES

AMERICANS ARE BECOMING more and more remote from the "real world." We read constantly that our lives are becoming dominated by our electronic media. For an entire generation, nature is something to be enjoyed vicariously in *National Geographic* specials. This may be less true in the Bay Area than in many parts of the country. Northern Californians are traditionally dedicated environmentalists and outdoorspeople. Enjoying butterflies is one reason to get out into the fresh air, if in fact one needs a reason. Here is a quick rundown of things you can do with butterflies. Some of them are purely personal; some have at least the possibility of providing new knowledge or enjoyment to others.

Butterfly Collecting

A couple of decades ago books like this devoted several pages to basic collecting techniques. In fact, collecting was the only recreational activity generally associated with butterflies. The cultural climate has changed, however, and nowhere more so than in our geographic area. The era of hobby collecting is effectively over. The general perception is that living organisms are not appropriate objects for collection, especially when so many species seem to be disappearing. There has been bitter controversy over whether collecting constitutes a threat to endangered populations or species. Whether it does or not, the psychic income derived from traditional collecting can be gotten from butterfly watching and butterfly photography—the joy of physical possession is redirected from the animal itself, dead on a pin, to one's notebooks or photos or video records of the animal in life. I think this is for the good.

That is not to say collecting is not sometimes necessary. Only a handful of professional scientists are working on butterflies. Most of what we know of butterfly distribution and variation is the work of amateurs, some of whom are exceedingly skilled and sophisticated. Our major institutional collections were built up very largely by the acquisition, by donation or purchase, of amateur collections; when professionals want to get the "big picture" of distribution and variation, those are the resources they need. When we seek to do molecular phylogeography on butterflies, we usually need to collect fresh material from the wild (though it is often possible to get usable DNA from old museum specimens). Similarly, to look for defensive chemicals or extract pigments for analysis or study thermoregulatory mechanisms, real animals are required.

If you do want to collect, information on techniques is readily available. See the "Resources" section at the end of this book. If you do collect, please obey all pertinent laws and regulations (when in doubt, ask!). And where there are no restrictions, please act responsibly and try to avoid contributing to the decline of our butterfly fauna. Thank you!

Butterfly Watching

Originally promoted as a nonconsumptive alternative to collecting, butterfly watching (often called "butterflying" by analogy to "birding") has taken off in a big way. Like birders, butterfly watchers often maintain a life list of species and let an element of competitiveness enter into what they do. On the other hand, as with birders, butterfly watchers may network on-line to let fellow enthusiasts know where a rare species can be seen right now. Virtually all the phenomena we associate with birding are duplicated in butterflying, except there are no songs or calls to imitate.

Butterfly watchers, who do not have the luxury of examining specimens in the hand, are forced to develop refined discriminatory skills to identify their beasts. This is especially true of difficult groups with many similar-looking species, for example, the blues, hairstreaks, branded skippers, or duskywings. Those of us who grew up with net in hand ultimately learned to do this sans net, and we should not be unduly skeptical of the claims of others. Concentrating on the living butterfly, performing its behavioral routines without apparent threat, allows the observer to internalize subtle cues that contribute to a "search image" for the species. To behavioral ecologists, a search image is a gestalt, a set of cues employed below the conscious level to look for and recognize something specific. That is what you are using when you say "That's a Sylvan Hairstreak!" without itemizing the reasons why you think it is one.

Butterfly watching is made much easier with a pair of close-focusing binoculars, which enable you to zoom in and observe an individual insect in great detail. The North American Butterfly Association Web site (www.naba.org) can steer you to reliable information on the best makes and models and the technical specifications that make them desirable. Once you become adept at butterfly watching, almost inevitably you will begin wishing you had preserved this or that moment photographically. A great many butterfly watchers become butterfly photographers.

If you are a butterfly watcher, you should keep just as detailed and accurate records as a skilled collector would. Some watchers record their voiced narratives and transcribe them later. Some who are "into" technology carry a global positioning system (GPS) unit to record their exact location any time they make an interesting observation.

Just as for birders, the pursuit of a life list can provide a powerful incentive for travel … and a boost to the recreational-vehicle industry. Butterfly watchers often get "into" wildflowers as well. Any trip has the promise of adding to one's lifetime trove of nature experiences.

Butterfly Photography

Nearly 30 years ago a newspaper photographer followed me around in the Sierra to document what I did for a personality profile article. He had never photographed a living butterfly and thought it would be extremely difficult. We went to a place where a very unusual skipper was flying and nectaring on rabbitbrush. I encouraged him to get in close with the most appropriate lens he had and fire away. The result was a whole roll of beautiful life shots and one very proud photographer. And that was before digital photography.

Most butterflies are remarkably easy to approach (at least if one is not trying to catch them), and there is no excuse but laziness or ignorance for the obviously faked photos one sees all too often in magazines. (Sometimes the hole is visible where the pin was removed from the thorax; the posture is generally one that does not exist in nature, and the compound eyes are brown or black, a real "dead" giveaway.) The number of stunning life portraits of butterflies available in print and on-line has exploded.

Although maintaining a butterfly garden maximizes your opportunities to shoot without leaving home, you will probably want to broaden your horizons by going afield. Take your equipment with you whenever you are likely to be in a butterfly habitat, because butterfly photography is very largely a matter of opportunity. Relatively isolated nectar sources tend to concentrate many individuals and species in a small area and to maximize the likelihood that a given individual, if scared away, will return. Clumps of milkweed, dogbane, giant hyssop, yerba santa, rabbit-brush, and similar butterfly magnets should provide you with more opportunities than you'll know what to do with! Away from flowers, keep an eye out for territorial perchers (if frightened, they usually return), mud puddle clubs (note how each species tends to cluster separately; is there anything more photogenic than 15 Pale Swallowtails and six Western Tiger Swallowtails in the middle of the road?), and pairs in copula, which are usually loath to move and provide fascinating glimpses into butterfly life. Torpid blues or ringlets clinging to dry grass at Point Reyes on a foggy day, with everything festooned with dew, are irresistible. And so it goes.

Traditional (film) photography may be in decline, but for many purists (or just us old-timers) it's still the way to go. We are talking about a 35 mm single-lens reflex (SLR) camera with interchangeable lenses. If you are unfamiliar with the technology, consult a trustworthy source (there are numerous on-line resources, and any good camera store can give you good advice—but keep in mind that they want to sell you things). Ninety, 100, and 105 mm macro lenses work very well for most butterfly work and can be combined with extension tubes to get you even closer. The 180 or 200 mm lenses generally should be used with a tripod, which is a pain to carry. Realistically speaking, great photo ops usually do not afford you the time to set up and calibrate a lot of equipment. Flexibility counts.

For digital photography, the equipment landscape is changing so fast and the quality improving so rapidly that any recommendation given here would already be out of date by the time you read it. Again, there are abundant on-line sources, and your trusty neighborhood camera store is ready and eager to help. Digital point-and-shoot cameras can focus at very close range, but if you want the background to be almost as clear as the butterfly (depth of field) you might be disappointed. Still, the images they provide are often extraordinary. (See the article by Rik Littlefield, "An introduction to extended depth of field digital photography," *News of the Lepidopterists' Society* 47 (2): 47–49, 2005.) Digital SLRs are now available that can work with the specialty lenses you may already have for film photography—isn't that great? You may ultimately want to rear your own subjects and photograph them (and their transformations) in a home studio where you can control the entire context. This is where tripods, light sources, and such really come into play. Studio photography presents a whole different set of challenges as compared to photography in the field; some people prefer one, some the other.

Butterfly Gardening

Butterfly gardening—more properly, gardening for butterflies— has become very popular. Books, articles, and on-line resources are plentiful. Unfortunately, some of these encourage unrealistic expectations, and others recommend plants that might work very well in other climates and with other butterfly faunas but are inappropriate here. What follows, then, is a discussion specifically keyed to the area served by this book.

First Principles

Be sure you understand these before you get started!

1. The principal function of a butterfly garden is to intercept individual butterflies as they move through an area, and detain them where they can be observed and enjoyed. Occasionally you can actually boost their overall numbers by planting nectar sources or larval hosts if these are in short supply. More often you are just shifting individuals around—from somewhere else into your garden, at least for a while. (We live immersed in a sea of moving butterflies, even in cities. Most people never notice them.)

2. You should not expect permanent establishment of breeding colonies on an ordinary residential lot, no matter what you plant. Most butterflies require a larger and more diverse resource base (larval host plants, adult food supply, pupation sites, mating territories) than an ordinary residential lot can provide. For territorial species, male territories are not very compres-

sible, and excess individuals will simply emigrate. Moreover, many of our species—including the ones most likely to be seen in a garden—are "weedy" or "fugitive"—they rarely raise two consecutive generations in the same place. If you do get breeding, it will be as part of a larger "metapopulation" whose shape is constantly changing.

3. Valuable natural history data can be obtained from a butterfly garden if you are a careful observer. What flowers are or are not visited by what species? Do the butterflies interact on flowers? Are they present only at particular times of the day? Keep a phenological calendar and compare the seasonal occurrence and abundance of the species you see, from year to year. And so on.

4. Skillful planting will enable you to maximize both the number of individuals and the number of species you see, but be realistic in your expectations; what you can get depends on what is available in your vicinity. Don't expect endangered species to breed in your backyard. And on that note, let's talk about...

Context

Your strategy in butterfly gardening, and the plant species you deploy, will depend on where you are: What is the surrounding landscape? What is its butterfly fauna? What is your local climate like?

Urban

In an urban setting, most or all of the vegetation is "artificial," in other words, made up of exotic plants selected for landscape value or self-selected as weeds. Most of the butterfly species are weedy, highly dispersive, and multivoltine and depend on a combination of introduced plants and irrigation for their continued presence. Most of these species go through a population bottleneck over the winter, then increase in density with each passing generation so that maximum densities are reached in fall. In the Bay Area this coincides with the period of sunny, warm weather in September and October when the air-pressure gradient between land and sea typically eases. In the Sacramento Valley you rarely see many butterflies in a garden before late July, and they peak in early October.

BUTTERFLIES IN THE CITY

Had there been an Endangered Species Act in the 1860s, San Francisco would be a very different place. The "Great Sand Bank" occupying the western third of the city would have been declared critical habitat for any number of plants and animals found nowhere else on Earth — including three butterflies that subsequently went extinct. We would still have the Xerces Blue *(Glaucopsyche xerces)*, the Pheres Blue *(Plebejus icarioides pheres)*, and the Sthenele Satyr *(Cercyonis sthenele sthenele)*, but there would be no Golden Gate Park and no Sunset District. The reclamation and stabilization of what was seen as a bleak, barren, fog-and-windswept wasteland were hailed at the time as triumphs of civilization. Now some environmentalists would like to turn the clock back and restore a little of that unique habitat. But some of its inhabitants, including those three butterflies, are gone, never to return.

Apart from the lost dune fauna, San Francisco has never been much of a butterfly place. The cool gray dampness of the summer is basically hostile to butterfly life. As the smallest county in the state, San Francisco still has 68 butterfly species recorded. That's 80 percent of the total for the British Isles! But by California standards it's a poor fauna, and strikingly, mostly a nonresident one.

The first collector to live in the city, H. H. Behr, arrived in 1851 and stayed until his death in 1904. An accomplished naturalist, he published intermittently on butterflies but never catalogued the fauna of the city, and his collection went up in flames with the California Academy of Sciences in the disaster of 1906. There were several other nineteenth-century collectors. Among them, they recorded about 36 species in San Francisco. The first (and only) published fauna for the city was prepared by Francis X. Williams, a policeman by trade, in 1910. It lists 43 species, but Williams cautions that "probably not more than thirty could be taken in several seasons," and his notes make it plain that he hardly found the place a Happy Hunting Ground. In 1956 J. W. Tilden published a lamentation for the lost dune fauna but did not update Williams's list. In an unpublished manuscript written in the 1970s, Harriet V. Reinhard listed 51 species she had seen attributed to San Francisco in various collections. Of course, she usually could not vouch for the accuracy of the data. She and her credible correspondents had seen some 36 species in the 1970s. The number of species thus has not changed greatly over the years. But how has the fauna itself changed?

The older papers, down through Tilden's, require interpretation because their taxonomy is often antiquated and sometimes badly misleading. For example, in Williams's day the specific epithet *piasus* was often thought to apply not to our Arrowhead Blue *(Glaucopsyche piasus)* but to the Echo Blue *(Celastrina ladon echo),* also called the Spring Azure. Once one figures out what the names used in the old papers really refer to, one can look for evidence of faunal change. And one finds it!

The Cabbage White *(Pieris rapae),* also called the European Cabbage Butterfly, arrived sometime in the late nineteenth century — after 1883, at least — but by 1910 was perhaps the commonest butterfly in San Francisco. The oldest record of the Gulf Fritillary *(Agraulis vanillae)* in the Bay Area is in Santa Clara County before 1908. It seems to have become established only in the 1950s. The Fiery Skipper *(Hylephila phyleus)* was unknown to Williams in 1910. Reinhard could find no records before 1937.

Several native species that were treated as scarce by Williams have become commoner due to introduced, weedy host plants. The Anise Swallowtail *(Papilio zelicaon)* and the Red Admiral *(Vanessa atalanta)* are prime examples. The West Coast Lady *(V. annabella),* most of whose hosts are weedy, was already very abundant in Williams's time.

More than half of the species recorded from San Francisco breed there only sporadically, if at all. With a rich resident fauna nearby — in places where the sun shines in the summer! — it should not be surprising that quite a few San Francisco butterflies are merely tourists passing through.

The first nature walk I led in the city was in Glen Canyon, and we happened to see a Buckeye *(Junonia coenia).* As a resident of the hot interior, I am used to seeing 50 Buckeyes a day. I was flummoxed when the "oohs" and "aahs" started and the cameras came whipping out. But, in fact, a Buckeye is a big deal in the city.

Even I find it thrilling to see the black-and-red flash of a Red Admiral amid the garbage along Sixth Street under the freeway, or when a West Coast Lady male sets up a mating territory in the bowels of the Tenderloin.

Hope springs eternal.

Suburban

Extensive subdivisions present the same panorama of unnatural vegetation as the inner city, but natural vegetation may be closer at hand, and nonweedy butterflies have a higher probability of appearance. The weedy species may show up later in the year, perhaps rather erratically.

Ranchettes

Isolated homes on large lots often have quite a bit of grassland, with scattered native trees and sometimes shrubs. The woody vegetation may be enough to sustain breeding populations of some butterflies, but much of the annual grassland is dominated by weeds not useful to butterflies, and its fauna is likely to be poor. Rocky areas have richer native floras and butterfly possibilities, as do creek bottoms. Hog wallows (ephemeral ponds in low areas between hills) may have good plants but do not support special butterflies. If your lot is dominated by yellow star-thistle *(Centaurea solstitialis),* whatever butterflies are there will have plenty of summer nectar without supplementation in a garden. What you see on the star-thistle is what you will get.

Seminatural Settings

In places such as the Oakland-Berkeley hills or Orinda, where homes are interdigitated into reasonably intact foothill landscapes, gardeners can tap into the resident fauna. The highest butterfly diversity occurs in riparian canyon situations. Typically, streams that flow east to west have richer butterfly faunas than those that flow north to south, because there is greater vegetational difference between north-facing and south-facing slopes and each plant community brings its own fauna. If you live in or near such a canyon you might see 40 or more species in your garden. If you live on a special soil (in our area, that usually means serpentine), you may have lower overall diversity but have some special species that do not occur elsewhere. When planting, keep in mind that the foothill fauna is adapted to the mediterranean climate and is mostly restricted to one or two broods a year in spring to early summer. So early-season plants are important in your garden, while in the city they will not bring in much if anything.

Choosing Plants for Your Butterfly Garden

Among the things you need to consider in choosing your plants are, Do I want adult nectar sources, larval host plants, or both? What butterflies occur in my area, what resources do they need, and when are they active? What are the mesoclimate and microclimates in my garden? What plants are butterfly friendly and well adapted to my climates, soil, sun, and shade? The lists that follow are based on experience in our area but are not meant to be exclusive. And you will encounter surprises. For example, the rose family (Rosaceae) in general are not very attractive to butterflies, but the larger species of cotoneaster *(Cotoneaster)* are very much so to migrating Painted Ladies *(Vanessa cardui)*, and the inconspicuous flowers of linden or lime *(Tilia)* also pull in Painted Ladies in huge numbers, but nothing else.

Some of the plants listed are not conventional garden subjects and may be available only from specialized native plant nurseries, if at all. There are now plenty of sources for native plants in the Bay Area and beyond, easily researchable on-line or in specialty magazines. Using a standard reference such as the *Sunset Western Garden Book* (Brenzel 2001), know your climatic zone and check out the description of any plant you are interested in. You can often find a color picture on-line. If you are lucky enough to live in a place with butterfly activity much of the year, you can plant a seasonal flowering succession to delight the eye as well as the proboscis. If butterfly activity in your area is strongly seasonal, there is no rule against planting off-season nonbutterfly flowers for purely esthetic reasons.

Because so much of our fauna is now adapted to exotic plants, you will probably get many more butterflies if you do not remove many of your weeds than if you do. Weeds have other virtues— some are edible, for example—but be mindful of what your neighbors think. If you want to grow weeds for the butterflies, make it look like you've had to work hard to get them to grow! (Tell them that milk thistle *[Silybum marianum]* is often grown as an ornamental in England. It is.)

Plants That Attract Adults

Trees

California buckeye *(Aesculus californica)*. Blooms April to June (later in fog belt). This tree is attractive to nearly everything flying during its flowering season but is most useful in foothill settings. (Remember that it dries out and turns brown in summer!)

Native Shrubs

Skunkbush *(Rhus trilobata)*, also called lemonade bush or squawbush. Blooms in spring. Best in foothills. Often mistaken by passersby for poison oak *(Toxicodendron diversilobum)!* Very attractive to checkerspots.

Rubber rabbitbrush *(Chrysothamnus nauseosus)*. Blooms July to October; extremely attractive to everything flying then. Tends to be short-lived on heavy clay soil, but fine on well-drained light sandy loam. Do not use as a cut flower (odd smell).

Shrubby buckwheats *(Eriogonum)*. Bloom spring to fall, depending on species and site. These are larval host plants in southern California but do not seem to attract a breeding fauna here.

Wild-lilacs *(Ceanothus)*, also called California-lilacs. Bloom in spring, occasionally again later. Useful in foothills, less so elsewhere due to seasonality of butterfly fauna. Host plants for several foothill species.

Coyote brush *(Baccharis pilularis* and *B. p. consanguinea)*. Blooms August to December. Both prostrate and erect forms are extremely attractive to all butterflies flying then; male plants are more attractive to butterflies than females but have a somewhat odd odor. The spring-blooming mule fat *(Baccharis salicifolia)* normally grows in creek bottoms but can be adapted to a garden and is also highly attractive to butterflies, especially hairstreaks.

Buttonbush *(Cephalanthus occidentalis)*. Blooms early summer.

Yerba santa *(Eriodictyon)*. Blooms spring to early summer.

Coffeeberry *(Rhamnus)*. Blooms late spring to early summer. Inconspicuous flowers attract inconspicuous butterflies, especially hairstreaks.

Nonnative Shrubs

Butterfly bush *(Buddleia)*. Blooms throughout late spring and summer and (if pruned) fall. The pink and light purple varieties are outstanding; blackish purple, white, and orange less so.

Lilac *(Syringa)*. Blooms in spring. The purple and pink varieties are excellent for swallowtails and nymphalids.

Lavender *(Lavandula)*. Blooms April to December. Excellent for hesperiids and the Cabbage White *(Pieris rapae)*.

Rosemary *(Rosmarinus officinalis)*. Blooms all year. Most visited in winter.

Escallonia *(Escallonia rubra, E. "exoniensis," etc.)*. Blooms all year. Most visited in winter.

Pride of Madeira *(Echium fastuosum)*. Blooms April to May inland, all year fog belt. Very attractive to many species.

Waxleaf privet *(Ligustrum japonicum)*. Blooms May to June. Attracts mainly smaller butterflies; *L. vulgare* and *L. ovalifolium* are somewhat less attractive.

Abelia *(Abelia)* and weigelia *(Weigelia)*. Bloom spring to summer, abelia into fall. Especially attractive to hesperiids.

Lantana *(Lantana* spp.*)*. Blooms all year (frost sensitive). Extremely attractive to butterflies generally; pink and orange varieties most attractive, pale yellow least.

Native Nonwoody Nectar Plants

Asters and Michaelmas daisies *(Aster)*. Bloom mostly July to December. Among the best butterfly flowers for fall planting. No difference in attractiveness between native western species and the eastern New England aster *(A. novae-angliae)* and others in cultivation; boltonia *(Boltonia asteroides)* also outstanding. Chinese aster *(Callistephus)*, however, and most fleabanes *(Erigeron)* are not butterfly plants. The native coastal seaside daisy *(Erigeron glaucus)* is attractive mainly in fall and near the coast.

Goldenrods *(Solidago)*. Bloom from July to November. Almost as outstanding as asters. Eastern species in cultivation are equally effective as California natives.

Gum plant *(Grindelia)*. Blooms May to November. Attracts some, mostly smaller, butterflies.

Salt marsh fleabane *(Pluchea)*. Blooms July to October. A native tule marsh plant that attracts a variety of butterflies.

Milkweeds *(Asclepias)*. Purple milkweed *(A. cordifolius)* blooms mostly in spring; narrow-leaf milkweed *(A. fascicularis)* late spring to fall. All are outstanding butterfly plants; most are invasive and need to be contained. The common tropical milkweed *(A. curassavica)*, as well as other exotic members of the group, work just fine.

Dogbanes *(Apocynum,* including the common tall Indian hemp [*A. cannabinum*]). Blooms early summer. Very good for many butterflies; also require containment.

Coyote mint *(Monardella)*. Blooms May to October. Good for everything, especially fritillaries and hesperiids. Needs light, loamy soil.

Giant hyssop *(Agastache nepetoides)*. Blooms June to September, longer if pruned. Attracts mostly larger species.

Golden fleece *(Ericameria arborescens)*. Blooms late summer. Resembles a giant rabbitbrush. Generally attractive.

Marsh baccharis *(Baccharis douglasii)*. A native tule marsh plant blooming intermittently, June to October. Generally attractive, especially to lycaenids and Buckeyes *(Junonia coenia)*.

Larger native umbels (Apiaceae), including angelica *(Angelica)* and hog fennel or cow parsnip *(Heracleum lanatum)*. Attract mainly lycaenids. (Also attract parasitoids and stinging hymenopterans!)

Native spring bulbs, including brodiaeas *(Brodiaea* and *Dichelostema)*, and wild onions *(Allium)*. Sorry, Mariposa lilies *(Calochortus)* and fritillarias *(Fritillaria)* are not butterfly flowers.

Larger lilies *(Lilium)*, both native and nonnative. Attractive to swallowtails only.

Wild buckwheat *(Eriogonum)*, herbaceous or semiwoody species.

California wallflower *(Erysimum capitatum)*, also called foothill or orange wallflower. Blooms late spring to early summer. Irresistible to large swallowtails, especially Pale Swallowtail *(Papilio eurymedon)*.

Nonnative, Nonwoody Nectar Plants

Sedum *(Sedum)*. Blooms summer to fall. Larger, showy species and varieties with pink or purple flowers, for example, *S. spectabile* (sometimes sold as "never-die"), are most attractive.

Horehound *(Marrubium vulgare)*. Blooms late spring. Outstanding for hairstreaks and coppers.

Mints *(Mentha)*. Bloom spring to fall. Mostly invasive but generally outstanding.

Onions, scallions, leeks *(Allium)*. Bloom mostly in summer. Large showy ones most attractive to butterflies, but also very attractive to wasps.

Gayfeather *(Liatris)*. Blooms in fall. All species attractive.

Verbena *(Verbena)*. Blooms most of year. Most species attractive.

Alfalfa *(Medicago sativa)*. Blooms most of year. Outstanding butterfly flower; light purple best, deep purple okay, yellow and white disfavored.

Heliotrope *(Heliotropium)*. Blooms midsummer to fall. All species excellent.

Marigolds (*Tagetes, Bidens,* etc.), zinnias *(Zinnia)*, other showy subtropical annuals. Bloom midsummer to fall. Excellent and especially attractive to skippers, Buckeyes *(Junonia coenia)*, and Monarchs *(Danaus plexippus)*.

Sunflowers *(Helianthus)*. Bloom midsummer to fall.

Ox-eye daisy *(Chrysanthemum leucanthemum)*. Blooms late spring to early summer.

Black-eyed Susans *(Rudbeckia)*. Bloom late summer to fall.

Pincushion flower *(Scabiosa)*. Blooms all year. Excellent.

Vetches *(Vicia and Lathyrus)*, but not garden annual or perennial sweet peas *(L. latifolius)*.

Ground Cover

Lippias *(Lippia,* also called *Phyla)*. Blooms repeatedly much of year. Excellent.

Vine

Salpichroa. Blooms in winter. An old-fashioned plant with inconspicuous flowers, much visited by West Coast Lady *(Vanessa annabella)* and Red Admiral *(V. atalanta)*.

Native Plants Used as Larval Hosts

California pipevine *(Aristolochia californica)*, also called Dutchman's Pipe: Pipevine Swallowtail *(Battus philenor)*.

Bush monkeyflower *(Mimulus aurantiacus)*; penstemon *(Penstemon* and *Keckiella)*, shrubby and herbaceous perennial species; bee plant *(Scrophularia)*: Variable Checkerspot *(Euphydras chalcedona)*. Also eaten by the Buckeye *(Junonia coenia)*.

Lippias *(Lippia)*: Buckeye *(Junonia coenia)* (which also eats numerous introduced plants, listed elsewhere).

Pearly and other everlastings *(Gnaphalium, Antennaria, Anaphalis)*: American Painted Lady *(Vanessa virginiensis)*.

Wild buckwheats *(Eriogonum)*, natives more likely to be used than south-state, shrubby species: Various lycaenids.

Native perennial vetches and sweet peas *(Vicia* and *Lathyrus)* and most perennial lupines (including bush lupine *[Lupinus albifrons]*): multiple species of blues, hairstreaks, and sulphurs.

Milkweeds *(Asclepias)*: Monarch *(Danaus plexippus)*.

Native carrot family (Apiaceae), including biscuitroot *(Lomatium)*, angelica *(Angelica)*, Tauschia, Oenanthe, yampah *(Perideridia)*, and so forth: Anise Swallowtail *(Papilio zelicaon)* (which also feeds on exotic carrot family species).

Wild-lilacs *(Ceanothus)*, in foothills only: California Tortoiseshell *(Nymphalis californica)*, Echo Blue *(Celastrina ladon echo)*, Pale Swallowtail *(Papilio eurymedon)*, Hedgerow Hairstreak *(Satyrium saepium)*, Pacuvius Duskywing *(Erynnis pacuvius pernigra)*.

Coffeeberry *(Rhamnus)*: Pale Swallowtail *(Papilio eurymedon)*.

Lotus *(Lotus)*, including deer weed *(Lotus scoparius)*, Spanish lotus *(L. purshianus)*, *L. crassifolius*, most other species: tailed blues, Acmon Blue *(Plebejus acmon)*, Persius Duskywing *(Erynnis persius)*, Northern Cloudy-wing *(Thorybes pylades)*, Silver-spotted Skipper *(Epargyreus clarus californicus)*, and Orange Sulphur *(Colias eurytheme)*.

False indigo *(Amorpha)*, also called California wild indigo or lead plant: California Dogface *(Zerene eurydice)* and Silver-spotted Skipper *(Epargyreus clarus californicus)*.

Turkey mullein *(Eremocarpus [Croton] setigerus)*: Common Hairstreak *(Strymon melinus)*.

Willows *(Salix)*: Mourning Cloak *(Nymphalis antiopa)*, Lorquin's Admiral *(Limenitis lorquini)*, Sylvan Hairstreak *(Satyrium sylvinum)*. (Of these, only the Mourning Cloak will use weeping willow *[S. babylonica]*.)

Oaks *(Quercus)*: California Sister *(Adelpha bredowii californica)*, several hairstreaks and duskywings.

Stinging nettle *(Urtica holosericea)*: Red Admiral *(Vanessa atalanta)*, West Coast Lady *(V. annabella)*, Satyr Anglewing *(Polygonia satyrus)*, Milbert's Tortoiseshell *(Nymphalis milberti)*.

Asters *(Aster)*: Field Crescent *(Phyciodes campestris)*.

Native thistles *(Cirsium)*: Mylitta Crescent *(Phyciodes mylitta)*, Painted Lady *(Vanessa cardui)*.

Bush mallow *(Malacothamnus)*: Northern White Skipper *(Heliopetes ericetorum)*.

Mountain-mahogany *(Cercocarpus)*: Mountain-mahogany Hairstreak *(Satyrium tetra)*.

Native mustard family (Brassicaceae) such as jewel flowers *(Streptanthus)* and rock cresses *(Arabis)*: Various whites and orange-tips.

Bleeding heart *(Dicentra)*: Clodius Parnassian *(Parnassius clodius)* (but you won't get it!)

Nonnative Plants Used as Larval Hosts

Mallows *(Malva)*: Painted Lady *(Vanessa cardui)*, West Coast Lady *(V. annabella)*, Common Hairstreak *(Strymon melinus)*, Common Checkered Skipper *(Pyrgus communis)*.

Annual vetches *(Vicia)*, birdfoot trefoil *(Lotus corniculatus)*; alfalfa *(Medicago sativa)*; nonnative clovers *(Trifolium)*: Eastern Tailed Blue *(Everes comyntas)*, Common Hairstreak *(Strymon melinus)*, Orange Sulphur *(Colias eurytheme)*.

Dock and sheep sorrel *(Rumex)*: Purplish Copper *(Lycaena helloides)*, Great Copper *(L. xanthoides)*.

Prostrate knotweed *(Polygonum aviculare* complex), also called yard grass: Purplish Copper *(Lycaena helloides)*, Acmon Blue *(Plebejus acmon)*.

Pigweeds *(Amaranthus)*, cockscomb *(Celosia)*: Common Sootywing *(Pholisora catullus)*.

Russian thistle *(Salsola)*, also called tumbleweed: Western Pygmy Blue *(Brephidium exile)*.

Weedy thistles *(Cirsium* and *Carduus)*, milk thistle *(Silybum marianum)*: Mylitta Crescent *(Phyciodes mylitta)*, Painted Lady *(Vanessa cardui)*.

Fluellin *(Kickxia)*: Buckeye *(Junonia coenia)*.

Plantain *(Plantago):* Buckeye *(Junonia coenia).*

Some mustards *(Brassica* and *Hirschfeldia),* wild radish *(Raphanus),* white-top *(Cardaria draba),* perennial peppergrass or tall white-top *(Lepidium latifolium):* whites and orange-tips.

Pellitory *(Parietaria judaica),* baby's tears *(Soleirolia,* also known as *Helxine):* Red Admiral *(Vanessa atalanta).*

Grasses, including Bermuda grass *(Cynodon dactylon),* Johnson grass *(Sorghum halepense),* Harding grass *(Phalaris aquatica),* orchard grass *(Dactylis glomerata),* Dallis grass *(Paspalum),* barnyard grass *(Echinochloa):* various branded skippers.

Passionflower *(Passiflora):* Sole host of the Gulf Fritillary *(Agraulis vanillae).*

Nonnative Trees Used as Larval Hosts

European and American elms *(Ulmus procera, U. Americana),* but not Chinese or Siberian elms; hackberry *(Celtis);* weeping willow *(Salix babylonica):* Mourning Cloak *(Nymphalis antiopa).*

Ash *(Fraxinus),* sycamore *(Plantanus),* sweet gum *(Liquidambar),* and Japanese tree privet *(Ligustrum lucidum):* Western Tiger Swallowtail *(Papilio rutulus),* Two-tailed Swallowtail *(P. multicaudatus).*

Cork oak *(Quercus suber)* and probably other exotic oaks: Mournful Duskywing *(Erynnis tristis).*

Some Final Admonitions

Sorry, but butterflies are not attracted to most of the evening primrose family (Onagraceae) including fireweed *(Epilobium angustifolium),* California fuchsia *(Zauschneria,* now put in *Epilobium),* or evening primroses *(Oenothera);* most of the rose family (Rosaceae) (including spireas *[Spirea]);* poppies *(Papaveraceae);* nightshades *(Solanum);* gardenias *(Gardenia);* camellias *(Camellia);* dahlias *(Dahlia);* pot-marigolds *(Calendula);* and most larger bulbs (except lilies *[Lilium]* and amaryllis *[Amaryllis]).*

Most California natives in cultivation are of no more butterfly interest than nonnatives, and most of the best butterfly flowers in our area are exotic.

If you want Great Purple Hairstreak *(Atlides halesus),* you have to tolerate mistletoe (to a point).

If you want butterflies at all, don't use insecticides. That includes Bt *(Bacillus thuringiensis)*, which is a bacterial agent specifically toxic to caterpillars.

It's helpful to provide a puddle or a moist pebbly site where butterflies can drink, but do not provide breeding habitat for disease-carrying mosquitoes!

So-called "butterfly houses" intended to provide overwintering sites for nymphalids are completely worthless.

Never under any circumstances try to augment your garden fauna by releasing nonnative butterflies!

Butterfly Rearing

Many people raised butterflies from caterpillars when they were in school. It is edifying and educational and awe-inspiring. Strangely enough, the early stages of many California butterflies are still poorly known. Rearing on a small scale at home provides an opportunity to contribute to science. It's also fun.

Eggs of many species are not hard to find. The Cabbage White *(Pieris rapae)* lays its yellowish white, spindle-shaped eggs mostly on the undersides of host leaves. The Gray-veined White *(P. "napi")* lays slightly larger, whiter eggs in similar places on its wildland hosts. The Sara Orange-tip *(Anthocharis sara)* lays pale green eggs that turn bright red on stems and in the inflorescences of slender mustards (Brassicaceae). The Large Marble *(Euchloe ausonides)* lays its bright orange red eggs singly on the terminal buds of mustards. The Pipevine Swallowtail *(Battus philenor)* lays its brick red eggs in clusters on tender young shoot tips of the host. Crescents and checkerspots lay their yellow eggs in large batches. The turban-shaped white eggs of the Sonoran Blue *(Philotes sonorensis)* are easy to find on its succulent gray host, dudleya *(Dudleya)*, even if no butterflies are present. And so forth. Rearing from wild eggs does not guarantee freedom from parasitoids, however; you may get minute trichogrammatid wasps instead! Keep in mind that the neonate larvae of quite a few species (all the true fritillaries, larger satyrs, and some hesperiids) go directly into diapause after hatching, having eaten only their eggshells.

Wild-collected caterpillars may be very difficult or impossible to identify unless you can rear them to the adult. The vast majority of lepidopteran species in California (and everywhere else) are moths, not butterflies—so, on average, a caterpillar you encounter in the wild is likely to be a moth. Because no photographs are available for the vast majority of California caterpillars—moth or butterfly—a field guide to California caterpillars is still well in the future, though such guides are beginning to appear for other areas. If you are "into" photography, photograph all the caterpillars you find, then rear them out. Yours could be the first pictures obtained of that species! (Photograph pupae and eggs too, of course. The sculpturing on the egg chorion requires microphotography or even scanning electron micrography. But even a coarse overall view of an egg or egg mass is helpful.)

Note carefully what plant your caterpillar is eating, and feed it that. If the plant is unlikely to be available close to home, collect enough to see the insect through to pupation and keep it cool—preferably in an ice chest—inside a plastic bag until you get home; then store it in the refrigerator and use it as needed. As noted elsewhere in this book, each species has specific hosts, which it usually recognizes by secondary plant chemistry. No matter how hungry, a Cabbage Worm will never eat lettuce. Because the digestive system of the individual becomes adapted to what it is eating, it may be very difficult to switch a caterpillar from its host afield to a different known host of that species. In fact, it may begin feeding but ultimately die. If your caterpillar is wandering rapidly across the landscape when you find it, either it has consumed its host and is seeking another (in which case it will probably be thin, and repeatedly raise its forequarters as if looking for something), or it is full-grown and looking for a pupation site. Mature larvae usually look fat, the head looks too small for the body, and the ground color may be dull or even change completely (the Small Marble [*Euchloe hyantis*], for example, turns from bright green to livid purple). If you do not know how your species pupates, provide some twigs and a couple of inches of dry uncompacted soil. (None of our butterflies burrows underground to pupate, but many moths do.) Densely hairy "woolly bear"–type caterpillars in our area are always moths, not butterflies. Caterpillars with rigid or branching spines that do not sting like nettles are usually butterflies. If the spines sting, it's a moth.

Caterpillars should not be crowded, and they require good ventilation and frequent removal of their droppings (frass). Poorly ventilated containers, especially glass, will "steam up" and provide a perfect environment for the incubation of disease. Do not allow the host material to deteriorate. Cuttings can be kept fresh by inserting them into a florist's water pick, or a pill bottle with a small hole cut in the lid. When using host material brought in from the wild, check carefully for predators (spiders, ants, true bugs) that might have hitchhiked in with it. If ants are present in the environment, place the rearing container in or over a dish of water to keep them out. In general, do not rear in direct sunlight. Your setup is likely to overheat, the insects may die, and at least, the plants will deteriorate faster than need be.

Most wild-collected caterpillars have already been found by a parasitoid. The parasitoid may leave the host to pupate (hymenopterans usually spin a cocoon on the host's back or actually incorporating its skin; dipterans drop to the ground and form a dark brown puparium that resembles a pine nut) or pupate within and first manifest as an adult. You should save all parasitoids you rear. They are often little known and may even be of economic importance. We know very little about the degree of host specialization (the range of hosts successfully attacked) for the vast majority of parasitoids, so all records are potentially interesting. Parasitoids may be "pickled" in alcohol for storage.

It is fairly unusual to find pupae afield, but if you do, remove them very carefully from the substrate to avoid injuring them. Diapausing caterpillars stop feeding and generally curl into a ball. If this persists for a week or more, the caterpillar should be refrigerated and allowed to work out its chilling requirement to resume development (do not freeze!). Overwintering larvae are prone to desiccation, so a lightly moistened bit of paper towel in the container is recommended. Diapausing pupae simply fail to develop into adults within a few weeks, as direct developers would. They should then be refrigerated too. The majority of species diapause for only one winter and can be brought out in January or February ready to resume development. But a few species (the Anise Swallowtail *[Papilio zelicaon]*, Small Marble, Spring White *[Pontia sisymbrii]*, and perhaps others) may diapause for a second year or even longer. Long-dormant pupae that have used up their food reserves and died usually signal this by stretching the abdomen, exposing the intersegmental membranes, which have turned blackish.

Diseased larvae generally stop feeding, may appear bloated, develop dark blotches, produce discharges from the mouth and/or anus, and ultimately may dissolve into a puddle of goo. Whatever the cause, the discharges and goo should be treated as infectious to healthy larvae and disposed of carefully. Diseased pupae often turn black and soft. Do not confuse the development of a healthy pharate adult with such symptoms!

Some butterflies lay eggs readily in captivity; others never do. In general a combination of warmth, light, and host plant must be provided. It is usually helpful to keep the female agitated by caging her with other butterflies, which need not be her own species (use males so you don't get eggs of more than one species!). A typical setup would be a mesh or screen cage 30 cm (12 in.) on a side, with sprigs of host plant and a good nectar source inserted in water picks inside, the whole cage placed 20 to 30 cm (8 to 12 in.) below a gooseneck lamp with a 100 W bulb. Fluorescent bulbs do not generate enough heat, and direct sunlight is guaranteed to overheat. Cabbage Whites, sulphurs, Anise Swallowtails, ladies, and Buckeyes *(Junonia coenia)* all oviposit readily in such circumstances. A single female Orange Sulphur *(Colias eurytheme)* may give you more than 500 eggs over a period of days. Most branded skippers will lay on any grass at all, and some will even lay without a host plant. Lycaenids are notoriously tricky, and some remain unconvinced to lay in the lab. Do not allow the plant material on which the eggs are laid to desiccate; this will usually suck moisture out of the eggs and kill them.

Butterflies as a rule do not mate readily in cages. Most require sunlight, which is difficult to provide without running the risk of overheating. At the University of California at Davis we obtain matings in a climate-controlled greenhouse, which is a luxury not usually available to amateurs. Remember that almost every female caught in the wild has already mated and is able to lay fertile eggs.

Species with large genitalia, such as swallowtails and whites, can be hand paired. This is a method in which the abdomens of the male and female are brought into contact, light pressure is applied to the male to cause him to open his claspers and attach to the female, and nature is allowed to take it from there. Skilled practitioners boast of success rates in excess of 50 percent. By circumventing all behavioral and pheromonal cues, this method facilitates experimental interspecific hybridization. Needless to say, it cannot be used on, say, Western Pygmy Blues *(Brephidium exile)!*

When rearing, handle your animals as little as possible—preferably not at all. Wear disposable rubber gloves and/or wash your hands thoroughly with soap and water before and after processing your culture. If caterpillars must be moved from a deteriorating host to a fresh one, it is best to remove the old host from water and let it dry out in contact with the new one; the larvae should move themselves. If it is necessary to "help" them, use a fine-point brush to lift very small larvae, or to prod larger ones into action.

Many rearing tips can be found in the book by William Winter, listed in the "Resources" section.

Amateurs Can Do Important Science!

Throughout this book, I have mentioned outstanding problems in our fauna that could be attacked profitably by amateurs.

Biology becomes more dependent on expensive high technology every day. What was cutting-edge technology 30 years ago—protein electrophoresis for population genetics, for example—is now regarded as a crude exercise suitable for freshman biology labs. The molecular revolution in biology has so expanded our knowledge and altered our ways of looking at the biosphere that the introductory biology curriculum has had to be redesigned to accommodate this brave new world. As a result, much of the systematics, morphology, and natural history has disappeared. Indeed, today's biology student gets appallingly little exposure to organisms.

And that means the natural history has been left to the amateurs.

At the University of California we are used to encountering brilliant young graduate students, well grounded in the conceptual structure of modern biology and eager to tackle important intellectual problems—but completely uninformed as to what kind of organism is appropriate for that. Or what species actually occur in their area. Or how to get them to lay eggs. Increasingly, they must turn to the experienced amateurs, who tend to be organism oriented rather than problem oriented, to learn how to proceed.

Some old-timers lament the end of the era of butterfly collecting. But in truth, the number of collectors was always small. While they got enormously savvy over years of experience, they usually published little or nothing, passed on their knowledge by personal networking and apprenticeship, and left their collections in institutions as their monuments. With the rising popularity of butterfly watching, photography, and gardening, we have the opportunity to enlist the eyes and brains of much larger numbers of eager observers than we ever had access to before. With instantaneous modes of communication, knowledge can be shared as it never was before. We may learn more about our butterflies in the next decade than we did in the previous century, thanks to people like you. If the butterflies don't go extinct first!

A Final Word about Conservation

Worldwide, amphibians (frogs, toads, and salamanders) are in precipitous decline. The reasons seem to differ from place to place and species to species, yet the overall pattern is global. When we see such a pattern, the implication is that a global process that we do not understand is actually at work, and the local causes are merely epiphenomena. We have seen in this book that our butterfly fauna seems to be in decline, but it is too early to tell whether the decline is cause for real alarm.

The main reason we have so many federally endangered or threatened butterflies in the Bay Area (table 2) is that our peculiar geography is predisposed to the fragmentation of populations—particularly in the coastal fog belt. These local evolutionary experiments may well have been dead ends in the long run ("In the long run," said the economist John Maynard Keynes, "we are all dead"). But they were so restricted to tiny chunks of habitat that even nineteenth-century development was enough to spell their doom.

Species/Subspecies	Status	As of Date
Mission Blue (*Plebejus icarioides missionensis*)	Endangered	June 1, 1976
Lotis Blue (*Lycaeides idas lotis*)	Endangered[a]	June 1, 1976
San Bruno Elfin (*Incisalia mossii bayensis*)	Endangered	June 1, 1976
Lange's Metalmark (*Apodemia mormo langei*)	Endangered	June 1, 1976
Bay Checkerspot (*Euphydryas editha bayensis*)	Threatened	Oct. 18, 1987
Myrtle's Silverspot (*Speyeria zerene myrtleae*)	Endangered[b]	June 22, 1992
Callippe Silverspot (*Speyeria callippe callippe*)	Endangered	Dec. 5, 1997
Behrens' Silverspot (*Speyeria zerene behrensii*)	Endangered	Dec. 5, 1997

[a] Presumed extinct. See text for explanation.
[b] Possibly extinct by interpretation. See text for explanation.

Data as of July 2005.

There is no doubt that habitat loss and fragmentation are the biggest known threats to butterflies. The role of collecting has been debated vigorously, but by and large it is a minuscule factor compared to habitat loss. Pesticides, predators, and diseases may all occasionally play a role but do not seem to be general problems. Habitat loss is easily seen as a peril for butterflies such as the San Bruno Elfin *(Incisalia mossii bayensis)* or the Mission Blue *(Plebejus icarioides missionensis)*. The decline of previously wide-ranging, common species adapted to naturalized host plants— for example, the Large Marble *(Euchloe ausonides)* or the Common Sootywing *(Pholisora catullus)*—is harder to understand. We have to try. They are trying to tell us something that may be very important for our own survival.

DISTRIBUTIONAL CHECKLIST

This list gives the known distributions of all species (subspecies are *not* distinguished) by counties for the Bay Area, for the Sacramento Valley in general, and for the Sutter Buttes. The Bay Area data are based on the most recent compilations by Ray Stanford and Paul Opler, and the Sutter Buttes data are from Walt Anderson. All are corrected to July 2005. Because the Sacramento Valley counties include nonvalley landscapes and vegetations, their species lists do not correspond to the Valley fauna, so the latter has been abstracted here based on the author's experience since 1972. We do not believe the Sutter Buttes fauna list is complete.

Obvious misidentifications and errors from the literature have been omitted. The single biggest source of these—fortunately, seldom encountered—is the butterfly section of the book *The Flora and Fauna of Solano County,* by W. J. Neitzel (Solano County Office of Education, 1965). This work contains eight entirely erroneous species records and much additional misinformation—beware!

● Regularly present
• Accidental
○ Occasional
▼ Migrates through without breeding
◆ Extinct
❖ Old, vague, or suspect record

Swallowtails and Parnassians (Papilionidae)

	Marin	Sonoma	Napa
☐ Clodius Parnassian (*Parnassius clodius*)	●		
☐ Pipevine Swallowtail (*Battus philenor*)	●	●	●
☐ Anise Swallowtail (*Papilio zelicaon*)	●	●	●
☐ Indra Swallowtail (*Papilio indra*)		●	●
☐ Western Tiger Swallowtail (*Papilio rutulus*)	●	●	●
☐ Two-tailed Swallowtail (*Papilio multicaudatus*)		●	●
☐ Pale Swallowtail (*Papilio eurymedon*)	●	●	●
☐ Lime Swallowtail (*Papilio demoleus*)			

Whites, Orange-tips, and Sulphurs (Pieridae)

	Marin	Sonoma	Napa
☐ Pine White (*Neophasia menapia*)		●	●
☐ Spring White (*Pontia sisymbrii*)	●	●	●
☐ Checkered White (*Pontia protodice*)	●	●	●
☐ Western White (*Pontia occidentalis*)			
☐ Gray-veined White (*Pieris* "napi complex")	●	●	●
☐ Cabbage White (*Pieris rapae*)	●	●	●
☐ Large Marble (*Euchloe ausonides*)	●		
☐ Small Marble (*Euchloe hyantis*)		●	●
☐ Sara Orange-tip (*Anthocharis sara*)	●	●	●
☐ Boisduval's Marble (*Anthocharis lanceolata*)		●	●
☐ Orange Sulphur (*Colias eurytheme*)	●	●	●
☐ Western Sulphur (*Colias occidentalis*)		●	●
☐ California Dogface (*Zerene eurydice*)	●	●	●
☐ Cloudless Sulphur (*Phoebis sennae*)	●		
☐ Mexican Yellow (*Eurema mexicana*)	•		
☐ Dainty Sulphur (*Nathalis iole*)			

Coppers, Hairstreaks, Blues, and Metalmarks (Lycaenidae)

	Marin	Sonoma	Napa
☐ Tailed Copper (*Lycaena arota*)	●	●	●
☐ Great Copper (*Lycaena xanthoides*)	●	●	●
☐ Gorgon Copper (*Lycaena gorgon*)	●	●	●
☐ Blue Copper (*Lycaena heteronea*)	●	●	
☐ Purplish Copper (*Lycaena helloides*)	●	●	●
☐ Golden Hairstreak (*Habrodais grunus*)	●	●	●

Solano	Contra Costa	Alameda	Santa Clara	Santa Cruz	San Mateo	San Francisco	Sacramento Valley	Sutter Buttes
				◆				
●	●	●	●	●	●	●	●	●
●	●	●	●	●	●	●	●	●
●								❖
●	●	●	●	●	●	●	●	●
●	●	●	●				○	●
●	●	●	●	●	●	●	○	●
			•					

Solano	Contra Costa	Alameda	Santa Clara	Santa Cruz	San Mateo	San Francisco	Sacramento Valley	Sutter Buttes
			❖					
●	●	●	●	●			○	
●	●	●	●	●	●	●	●	
			●	●			○	
●	●	●	●	●	●			
●	●	●	●	●	●	●	●	●
●	●	●	●	●	●	●	●	●
●								
●	●	●	●	●	●	●	○	●
	●					❖		
●	●	●	●	●	●	●	●	●
❖								
●	●	●	●	●		❖	●	●
●		●	●	●	●	●		
							•	
●	●	●			●			

Solano	Contra Costa	Alameda	Santa Clara	Santa Cruz	San Mateo	San Francisco	Sacramento Valley	Sutter Buttes
●	●	●	●	●	●	●		
●	●	●	●	●	●	●	●	●
●	●	●	●	●	●		○	
●	●	●	●	●	●	●	●	●
●	●	●	●	●	●			

continued ➤

Legend:
- ● Regularly present
- • Accidental
- ○ Occasional
- ▼ Migrates through without breeding
- ◆ Extinct
- ❖ Old, vague, or suspect record

Coppers, Hairstreaks, Blues, and Metalmarks (Lycaenidae) *cont.*

	Marin	Sonoma	Napa
☐ Great Purple Hairstreak *(Atlides halesus)*	●	●	●
☐ California Hairstreak *(Satyrium californicum)*	●	●	●
☐ Sylvan Hairstreak *(Satyrium sylvinum, incl. S. dryope)*	●	●	●
☐ Gold-hunter's Hairstreak *(Satyrium auretorum)*	●	●	●
☐ Mountain-mahogany Hairstreak *(Satyrium tetra)*		●	●
☐ Hedgerow Hairstreak *(Satyrium saepium)*	●	●	●
☐ Bramble Hairstreak *(Callophrys "dumetorum complex")*			
☐ "Inland" entity	●	●	●
☐ "Coastal" entity	●	●	
☐ Thicket Hairstreak *(Mitoura spinetorum)*		●	●
☐ Johnson's Hairstreak *(Mitoura johnsoni)*			●
☐ John Muir's Hairstreak *(Mitoura gryneus muiri)*	●	●	●
☐ Nelson's Hairstreak *(Mitoura gryneus nelsoni)*			●
☐ Western Brown Elfin *(Incisalia augustinus iroides)*	●	●	●
☐ Moss's Elfin *(Incisalia mossii)*	●	●	●
☐ Western Pine Elfin *(Incisalia eryphon)*	●	●	●
☐ Common Hairstreak *(Strymon melinus)*	●	●	●
☐ Eastern Tailed Blue *(Everes comyntas)*	●	●	●
☐ Western Tailed Blue *(Everes amyntula)*	●	●	●
☐ Silvery Blue *(Glaucopsyche lygdamus)*	●	●	●
☐ Xerces Blue *(Glaucopsyche xerces)*			
☐ Arrowhead Blue *(Glaucopsyche piasus)*	●	●	●
☐ Marine Blue *(Leptotes marina)*			
☐ Reakirt's Blue *(Hemiargus isola)*			
☐ Ceraunus Blue *(Hemiargus ceraunus)*			
☐ Western Pygmy Blue *(Brephidium exile)*	●	●	●
☐ Sonoran Blue *(Philotes sonorensis)*		❖	
☐ San Bernardino Blue *(Euphilotes bernardino)*			
☐ Dotted Blue *(Euphilotes enoptes)*	●	●	
☐ Echo Blue *(Celastrina ladon echo)*	●	●	●
☐ Acmon Blue *(Plebejus acmon)*	●	●	●
☐ Lupine Blue *(Plebejus lupini)*			

Solano	Contra Costa	Alameda	Santa Clara	Santa Cruz	San Mateo	San Francisco	Sacramento Valley	Sutter Buttes
●	●	●	●		●		●	●
●	●	●	●	●	●		●	●
●	●	●	●	●	●		●	
●	●	●	●				●	
●	●	●	●		●			
●	●	●	●	●	●	●		
●	●	●	●	●	●		●	●
			●	●	●	●		
●	●	●	●					
●								
	●	●	●					
●	●	●	●	●	●	●	●	●
●	●			●	●	❖		
	●	●			◆	◆		
●	●	●	●	●	●	●	●	●
●	●	●	●		●	❖	●	●
	●	●	●	●	●	●		
●	●	●	●	●	●	●	●	
						◆		
			●		●			
●	●	●			●	●		●
	•				•		•	
	•	•						
●	●	●	●	●	●	●	●	
	●	●	●	●				
			❖					
●	●		●	●		●		
●	●	●	●	●	●	●	○	
●	●	●	●	●	●	●	●	●
			❖	❖				

continued ➤

- ● Regularly present
- • Accidental
- ○ Occasional
- ▼ Migrates through without breeding
- ◆ Extinct
- ❖ Old, vague, or suspect record

	Marin	Sonoma	Napa
Coppers, Hairstreaks, Blues, and Metalmarks (Lycaenidae) *cont.*			
☐ Greenish Blue (*Plebejus saepiolus*)	●	●	
☐ Boisduval's Blue (*Plebejus icarioides*)	●	●	●
☐ Melissa Blue (*Lycaeides melissa*)			
☐ Lotis Blue (*Lycaeides idas lotis*)		❖(◆)	
☐ Mormon Metalmark (*Apodemia mormo*)	●		●
Brushfoots (Nymphalidae)			
☐ Gulf Fritillary (*Agraulis vanillae*)	●	●	●
☐ Variegated Fritillary (*Euptoieta claudia*)			
☐ Western Meadow Fritillary (*Boloria epithore*)			
☐ Crown Fritillary (*Speyeria coronis*)	●	●	●
☐ Zerene Fritillary (*Speyeria zerene*)	●	●	●
☐ Callippe Fritillary (*Speyeria callippe*)	●	●	●
☐ Hydaspe Fritillary (*Speyeria hydaspe*)			❖
☐ Unsilvered Fritillary (*Speyeria adiaste*)			
☐ Field Crescent (*Phyciodes campestris*)	●	●	●
☐ California Crescent (*Phyciodes orseis*)	◆	◆	◆
☐ Mylitta Crescent (*Phyciodes mylitta*)	●	●	●
☐ Northern Checkerspot (*Chlosyne palla*)	●	●	●
☐ Bordered Patch (*Chlosyne lacinia*)			
☐ Leanira Checkerspot (*Thessalia leanira*)	●	●	●
☐ Variable Checkerspot (*Euphydryas chalcedona*)	●	●	●
☐ Edith's Checkerspot (*Euphydryas editha*)		●	●
☐ Satyr Anglewing (*Polygonia satyrus*)	●	●	●
☐ Rustic Anglewing (*Polygonia faunus*)	●	●	●
☐ Zephyr Anglewing (*Polygonia zephyrus*)	●	●	●
☐ Oreas Anglewing (*Polygonia oreas*)	●	●	●
☐ California Tortoiseshell (*Nymphalis californica*)	●	●	●
☐ Compton Tortoise (*Nymphalis j-album*)			
☐ Milbert's Tortoiseshell (*Nymphalis milberti*)		●	
☐ Mourning Cloak (*Nymphalis antiopa*)	●	●	●
☐ American Painted Lady (*Vanessa virginiensis*)	●	●	●
☐ Painted Lady (*Vanessa cardui*)	●	●	●

Solano	Contra Costa	Alameda	Santa Clara	Santa Cruz	San Mateo	San Francisco	Sacramento Valley	Sutter Buttes
	●				❖			
●	●	●	●	●	●	●	○	
				❖				
●	●		●	●	●			
●	●	●	●	●	●	●	○	
							•	
			●	●	●	●		
●	●	●	●	●	●	●	○	
					❖			
●	●	●	●	●	●	●	○	
			◆	●	●			
●	●	●	●	●	●	●	●	●
						◆		
●	●	●	●	●	●	●	●	●
●	●	●	●	●	●	●	○	
							•	
●	●	●	●	●	●	●		
●	●	●	●	●	●	●	○	
	●	●	●		●	●		
●	●	●	●	●	●	●	●	●
	●			●	●			
●				●	●			
●	●	●	●	●	●	●		
●	●	●	●	●	●	●	●	●
					•			
●		●					●	●
●	●	●	●	●	●	●	●	●
●	●	●	●	●	●	●	●	●
●	●	●	●	●	●	●	●	●

continued ➤

	Marin	Sonoma	Napa
Brushfoots (Nymphalidae) *cont.*			
☐ West Coast Lady *(Vanessa annabella)*	●	●	●
☐ Red Admiral *(Vanessa atalanta)*	●	●	●
☐ Buckeye *(Junonia coenia)*	●	●	●
☐ Lorquin's Admiral *(Limenitis lorquini)*	●	●	●
☐ California Sister *(Adelpha bredowii)*	●	●	●
☐ California Ringlet *(Coenonympha tullia californica)*	●	●	●
☐ Ox-eyed Satyr *(Cercyonis pegala boopis)*	●	●	●
☐ Great Basin Wood Nymph *(Cercyonis sthenele)*	●	●	●
☐ Great Arctic *(Oeneis nevadensis)*		●	
☐ Monarch *(Danaus plexippus)*	●	●	●
☐ Queen *(Danaus gilippus)*			
Skippers (Hesperiidae)			
☐ Silver-spotted Skipper *(Epargyreus clarus)*	●	●	●
☐ Dorantes Skipper *(Urbanus dorantes)*		•	
☐ Northern Cloudy-wing *(Thorybes pylades)*	●	●	●
☐ Dreamy Duskywing *(Erynnis icelus)*		●	
☐ Sleepy Duskywing *(Erynnis brizo lacustra)*	●	●	●
☐ Propertius Duskywing *(Erynnis propertius)*	●	●	●
☐ Mournful Duskywing *(Erynnis tristis)*	●	●	●
☐ Funereal Duskywing *(Erynnis funeralis)*			
☐ Pacuvius Duskywing *(Erynnis pacuvius)*	●	●	●
☐ Persius Duskywing *(Erynnis persius)*	●	●	●
☐ Two-banded Skipper *(Pyrgus ruralis)*	●	●	●
☐ Least Checkered Skipper *(Pyrgus scriptura)*		●	●
☐ Common Checkered Skipper *(Pyrgus communis)*	●	●	●
☐ Northern White Skipper *(Heliopetes ericetorum)*	●	●	●
☐ Common Sootywing *(Pholisora catullus)*		●	●
☐ Arctic Skipper *(Carterocephalus palaemon)*	●	●	
☐ Fiery Skipper *(Hylephila phyleus)*	●	●	●
☐ Yuba Skipper *(Hesperia juba)*		●	●
☐ Comma Skipper *(Hesperia colorado* complex)	●		●
☐ Columbia Skipper *(Hesperia columbia)*	●	●	●

Solano	Contra Costa	Alameda	Santa Clara	Santa Cruz	San Mateo	San Francisco	Sacramento Valley	Sutter Buttes
●	●	●	●	●	●	●	●	●
●	●	●	●	●	●	●	●	●
●	●	●	●	●	●	●	●	●
●	●	●	●	●	●	●	●	●
●	●	●	●	●	●	●	●	●
●	●	●	●	●	●	●	●	
●	●	●	●	●	●	●	○	
○	●	●	●	●		◆		
●	●	●	●	●	●	●	●	●
	•							

Solano	Contra Costa	Alameda	Santa Clara	Santa Cruz	San Mateo	San Francisco	Sacramento Valley	Sutter Buttes
●							○	
●	●	●	●	●	●			
●	●	●	●	●				
●	●	●	●	●	●	●	○	●
●	●	●	●	●	●	●	●	●
•		•	•				•	
●			●	●	●			
●	●		●	●			●	
		●	●	●	●	●		
●	●	●	●	●			●	
●	●	●	●	●	●	●	●	●
●	●	●	●	●			▼	●
●	●	●	●	●		●	●	●
●	●	●	●	●	●	●	●	●
	●	●	●	●	●			
	●	●	●	●	●			
●	●	●	●	●				

continued ➤

- ● Regularly present
- • Accidental
- ○ Occasional
- ▼ Migrates through without breeding
- ◆ Extinct
- ❖ Old, vague, or suspect record

	Marin	Sonoma	Napa

Skippers (Hesperiidae) *cont.*

	Marin	Sonoma	Napa
☐ Lindsey's Skipper *(Hesperia lindseyi)*	●	●	●
☐ Sandhill Skipper *(Polites sabuleti)*	●	●	●
☐ Dogstar Skipper *(Polites sonora)*	●	●	
☐ Sachem *(Atalopedes campestris)*	●	●	●
☐ Woodland Skipper *(Ochlodes sylvanoides)*	●	●	●
☐ Farmer *(Ochlodes agricola)*	●	●	●
☐ Yuma Skipper *(Ochlodes yuma)*			
☐ Umber Skipper *(Poanes melane)*	●	●	●
☐ Dun Skipper *(Euphyes vestris)*		●	
☐ Roadside Skipper *(Amblyscirtes vialis)*		●	●
☐ Eufala Skipper *(Lerodea eufala)*	●	●	●

	Marin	Sonoma	Napa
Totals (all)	92	108	100
Extinct	1	2	1

Solano	Contra Costa	Alameda	Santa Clara	Santa Cruz	San Mateo	San Francisco	Sacramento Valley	Sutter Buttes
	●	●	●			●		
●	●	●	●	●	●	●	●	●
		●						
●	●	●	●	●	●	●	●	●
●	●	●	●	●	●	●	●	●
●	●	●	●	●	●	●	○	●
●	●						○	
●	●	●	●	●	●	●	●	●
				●				
		●						
●	●	●	●		●		●	●
90	94	90	96	88	84	68	71	45
0	0	0	1	1	1	5	0	0

GLOSSARY

Aberration An abnormal phenotype, usually in wing color or pattern; may be caused by genetic, environmental, or developmental factors.

Alkaloids A large class of secondary plant compounds, many of which are produced by specific plant families or genera and which may be important in host-plant selection by butterflies.

Allochronic Active at different seasons or times. Compare *synchronic*.

Allopatric Occurring in different places. Compare *sympatric*.

Androconia Specialized wing scales that secrete or disperse pheromones, as in courtship.

Anthropic landscape The landscape as modified or affected by human activity, including habitat conversion and the introduction and extirpation of species.

Aposematism See *warning coloration*.

Automimicry Protective resemblance afforded palatable members of a species (e.g., the Monarch) by their perfect resemblance to unpalatable individuals. See *mimicry*.

Basad Toward the base (as on wings).

Batesian mimicry The mimicry of an unpalatable species (model) by a palatable one (mimic), named for Henry Walter Bates, nineteenth-century naturalist. See *mimicry*. Compare *Mullerian mimicry*.

Biogeography The scientific discipline that studies the geographic distribution of organisms and its causes, both historic and contemporary.

Bivoltine Having two flights per year.

Body-basking A mode of thermoregulation in which the wings are opened just enough to expose the body to incoming solar radiation. See *thermoregulation*.

Cardenolides A class of secondary plant compounds found mostly in the milkweed and dogbane families, involved in host selection by species feeding on these plants and commonly sequestered by them for their own defense against predators. Cardenolides are generally very bitter tasting, induce vomiting, and can cause problems with the heartbeat. Also called *cardiac glycosides*.

Cervical shield A heavily pigmented, sclerotized area behind the head of hesperiid caterpillars.

Chaparral A plant formation dominated by evergreen shrubs; characteristic of mediterranean climates. See *formation*.

Character An attribute of an organism that is used in definition and discrimination in taxonomy.

Chill coma Torpor induced by low temperatures above the lethal level. See *torpor*.

Chorion The eggshell, often ornately sculptured.

Chrysalis A butterfly pupa; the word is from a root alluding to "gold," because many nymphalid pupae are marked with iridescent, metallic patches.

Cladistics A philosophy and methodology in systematics in which strict logical methods are used to make classification correspond to the evolutionary history of a lineage. See *systematics*.

Community A unit consisting of all the species populations in a defined area, often visualized by ecologists as functionally integrated and repeating in space and time; in vegetation ecology defined by species composition (floristics) rather than growth form (physiognomy). See *floristics, physiognomy*.

Congeneric Belonging to the same genus. A congener is another species of a genus under consideration.

Conspecific Belonging to the same biological species.

Costa The heavily reinforced vein forming the leading edge of a butterfly wing.

Cremaster A hook-bearing structure at the tail end of a pupa, used to anchor the animal to a silken pad.

Crypsis Concealing coloration or "camouflage," which may entail a generalized or highly specific resemblance to the background or to an object of no interest to a visual predator.

Density dependence The extent to which the effectiveness of a mortality factor acting on a population is a function of the density of that population. Biotic factors such as disease and parasitism are normally density dependent, while abiotic ones such as weather are density independent.

Developmental zero The temperature below which growth ceases.

Diapause A type of dormancy characterized by developmental arrest; it may entail overall development or only (in adults) reproductive capability. True diapause is under hormonal control and is not simply a direct response to current environmental conditions.

Disjunct Occurring in widely separated localities.

Dorsal Pertaining to the back (the upper surface of a butterfly).

Dorsal-basking A mode of thermoregulation in which the wings are opened broadly to incoming solar radiation. See *thermoregulation*.

Dyar's law An empirical generalization attributed to Harrison Dyar that successive larval instars in Lepidoptera are related by a constant size multiplier; this enables one to use the shed head capsules to determine the number of molts.

Ecdysis Molting, from a root meaning "to disrobe"; "ecdysone" is molting hormone.

Ecotone An area transitional between ecological communities.

Ecotype An ecological "race," generally unrecognized taxonomically but adapted to local ecological conditions in its host selection, phenology, etc. See *race*.

Edaphic Pertaining to the soil. Edaphic endemism is restriction to a particular soil type.

Electrophoresis A technique used to visualize genetically controlled protein variation at the level of individuals and populations, much used in studies of population genetics and to detect gene flow or hybridization.

Encapsulation The suppression of parasitoid eggs by surrounding them with host cells.

Endemism Restriction of an organism, population, species, and so on to a well-defined, limited area or type of environment.

Entomology The scientific study of insects and their biology.

Epigamic behavior Behavior that functions to bring the sexes together, such as hilltopping and territoriality. See *hilltopping, territoriality.*

Estivation Dormancy in summer, which may or may not involve diapause.

Evolutionary systematics A nonideological approach to classification in wide use in the second half of the twentieth century, now being replaced by cladistics. See *cladistics.*

Facultative Optional, as in diapause, which is induced by environmental cues only at certain seasons.

Floristics The taxonomic composition of vegetation; used in ecological classification.

Form A broad, informal term referring to various kinds of phenotypic variation, both polymorphic and polyphenic. See *polymorphism, polyphenism.*

Formation A unit of vegetation defined by its growth form (trees, shrubs, etc.) and thus by physiognomy rather than by floristics. Compare *community.*

Frass Caterpillar droppings.

Gabbro A crystalline rock type weathering to an unusual soil supporting a distinctive plant community. Compare *serpentine.*

Genitalia The reproductive structures of adults, which are often of diagnostic value in butterfly classification.

Genome The entire body of genetic information of an organism (individual or species).

Genotype The genetic information contained in an individual for a specific character or gene. Compare *phenotype.*

Glucosinolates A class of secondary plant compounds restricted largely to the mustard family and its close relatives, involved in host selection by the Cabbage White butterflies and their kin. Also called *mustard oil glycosides*.

Granulosis A common inclusion-body disease of butterfly larvae. See *inclusion-body diseases*.

Gynandromorph A type of sexual mosaic in which parts of the animal are male and other parts female; a type of aberration, usually rare and much prized by collectors. When one side (right or left) is male and the other female, it is known as a bilateral gynandromorph. See *sexual mosaic*.

Heliotherm An animal dependent on incoming solar radiation to elevate and maintain its body temperature above ambient.

Hemolymph The "blood" of insects. Insect blood does not usually carry oxygen and thus does not have a respiratory pigment like hemoglobin.

Hibernation Dormancy over winter, which may or may not involve diapause.

Hilltopping A type of epigamic behavior in which males and mateable females rendezvous at open summits; common in butterflies of mountainous terrain. See *epigamic behavior*.

Holocene The current geologic epoch, since the end of the last glaciation 10,000 to 20,000 years ago. Also called *Recent*.

Holometabolous Having complete metamorphosis. See *metamorphosis*.

Imaginal discs Clusters of embryonic cells that give rise to the structures of the adult during the late larval and pupal stages.

Inclusion-body diseases A group of important diseases of caterpillars and other insects, in which infective virus is shed within characteristic protein crystals ("inclusion bodies") that are very resistant to degradation in the environment.

Inflorescence The flowering part of a plant.

Infructescence The fruiting part of a plant.

Infuscated Washed or densely shaded with black or gray.

Instar The period between larval molts. In most species the number of instars is genetically fixed.

Intersex An individual that is, overall, intermediate between the sexes. Compare *sexual mosaic.*

Juvenile hormone A hormone with profound effects on insect maturation and metamorphosis.

Key-factor analysis A method for identifying the factors most responsible for population dynamics, using empirical data. Also called *k-factor analysis.*

Larva Caterpillar; from a root meaning "mask," since the adult is "concealed" therein.

Lateral-basking A mode of thermoregulation in which one side of the body and wings is exposed to incoming solar radiation. The wings are closed and the animal leans so that its plane is perpendicular to the rays. See *thermoregulation.*

Lek A site where males of a species congregate to attract females, or where females are instinctively programmed to seek them. Within the lek, males may be territorial. Compare *epigamic behavior.*

Lepidoptera Literally, "scaly wings"; the order comprising butterflies and moths.

Lepidopterist One who studies Lepidoptera.

Life table A concise statement of the vital statistics of a population (age-specific probabilities of death and reproduction), much used in studies of population dynamics and regulation.

Lineage A group of organisms connected by common descent.

Lumper A taxonomist who tends to emphasize resemblances rather than differences among organisms, thus creating or recognizing few taxa. Compare *splitter.*

Mark-release-recapture A standard methodology for quantifying population size, survivorship, and movements; often used with butterflies.

Meconium The waste products of metamorphosis, often red or pink, voided by the recently emerged adult, usually prior to flight.

Mediterranean climate The type of climate found in most of California, characterized by hot, sunny, dry summers and cool, cloudy, wet winters.

Mesic Referring to a temperate climate or vegetation: one with plenty of moisture and no extreme conditions.

Metamorphosis Individual development (ontogeny) characterized by passage through a series of distinctive stages differing conspicuously in morphology. Butterflies have complete metamorphosis (egg, larva, pupa, and adult).

Metapopulation A set of local populations interconnected by exchange of individuals and genes.

Micropyle An opening at the top of an egg, through which sperm enter to accomplish fertilization.

Migration Systematic dispersal. In butterflies, migrations may be latitudinal, altitudinal, or both and usually have a seasonal component; they may involve environmentally induced hormonal changes affecting development, phenotype, behavior, and/or reproduction.

Mimicry Resemblance between organisms that are not very closely related, in which adaptive benefit results to one or both (all). See *automimicry, Batesian mimicry, Mullerian mimicry.*

Monophagous Narrowly specialized in host choice, feeding on just one family or even one genus or species of plants.

Monophyletic Descended from a single common ancestor.

Mullerian mimicry A type of mimicry in which two or more unpalatable species display very similar phenotypes, which allows predators to generalize from experience with any of them, named for Fritz Muller, nineteenth-century naturalist. See *mimicry.* Compare *Batesian mimicry.*

Multivoltine Having three or more flights per year.

Naturalized Self-perpetuating in a new environment without human assistance. The introduced Cabbage White butterfly is a good example.

Neonate A newly born or hatched individual.

Nominate subspecies The subspecies containing the type specimen of a species and thus carrying the name of the species, for example, *Hesperia colorado colorado.*

Numerical taxonomy See phenetics.

Obligate Invariant, as in species that always diapause.

Obtect Rounded, without prominences.

Oedeagus The male intromittent organ in butterflies.

Oligophagous Moderately specialized in host selection, feeding on plants of only a few families.

Ommatidia The individual "facets" in compound eyes.

Ontogeny Individual (embryonic, etc.) development.

Osmeterium An eversible Y- or V-shaped organ behind the head of papilionid larvae, generally producing a defensive secretion that may have an offensive (to humans) odor.

Oviposition Egg laying.

Paleoclimatology The study of ancient climates and their characteristics.

Paleoecology The study of ancient environments.

Parasitoid An organism that is free-living as an adult but whose larva feeds on or in a single host organism. Parasitoids are among the most important natural enemies of butterflies and include many species of small wasps and true flies.

Patrolling Epigamic behavior in which males fly back and forth or describe a circuit, rather than remaining in a fixed position. See *epigamic behavior*. Compare *perching*.

Pelage Fur, as on a butterfly's body.

Perching Epigamic behavior in which males remain in a fixed position, often centrally located within a territory, and investigate passing organisms. See *epigamic behavior*. Compare *patrolling*.

Pharate adult The completely developed adult as visible within the pupal case before emergence.

Phenetics A philosophy and methodology in systematics in which all definable characters are weighted equally and classification is based on quantitative overall similarity or dissimilarity; largely replaced by cladistics. Also called *numerical taxonomy*. See *systematics, cladistics*.

Phenology The study of the timing (seasonality) of biological events (phenophases); biological seasonality.

Phenotype The appearance of an organism, particularly with reference to a particular character. Compare *genotype*.

Pheromone A chemical secreted by one organism and that induces specific developmental, physiological, or behavioral changes in another of the same species. Sexual attractants and aphrodisiacs are pheromones.

Photoperiod Daylength. Many physiological and ecological processes relating to phenology are regulated by photoperiod in temperate latitudes. See *phenology*.

Photophase The day part of a 24-hour cycle. Compare *scotophase*.

Phylogenetic systematics See *cladistics*.

Phylogeny Relationship by descent; an ancestor-descendant relationship.

Phylogeography A method using genetic (usually molecular genetic) data in a geographic context to reconstruct the history and historical distribution patterns of populations.

Physiognomy The growth form of vegetation (trees, shrubs, bunchgrasses, etc.); used in ecological classification.

Pleistocene The geologic epoch corresponding to the Ice Age, ending 10,000 to 20,000 years ago.

Pollination syndrome A set of phenotypic characteristics (form, color, odor, etc.) maximizing floral attractiveness to specific classes of pollinators.

Polyembryony Repeated fission of a fertilized egg, resulting in multiple instances of twinning; common in parasitoids.

Polyhedrosis A common group of inclusion-body diseases of caterpillars. See *inclusion-body diseases*.

Polymorphism In the broad sense, the occurrence within a population or species of multiple phenotypes or "forms"; today usually restricted to cases where the "forms" are under direct genetic control. Compare *polyphenism*.

Polyphagous Generalized in host-plant choice, feeding on many families of plants.

Polyphenism The occurrence within a population or species of multiple phenotypes or "forms" that are induced by environmental influences such as photoperiod or temperature. Seasonal variation in butterflies is polyphenic; usually all "forms" can be induced in the progeny of a single mating by rearing in different environments. Compare *polymorphism.*

Population A group of organisms belonging to the same species and occupying a defined area. See *metapopulation.*

Population viability analysis A technique used in conservation biology to assess the degree of risk to the continued existence of a population in a given time frame.

Preference/performance studies Research that assesses the degree to which female preference for oviposition substrate correlates with the suitability of the host for larval growth and development.

Proboscis The feeding tube of adult Lepidoptera, kept coiled beneath the head when not in use.

Puddling The aggregation of butterflies (usually young males) at mud, thought to represent a method of harvesting mineral salts for physiological use.

Pupa The stage after the caterpillar, in which the major transformation to the adult takes place. Compare *chrysalis.*

Quaternary The period on the geologic timescale embracing the Pleistocene (Ice Age) and Holocene; about the last two million years.

Race An informal term, not recognized in taxonomy, for subdivisions of a species recognizable on phenotypic, ecological, or similar bases. Compare *ecotype.*

Relict A species or population "left behind" in an area during an episode of geologic or climatic change. Many disjunct distributions are relictual. See *disjunct.*

Residence time The length of time a marked individual remains in a population, as determined by the dates of its first and last capture or record. See *mark-release-recapture.*

Riparian Of the river; refers to streamside habitats, also called gallery forests.

Sclerotin The hardened material of the insect exoskeleton. Structures such as the male genitalia are described as sclerotized.

Scotophase The night part of a 24-hour cycle. Compare *photophase*.

Secondary plant substances Chemical compounds characteristic of particular plant lineages, generally thought to have defensive functions, and usually important in insect-host plant relationships. Also called *secondary plant metabolites*. See *alkaloids, cardenolides, glucosinolates, tannins*.

Semivoltine Having one flight every other year.

Sensu lato "In the broad sense"; an inclusive sense of a scientific name.

Serpentine The state rock of California. Its unusual chemistry gives rise to soils that are toxic or nutritionally inadequate for many plants, resulting in distinctive vegetation with special (endemic) butterfly associates.

Sexual dimorphism Conspicuous phenotypic differences between the sexes of a species.

Sexual mosaic An individual that is part male and part female. Mosaics are easiest to recognize in species with strong sexual dimorphism. See *gynandromorph*.

Sibling species Species that are so similar phenotypically as to be nearly or quite unrecognizable as such without detailed study; usually first recognized based on behavioral or ecological differences. See *species*.

Sister taxon The most closely related taxon at the same rank (species, genus, etc.).

Source-sink dynamics A type of metapopulation dynamics in which some populations (sinks) are maintained only by outflow from others (sources). See *metapopulation*.

Species Under the biological species concept (which applies only to sexual, outcrossing organisms such as butterflies), a group of interbreeding individuals reproductively isolated from other such groups. Species are what we name with Latin binomials, such as *Pieris rapae*. Because the definition is based on reproductive compatibility, species need not differ in phenotype. See *sibling species*.

Spermatogenesis Sperm formation, which in butterflies begins in the late larval stage of males.

Spermatophore A sperm-containing package deposited by the male during copulation, usually including nutrients that can be mobilized by the female.

Sperm precedence A mechanism whereby the most recent male to copulate with a female inactivates any sperm left over from the last mating, apparently quite common in butterflies.

Sphragis A vaginal plug formed by a male during copulation, thus preventing the female from mating again; found in parnassians. See *sperm precedence* for the functional significance of the sphragis.

Splitter A taxonomist who tends to emphasize differences rather than similarities among organisms, and thus creates many new taxa. Compare *lumper*.

Stigma A pheromone gland on the wings of male butterflies, in locations characteristic of specific lineages. Also called *sex patch*.

Subspecies A taxon below the species level, allegedly recognizable phenotypically and with a definite geographic distribution. Unlike species, subspecies have no "biological" criterion and are often considered arbitrary, but some are much more well marked than others. Visible subspecific differences often do not reflect major differences at the genomic level.

Succession The process of more or less predictable species turnover over time in an ecological community.

Sympatric Occurring in the same location. Compare *allopatric*.

Synchronic Active at the same times or seasons. Compare *allochronic*.

Systematics The science embracing the naming of organisms (taxonomy) and the evolutionary, ecological, and philosophical aspects of classification.

Tannins A class of secondary compounds found particularly in woody plants, thought to have multiple functions including defense against herbivory, and strongly affecting the phenology of insects feeding on these plants.

Taxon Things that are named by taxonomists at any level in the hierarchy (subspecies, species, genera, families, etc.). In this book the biological objects themselves are referred to as "entities." The plural is *taxa*.

Taxonomy Biological classification; a narrower term than systematics. See *systematics*.

Teneral Of a butterfly, newly emerged from the pupa and not fully hardened and capable of flight.

Territoriality The association of individual organisms with specific sites, which they are typically visualized as "defending" against others. In butterflies, territoriality appears to be purely epigamic, and it is doubtful that true "defense" occurs. See *epigamic behavior, lek*.

Tertiary The geologic period before the Quaternary, beginning roughly 65 million years ago. Many distributional patterns of organisms reflect Quaternary disruption of previous Tertiary distributions.

Thermoregulation A collection of mechanisms whereby organisms control their own body temperatures; most butterflies rely on distinctive postures to maximize or minimize solar radiation.

Tomentose Densely covered with short hairs; woolly.

Torpor Inactivity induced by environmental conditions and quickly reversible by them. See *chill coma, diapause*.

Tule marsh A characteristic freshwater wetland of lowland California, dominated by tules (bulrushes) and other grasslike plants, believed to be the historic source of much of today's lowland butterfly fauna.

Univoltine Having one flight per year.

Ventral Referring to the underside (corresponding to the belly).

Vernal Of or pertaining to spring.

Voltinism The number of flights per year, which usually but not always corresponds to the number of generations.

Warning coloration Bright, contrasting colors that "advertise" the unpalatability of an organism to predators; often such species are gregarious. This is the basis for mimicry. Also called *aposematism*. See *mimicry*.

RESOURCES

Books

There are hundreds of butterfly books out there, and with the popularity of the subject, new ones appear regularly. I list only a handful that are available at this writing (2005) and that I have found generally useful and reliable. Unless otherwise indicated, they are available through bookstores and on-line booksellers. As to taxonomic inconsistencies, you're on your own!

ABOUT BUTTERFLY BIOLOGY GENERALLY

Schappert, Phil. 2005. *A World for Butterflies: Their Lives, Behavior, and Future.* Buffalo, NY: Firefly Books.

ABOUT DOING THINGS WITH BUTTERFLIES

Winter, William D., Jr. 2000. *Basic Techniques for Observing and Studying Moths and Butterflies.* Lepidopterists' Society. Available from the society; see www.lepsoc.org for ordering information.

ABOUT BUTTERFLY PHOTOGRAPHY

Folsom, William B. 2000. *Art and Science of Butterfly Photography.* Buffalo, NY: Amherst Media. Order by calling 1-800-622-3278 or check www.wfolsom.com/book.html for availability information.

ABOUT BUTTERFLY GARDENING

Xerces Society and Smithsonian Institution. 1990. *Butterfly Gardening: Creating Summer Magic in Your Garden.* San Francisco: Sierra Club Books.

FIELD GUIDES

Brock, Jim P., and Kenn Kaufman. 2003. *Butterflies of North America.* Kaufman Focus Guides. Boston: Houghton Mifflin.

Glassberg, Jeffrey. 2001. *Butterflies through Binoculars: The West.* New York: Oxford University Press.

Opler, Paul A., and Amy Bartlett Wright. 1999. *A Field Guide to Western Butterflies.* Boston: Houghton Mifflin.

Scott, James A. 1986. *The Butterflies of North America: A Natural History and Field Guide.* Palo Alto, CA: Stanford University Press. 1986.

SPECIALTY TOPICS

The Monarch Butterfly

Halpern, Sue. 2001. *Four Wings and a Prayer: Caught in the Mystery of the Monarch Butterfly.* New York: Pantheon Books.

Pyle, Robert M. 1999. *Chasing Monarchs: Migrating with the Butterflies of Passage.* Boston: Houghton Mifflin.

Schappert, Phil. 2004. *The Last Monarch Butterfly: Conserving the Monarch Butterfly in a Brave New World.* Buffalo, NY: Firefly Books.

Sutter Buttes

Anderson, Walt. 2004. *Inland Island: The Sutter Buttes.* Prescott, AZ: Natural Selection and Middle Mountain Foundation. Visit www.geolobo.com/sutterbuttes.

Serpentine

Kruckeberg, Arthur R. 1984. *California Serpentines: Flora, Vegetation, Geology, Soils, and Management Problems.* Berkeley and Los Angeles: University of California Press.

Kruckeberg, Arthur R. 2006. *Introduction to California Soils and Plants: Serpentine, Vernal Pools, and Other Geobotanical Wonders.* Berkeley and Los Angeles: University of California Press.

BASIC REGIONAL BACKGROUND

Gilliam, Harold. 2002. *Weather of the San Francisco Bay Region,* 2nd ed. Berkeley and Los Angeles: University of California Press.

Ornduff, Robert, Phyllis M. Faber, and Todd Keeler-Wolf. 2003. *Introduction to California Plant Life,* rev. ed. Berkeley and Los Angeles: University of California Press.

PREVIOUS REGIONAL BUTTERFLY BOOKS

Garth, John S., and James W. Tilden. 1986. *California Butterflies.* Berkeley and Los Angeles: University of California Press.

Tilden, James W. 1965. *Butterflies of the San Francisco Bay Region.* Berkeley and Los Angeles: University of California Press.

Collections

Our area has three major institutional collections. They are intended as research resources for advanced students and professionals and are not generally open on a drop-in basis, but all conduct programs and offer other services to the public. For details on holdings, access, location, public programs, and so on, use the contact information below.

The California Academy of Sciences is currently in temporary quarters in downtown San Francisco while its new home is under construction in Golden Gate Park. Visit www.calacademy.org. The collections manager is Norman D. Penny, phone 415-321-8640, fax 415-321-8640; npenny@calacademy.org.

The Essig Museum of Entomology is located in Wellman Hall on the University of California Berkeley campus. Its Web site is http://essig.berkeley.edu; the collections manager is Cheryl Barr, phone 510-643-0804, fax 510-642-7428, email cbarr@nature .berkeley.edu.

The Bohart Museum of Entomology is in the Academic Surge Building on the University of California Davis campus; visit www.bohart.ucdavis.edu. The collections manager is Steve Heydon, phone 530-752-0493, slheydon@ucdavis.edu.

Online Resources

There are plenty of chat rooms, bulletin boards, and so forth for butterfly enthusiasts. A search of almost any butterfly keyword will turn up many links. A general word of caution is in order! The Web has no quality control; anyone can post anything. For example, misidentified photos are routinely posted. As I write this, a search of "Mission Blue Butterfly," an endangered Bay Area subspecies, will among other things take you to a lovely photograph of some (very unendangered) Echo Blues. Beware!

This qualification even extends to the most-used, and most useful, butterfly Web site, www.butterfliesandmoths.org. Although mounted by four extremely knowledgeable lepidopterists—Ray Stanford, Paul Opler, Michael Pogue, and Harry Pavulaan—it attempts to survey the entire continent and thus pools information from a great variety of sources. Those sources cannot easily be traced, and the natural history information (number of broods, seasonality, host plants, etc.) may not be accurate for our area and may even be grossly misleading. Much of the geographically inappropriate butterfly information that appears in the media (for example, during the 2004 California Tortoiseshell or 2005 Painted Lady mass migrations in California) results from the uncritical repetition of material from this site. Again, beware! Efforts are being made to refine coverage on a state-by-state basis.

A modified (and very usefully organized) version of this site has been incorporated in the North American nature encyclopedia site "Nearctica" (www.nearctica.com)—recommended.

Also check out the author's own web site at http://butterfly.uc davis.edu.

Societies

The Lepidopterists' Society (www.lepsoc.org) was founded in 1947 to promote the science of lepidopterology in all its branches. Its membership is international. It publishes the *Journal of the Lepidopterists' Society* for research articles, a more informal *News of the Lepidopterists' Society* with color photography, and occasional memoirs and special publications; facilitates networking among both professionals and amateurs; and sponsors an annual meeting. The Pacific Slope Section sponsors its own annual meeting, and an informal annual get-together that alternates between the University of California at Berkeley and Davis campuses. As of 2006 the national secretary is David Lawrie, 307-10820 78th Avenue, Edmonton, Ontario, Canada T6E 1P8 (dlawrie@phys.u alberta.ca).

The North American Butterfly Association (NABA) (www.naba.org) promotes nonconsumptive forms of public enjoyment, awareness, and conservation of butterflies, including watching, gardening, and photography. It publishes a full-color magazine, *American Butterflies,* and a newsletter, *Butterfly Gardening News.* It sponsors the annual Fourth of July Butterfly Count, an event modeled on the Audubon Christmas Bird Count. It also has a national meeting and local chapters. NABA, 4 Delaware Road, Morristown, NJ 07960.

The Xerces Society (www.xerces.org), named for the extinct Xerces Blue from San Francisco, promotes invertebrate conservation and offers small research grants. Xerces, 4828 S.E. Hawthorne Boulevard, Portland, OR 97215; phone 503-232-6639.

Monarch Watch (www.monarchwatch.org) is devoted specifically to tracking, studying, and conserving the Monarch butterfly and maintains a continentwide observer network. Monarch Watch, c/o Entomology Program, University of Kansas, 1200 Sunnyside Avenue, Lawrence, KS 66045; phone 1-888-TAGGING.

A Few Print Classics

Here are just a handful of foundational papers on Bay Area lepidopterology. There are, of course, dozens more. The journals in question are available at major university libraries. As of this date, some have not been digitized (alas). Keep in mind that the taxonomy used is antiquated, and sometimes just plain wrong!

Arnold, R. A. 1980. Ecological studies of six endangered butterflies (Lepidoptera: Lycaenidae): Island biogeography, patch dynamics, and the design of habitat reserves. *University of California Publications in Entomology* 99:1–161.

Coolidge, K. R. 1908. The Rhopalocera of Santa Clara County, California. *Canadian Entomologist* 40:425–431.

Coolidge, K. R. 1909. Further notes on the Rhopalocera of Santa Clara County. *Canadian Entomologist* 41:187–188.

Langston, R. L. 1974. Extended flight periods of coastal and dune butterflies in California. *Journal of Research on the Lepidoptera* 13:83–98.

Opler, P. A., and R. L. Langston. 1968. A distributional analysis of the butterflies of Contra Costa County, California. *Journal of the Lepidopterists' Society* 22:89–107.

Shapiro, A. M. 1974. The butterfly fauna of the Sacramento Valley, California. *Journal of Research on the Lepidoptera* 13:73–82, 115–122, 137–148.

Tilden, J. W. 1956. San Francisco's vanishing butterflies. *The Lepidopterists' News* 10:113–115.

Williams, F. X. 1910. The butterflies of San Francisco, California. *Entomological News* 21:30–41.

Williams, F. X., and F. Grinnell, Jr. 1905. A trip to Mt. Diablo in search of Lepidoptera. *Entomological News* 16:235–238.

References

Austin, G. T. 1998. New subspecies of Hesperiidae (Lepidoptera) from Nevada and California. In *Systematics of Western North American Butterflies,* ed. T. C. Emmel, 523–532. Gainesville, FL: Mariposa Press.

Behr, H. H. 1863. On California Lepidoptera, no. 3. *Proceedings of the California Academy of Natural Sciences* 3:84–93.

Brenzel, Kathleen Norris. 2001. *Sunset Western Garden Book,* rev. ed. Menlo Park, CA: Sunset Publishing.

Emmel, J. F., T. C. Emmel, and S. O. Mattoon. 1998. The types of California butterflies named by Jean Alphonse Boisduval. In *Systematics of Western North American Butterflies,* ed. T. C. Emmel, 3–78. Gainesville, FL: Mariposa Press.

MacNeill, C. D. 1964. The skippers of the genus Hesperia in western North America with special reference to California (Lepidoptera: Hesperidae). *University of California Publications in Entomology* 35:1–230.

Steiner, J. 1990. Bay Area butterflies: The distribution and natural history of San Francisco Bay Region *Rhopalocera.* Unpublished master's thesis, California State University, Hayward.

INDEX

Page numbers in **boldface type** refer to main discussions of butterfly species.

Series Design:	Barbara Jellow
Design Enhancements:	Beth Hansen
Design Development:	Jane Tenenbaum
Indexer:	Jean Mann
Composition:	Jane Rundell
Text:	9/10.5 Minion
Display:	ITC Franklin Gothic Book and Demi

ABOUT THE AUTHOR

Photo by Sam W. Woo

Arthur M. Shapiro is professor of evolution and ecology at the University of California, Davis. As a boy, he spent long hours walking alone in the woods and fields northwest of Philadelphia and in the New Jersey Pine Barrens. During these outings he became fascinated with natural history, and by fifth grade he had narrowed his focus to insects, especially Lepidoptera. Arthur received his BA in biology from the University of Pennsylvania in 1966 and his PhD in entomology from Cornell University in 1970. After a brief stint teaching in the City University of New York, he came to Davis in 1971. His entire research career has focused on aspects of butterfly biology, including ecology, evolution, host-plant relationships, environmental physiology, and biogeography, and he has published roughly 275 scientific papers. He spends some 200 days a year afield in western North America and has a parallel research program in the Andes and Patagonia.

Timothy D. Manolis is an artist, illustrator, and biological consultant who received his PhD in biology from the University of Colorado. A former editor and art director of *Mainstream* magazine, he has written articles about birds for many journals, and he is the author and illustrator of *Dragonflies and Damselflies of California* (University of California Press, 2003).